Peatlands and Environmental Change

Peatlands and Environmental Change

Dan Charman

University of Plymouth, UK

JOHN WILEY & SONS, LTD

Other Wiley Editorial Offices

John Wiley & Sons, Inc., 605 Third Avenue,
New York, NY 10158-0012, USA

WILEY-VCH Verlag GmbH, Pappelallee 3,
D-69469 Weinheim, Germany

John Wiley & Sons Australia, Ltd, 33 Park Road, Milton,
Queensland 4064, Australia

John Wiley & Sons (Asia) Pte Ltd, 2 Clementi Loop #02-01,
Jin Xing Distripark, Singapore 129809

John Wiley & Sons (Canada) Ltd, 22 Worcester Road,
Rexdale, Ontario M9W 1L1, Canada

Library of Congress Cataloging-in-Publication Data

Charman, Daniel.
    Peatlands and environmental change / by Daniel Charman.
        p.   cm.
    Includes bibliographical references (p.).
    ISBN 0-471-96990-7 (alk. paper) – ISBN 0-471-84410-8 (alk. paper)
    1. Peatlands.   2. Peatland ecology.   3. Peatland conservation.   I. Title.

GB621.C43   2002
577.68′7 – dc21   2001046744

British Library Cataloguing in Publication Data

A catalogue record for this book is available from the British Library

ISBN 0-471-96990-7 (HB)   0-470-84410-8 (PB)

Typeset in 9/11 pt Times by C.K.M. Typesetting, Salisbury

Contents

Preface and acknowledgements

'Peat? You're writing a book about peat?' has been the incredulous response of a number of my less-than-peaty colleagues and friends to my work on this book over the past few years. 'Is there that much to say about peat?' they quite reasonably ask. A straight answer is undoubtedly 'yes', there certainly is more than enough to fill a book on the subject of peatlands. However, to answer the question on why anyone would want to, perhaps requires a bit more amplification!

This book was conceived for a number of reasons. First, it is now more than 25 years since the publication of Peter Moore and David Bellamy's book *Peatlands* (1973). Although there have been a number of edited works and collected papers since that time, there has been no affordable balanced and up-to-date synthesis on these fascinating environments. Second, we felt that a treatment of peatlands that emphasised their functioning as whole systems was needed to help in the development of peatland science to tackle contemporary issues in peatland management and environmental change. Finally, we wanted to provide a text that would be relevant to peatlands throughout the world. So much of the research has been on northern peatlands in Europe and North America, but many of the principles are now being applied and tested on other areas such as the tropical peat swamps and temperate Southern Hemisphere systems. Findings on these other systems in turn cause us to question some of the supposed generally applicable ideas that were developed on the better-known northern peatlands.

Peatlands cover around 400–500 million ha so they are a significant fraction of the world's land surface – as important as tropical forests or hot deserts in terms of area. Hovering halfway between terrestrial and aquatic ecosystems they provide an unusual habitat for unique assemblages of plants and animals. The combination of a saturated surface and organic soil, sometimes with difficult nutrient balance and acidity conditions, encourages specialist adaptations in the flora and fauna. Slight microtopographic variations often create patterns of pools and different plant communities that

are spectacular from the air and frustrating for navigation at ground level. At a larger scale, variations in the morphology, plant communities and hydrology of peatlands give rise to a surprising variety of peatland types throughout the world. However, to discuss only the spatial extent and variability of peatlands misses an important point (probably the most important point) – peatlands have a third dimension – depth. Peat accumulates to a very great depth in many places, creating an important economic raw material, store of carbon and archive of past environmental change. Peatlands are therefore of importance not only because they are unique and interesting ecosystems, but also because they perform a variety of other functions beyond their value as unique wildlife habitats and landscapes. The economic value of peat is very high as a fuel and growing medium for horticulture, yet peatlands currently store very large amounts of carbon that can be released to the atmosphere by exploiting this resource, contributing to the anthropogenic greenhouse effect. Managing such conflicting values is not easy and becomes a political and economic debate as much as a scientific issue.

However, an adequate scientific understanding of the nature and functions of peatlands is critical to informing this debate, and this book is intended to contribute to this by providing a comprehensive summary for students, researchers and those involved in peatland management throughout the world. The attributes of peatlands require us to be truly multidisciplinary in outlook and to acquire knowledge and skills in ecology, geology, hydrology and geochemistry if we are really to understand how the peatland system functions. The book is divided into four parts. Part 1 is an introduction to key concepts, classification, distribution and structure of peatlands. Part 2 moves on to consider processes in peatland ecology and hydrology (Chapter 3), as well as looking at the reasons for, and processes involved in, peat growth and accumulation (Chapters 4 and 5). Part 3 explores different aspects of change in peatlands. This includes Chapter 6, which looks at the techniques used for reconstructuring change over long time periods, Chapter 7 which examines changes that occur as a result of internal processes in the system and Chapter 8 which discusses the effect of external natural changes such as fire and climate on peatlands. Part 3 concludes by investigating the ways in which peatlands affect the surrounding local and global environment through hydrological processes and the carbon cycle (Chapter 9). Part 4 turns to the management of the peatland resource, exploring values and impacts of modern human activity (Chapter 10), with the final chapter (11) discussing the various options for management and restoration of damaged peatlands, before concluding with a look to the future of peatlands in the twenty-first century.

The book is inevitably strongly coloured by my own experiences of peatlands in Europe, North America and New Zealand in particular, as well as by the areas of peatland science where I have undertaken research myself. However, I hope that I have managed to compensate for these biases in the text and that the result is a useful balance in coverage of subject matter, as well as depth, for both newcomers to the world of peatlands and those wishing to broaden their knowledge of peatland science. Some may feel I have over-simplified particular aspects but I hope this has not resulted in gross inaccuracy or error! I have illustrated general ideas with specific examples from many sources; the choice of examples does not suggest that other work is less important but is a result of what has had to be a highly selective process, given the sheer amount of available material and considerations of balance. Almost 1000 separate references are included here, yet being more inclusive could easily double this number.

I would like to acknowledge the contributions of a number of people in the development of my own interest in peatlands, and in the production of this book. There is not room to mention everyone individually but I owe a particular debt to several colleagues and friends. First to those who encouraged and stimulated my interest in peatlands at an early stage: Roger Smith, Richard Lindsay, Terry Rowell and Keith Barber in particular. Second to those who helped in providing some of the material for the book, especially the illustrations; Cheryl Hayward and Brian Rogers redrew many of the figures, and Paul Glaser, Linda Halsey, Richard Lindsay, Edgar Karofeld, Päivi Lundvall, Jim Aber, Sue Page and Jack Reiley provided additional illustrations. Also thanks to Peter Moore, Dicky Clymo and other authors and publishers whose original material I have reproduced. Third, to the staff at John Wiley Publishers, especially Lyn Roberts for

encouragement in the final stages of writing and help with the production process. Finally, my greatest debt is to Barry Warner of the Wetlands Research Center, University of Waterloo, Canada. I worked with Barry during a period of postdoctoral research in 1990/91 and subsequently on other projects that opened up new areas of peatland science to me. Barry and I conceived the idea for this book together on a train journey some time ago, when we drafted the initial structure and contents. Barry graciously stepped aside from further involvement in writing the book when it became clear that he had too many other demands on his time. Although I wrote the bulk of the text as it appears here, Barry commented on and added to the first two chapters. I am very grateful to Barry for his input, encouragement and generosity; the book would never have got to this stage without him.

Dan Charman
Plymouth, England
August 2001

Publisher's note
Whilst every effort has been made to trace the owners of copyright material, in a few cases this has proved impossible and we take this opportunity to offer our apologies to any copyright holders whose rights we may have unwittingly infringed.

PART 1

INTRODUCTION

Peat (the substance) and peatlands (the ecosystems) mean many different things to different people, depending on their particular interest and background. In this book the aim is to provide a comprehensive text on peatlands in terms of their ecology, hydrology, development and their interactions with the wider environment, including people. The first part of the book therefore provides a summary of some of the main definitions and terminology used in peatland science, the classification and the world distribution of peatlands. Other texts provide much more detailed regional accounts, and only a very generalised overview of these last aspects is given in Chapter 1 here. Chapter 2 moves on to establish one of the central ideas in this book, which is a consideration of the whole peat system, in discussing peatlands as landforms. While peatlands show great spatial variability, they also develop vertically over time so that peatland typology has to be considered in three dimensions rather than two. Growth over time also involves lateral spread and diversification of mire type in response to hydrology, morphology of the underlying mineral landscape and climate conditions. We are accustomed to seeing landforms composed of bedrock and unconsolidated deposits moulded through physical processes; peat landforms are similar except that they have been moulded by biological processes.

1

Peat and Peatlands

1.1 INTRODUCTION: WETLANDS AND PEATLANDS

Wetlands are a widespread feature of the landscape throughout the world. Three distinguishing features set wetlands apart from other landscape units: (i) the presence of water at or near the land surface, (ii) unique soil conditions that are most often characterised by low oxygen content, and (iii) specialised biota (most obviously plants) that are adapted to growing in these environments (Etherington, 1983; Keddy, 2000; Mitsch and Gosselink, 2000). Peatlands, also referred to as 'organic wetlands', are a large subset, and probably the most widespread, group of wetlands. The dominance of a living plant layer and thick accumulations of preserved plant detritus from previous years' growth sets peatlands apart from mineral wetlands, which lack any substantial thickness and accumulation of organic remains. Not surprisingly, it is therefore the presence of peat that is the key feature of peatlands. Lakes and other non-wetland open water bodies may contain peat, but this only occurs in limited areas or where organic remains have been carried in by water movement.

Most peatlands begin their existence as mineral wetlands and become organic wetlands over time, either naturally through successional processes or with the assistance of humans or some other external factor. It is also possible for a peatland to become a mineral wetland, again by natural causes, human disturbance and other factors. Some of these topics will form the basis of large parts of this book. As we shall see, one of the great advantages of peatlands is that they store an archive of information about the history of the peatland in the peat itself. This archive can be unearthed and used to reconstruct the development of the peatland through past millennia.

1.2 PEAT AND PEATLAND DEFINITIONS AND TERMINOLOGY

The term 'peatland' has different meanings to different people in many disciplines, which has led to confusion and misunderstanding of the nature of peat and peatlands. Table 1.1 introduces some common key terms, some of which we have already begun to use above. Peat, the material that peatlands are composed of, can be thought of as a fresh material or as a source of various chemical and biotic compounds extracted from it, which have a number of applications in horticulture, energy generation, medicine, mining and specialised product industries (Fuchsman, 1986). Peatlands are of interest to ecologists, soil scientists, biogeochemists, hydrologists, archaeologists, cultural historians, agricultural and horticultural scientists, foresters and engineers. All of these have developed, to varying degrees, a discipline-specific and sometimes conflicting vocabulary to describe peat and peatlands.

It is not possible to consider peatlands without considering functional criteria, especially hydrological conditions and associated biological features that give rise to peat and peatland formation (Ingram, 1978; Chapter 4). Low oxygen content and waterlogging are key factors required for peatlands to form. The term 'peatland' originates from early writings where 'peat land' was used to describe any land area covered by peat (Gorham, 1953). It seems that early historians were differentiating, without consciously realising it, the special attributes of peat-dominated wetlands, as opposed to other wet areas. Peatland is also a shortened version of 'peat landform' in geomorphology that refers to the biotic landforms formed of peat.

Peat landforms generally have two layers, a near-surface oxygen-rich layer and a deeper oxygen-poor layer which are necessary for active formation and

Table 1.1   Main terms used in the English-language peatland literature. The definitions are those used throughout this book and reflect general usage in the recent literature. It should be noted that all of these terms are sometimes used more specifically within certain national systems of ecosystem classification. There are also many additional national or regional terms such as 'moss' (British Isles), 'muskeg' (North America) and 'aapa mire' (Scandinavia) which are explained in the text.

Wetland	Land with the water table close to or above the surface or which is saturated for a significant period of time. Includes most peatlands but also ecosystems on mineral substrates, flowing and shallow waters
Peatland	Any ecosystem where in excess of 30–40 cm of peat has formed. Includes some wetlands but also organic soils where aquatic processes may not be operating (e.g. drained or afforested peatlands)
Mire	All ecosystems described in English as swamp, bog, fen, moor, muskeg and peatland (Gore, 1983c), but often used synonymously with peatlands (Heathwaite et al, 1993b). Includes all peatlands, but some mires may have a mineral substrate
Fen	A mire which is influenced by water from outside its own limits[1]
Bog	A mire which receives water solely from rain and/ or snow falling on to its surface[1]
Marsh	Loose term usually referring to a fen with tall herbaceous vegetation often with a mineral substrate
Swamp	Loose term with very wide range of usage. Usually referring to a fen and often implying forest cover

[1] These terms are described more fully in section 1.4.

accumulation of peat. Such two-layered landforms have been referred to as *mires*. In Europe and Russia, there is a tendency to differentiate actively peat-accumulating *mires* from the more generic *peatland* term, which applies to all peatland regardless of whether or not the system is actively accumulating peat. In North America the term 'mire' is not widely recognised, and instead the more generic *peatland* term is used in the broadest sense to denote any peat land-form irrespective of whether it is two-layered and actively forming and accumulating peat or not (Warner, 2001). There are instances where the near-surface oxygen-rich layer is less well developed or absent and only the deeper oxygen-poor layer remains (Ingram, 1978). This most often is the case where erosion or human alteration has eliminated the near-surface layer as in the case of managed peatlands. Whether natural ageing of the peat accumulating system can also contribute to disappearance of the

near-surface layer is less clear. For example, polygon peatlands in Arctic Canada may have very old surface layers of peat that suggest that these sites are no longer actively accumulating peat and have not been doing so for several thousand years (Vardy et al, 2001). Sometimes it can be difficult to differentiate between pristine and once managed peatlands that have been abandoned and have undergone natural processes of regeneration, because they can resemble pristine peatlands to the untrained eye.

Peat is a highly variable substance (Clymo, 1983; Warner, 2001). It is possible to define it in two ways. The first, and most common, is to define peat as a substance that is composed of the partially decomposed remains of plants with over 65% organic matter (dry weight basis) and less than 20–35% inorganic content (Clymo, 1983; Heathwaite et al, 1993a). It is also possible to define peat in its intact or natural state (i.e. in the peatland) as 88–97% water, 2–10% dry matter and 1–7% gas (Heathwaite et al, 1993a). Botanical composition and state of decomposition are generally agreed to be the two most important characteristics. It is the plant communities that will dictate the litter source for the formation of peat. This has led to first-order classifications that differentiate peat into four broad categories: moss, herbaceous, wood and detrital or humified peat. Moss is often subdivided into *Sphagnum* (i.e. Sphagnaceae – the peat mosses) and brown mosses (i.e. Amblystegiaceae) because both groups are common peat formers. *Sphagnum* is the single most important moss genus. Few other moss groups, except members of the Polytrichaceae, are important peat formers, for example, in the Southern Hemisphere (Clymo, 1983). Identification of the moss remains to family and genus is often possible in the field with a hand lens, and where preservation is good it is possible to identify to species level with the aid of a microscope. The same is true for flower and seed parts, leaves, stems and roots of sedge (i.e. Cyperaceae), grass (i.e. Poaceae) and rush (i.e. Juncaceae) and restiad (Restionaceae) remains, which are probably the most important groups comprising herbaceous peats. It is easy to differentiate *Carex*, *Eriophorum*, *Phragmites*, *Cladium* and *Juncus* peat, and other more unusual peats such as the *Oxychloe* peats in the Andes, for example.

Woody peats are common in forested and shrub-dominated peatlands. Wood is common in peat formed under these vegetation communities. Much of the wood is identifiable depending upon the state of preservation. Stumps, logs or large branches can be sectioned and identified using standard wood identification techniques. Seeds, nuts, strobili, cones, leaves, needles, buds and bud scales, and bark can also be identified, almost certainly to family, and often to genus and even species.

The last major category refers to peat where the plant remains are so disarticulated (i.e. detrital peat) or decomposed (i.e. humified peat) that the bulk of the plant is no longer recognisable and identifiable.

The maximum depth for peatlands is unbounded. Average depth globally has been estimated to be 1.3–1.4 m (Lappalainen, 1996b) and a maximum of 15 m is rarely exceeded (Clymo, 1983). At the other end of the spectrum, it has been customary to take an arbitrary minimum depth of 30, 40 or 50 cm. In two-layered systems, the deeper layer is usually thicker, being several metres compared to the thinner near-surface layer, which is usually not much more than 0.5–1.0 m at the most.

1.3 SCIENTIFIC CLASSIFICATION SYSTEMS

Scientific classification of ecosystems is the recognition and arrangement in a logical framework of different types, which is an essential underpinning to understanding their ecology and functioning, particularly when concepts and information have to be communicated. Scientific classifications are based on scientific principles and characteristics as opposed to other systems such as regulatory and legal classifications, which are based on more practical societal needs, economic values or jurisdictional requirements. While all such taxonomic systems are subject to discussion, change and dispute, the classification of peatlands is probably one of the most fraught and misunderstood systems of all. As Moore (1984b) points out, this is probably because there are so many possible criteria that can be used, and almost all of them are continuous rather than discrete variables. Most countries with appreciable areas of peatland have developed classification systems of their own and few are identical or even similar in detail. However, there are common elements between most systems and this section attempts to provide a concise summary of the main systems in usage around the world and to explain clearly what the relationships between them are. Moore (1984b) lists the main criteria used in peatland classification as:

1. *Floristics*: The vegetational composition of the plant communities and often used as a proxy measure of other criteria such as chemistry and hydrology.

2. *Vegetation physiognomy*: The structure of the dominant plants on the surface. Used particularly in Russian and Fennoscandian schemes.

3. *Morphology*: The three-dimensional shape of the peat deposit itself and of the smaller-scale features on the peatland surface.

4. *Hydrology*: The source and flow regime of the water supply.

5. *Stratigraphy*: The nature of the underlying peat deposits and their implications for development of the peatland.

6. *Chemistry*: The chemical attributes of the water at the surface.

7. *Peat characteristics*: Usually specifically for energy, agricultural or horticultural use and based on simple assessments of botanical composition, nutrient content and structure.

However, no single factor is the basis for any national system of peatland classification and most use a 'mixed hierarchy' to attempt to encompass all the variability which occurs in a given region. In addition to national perspectives, a variety of classification systems exist for particular functional purposes. For example, while ecologists are primarily interested in classifying according to the full range of ecological features, the peat extraction industry will concentrate on botanical composition, decomposition and nutritional state of the peat itself (Kivinen, 1977), and forest managers may require a simple key combining soil and vegetation characteristics (e.g. Jones *et al*, 1983). Clymo (1983) has compared the search for an all-encompassing peatland classification to the search for the Holy Grail(!) and reminds us that all such systems are essentially for convenience and are subject to change over time. Here we restrict the description to the principal ecological factors that are most relevant to the understanding of the functioning of peatland systems. In terms of identifying peatland types, a hydromorphological classification is probably the only universally applicable system. This uses morphology and the way in which this is related to hydrology as its principal focus, but peat stratigraphy may also be relevant. Water chemistry is determined by these features, as is vegetation within a specific biogeographic region. Vegetation is widely used within specific regions to further subdivide peatland types but is really only useful at this more local scale. Vegetation physiognomy is also useful at this scale but has not been universally accepted as a criterion around the world.

1.4 FENS AND BOGS: A KEY CONCEPT

A fundamental element of almost all peatland classification systems is the gradient from relatively nutrient-rich conditions to very poor conditions (the fen–bog gradient). There are two senses in which this gradient is used, either referring to the present trophic status (suffix 'trophic') or the origin of the peatland (suffix 'genous'). Although terms are often (incorrectly) used interchangeably, the distinction between present conditions and origins is important to make.

1.4.1 Trophic status (nutrient status)

This refers to the present status of the mire and no inference should be made regarding the origins of the system. As mentioned above, 'bogs' are strictly *ombrotrophic* systems which receive all their water and nutrients from the atmosphere and are therefore acid and low in plant nutrients. *Minerotrophic* peatlands ('fens') receive inputs from outside their confines, from groundwater or surface runoff, and therefore tend to be more nutrient-rich and more alkaline. However, there is a great range in trophic status of fens and the terms *oligotrophic* and *eutrophic* describe rather nutrient-poor fens and more alkaline, calcium-rich fens respectively. While this division seems straightforward and theoretically easy to apply, there are complications in defining the boundary between types of system and no *precise* geochemical criteria can be used to differentiate systems. Recently it has been suggested that the primary division between peatlands on the basis of trophic status should be between bogs and poor fens on the one hand and mesotrophic and eutrophic fens on the other (Wheeler and Proctor, 2000; see Chapter 3). One particular area of confusion over terminology arises from the oceanic influence in many of the peatlands on the coastal fringes of Europe and North America. Oceanic ombrotrophic peatlands tend to receive more nutrients than those in continental areas, because precipitation is often much greater and it has a slightly higher ionic content. As a result, vegetation on oceanic ombrotrophic peatlands may be similar to that on oligotrophic fens in nearby continental regions or contain species normally regarded as 'fen species'. This is bound to cause confusion among scientists trained in different regions, because most assessments of trophic status are based on using plant indicators rather than actual measurements of water chemistry. A well-known example is the occurrence of *Schoenus nigricans* (the black bog rush) on ombrotrophic blanket mires in western Ireland (Sparling, 1962). This can be further complicated in extreme cases such as peatlands in areas where precipitation is especially high in other substances such as industrial pollution or volcanic ash. A good example of the latter are the sloping 'halamyri' in Iceland (Steindorsson, 1975) which have a high ash content and are characterised by *Carex nigra*. A further term which is still in use is *rheophilous* ('flow loving' – Kulczynski, 1949) or sometimes *rheotrophic* (Moore and Bellamy, 1973) and refers to a flowing minerotrophic water supply. Kulczynski also used the term ombrophilous ('rain loving'), which appears to be synonymous with ombrotrophic.

1.4.2 Origin

These terms refer to the origin of the mire but also define this in terms of the water source at the time of formation. *Ombrogenous* peatlands formed solely as a result of an atmospheric water supply. In contrast, the terms *geogenous* or *minerogenous* suggest that the water supply is influenced by soils or rocks. Subdivisions of the last two terms are *topogenous*, meaning the water is static and a result of topographic position, such as in a basin or on a floodplain, and *soligenous*, meaning the water is flowing, such as in a spring or sloping fen. A further term sometimes used is *limnogenous*, referring to a mire forming as a result of the influence of water from rivers or lakes, such as at a lake shore. In practice it is hard to see how these can be easily differentiated from topogenous mires in all cases. As these terms are not always used precisely, in practice they tend to cause confusion and add little to the trophic terms discussed above.

Much of the confusion over terminology has been caused by the proliferation of terms such as those discussed above. The early literature is littered with such terms, some of which are synonymous and some of which are used in different ways at different times. This has left a legacy of confusion even at this gross scale of classification. The simplest common terms to use are 'bog' for ombrotrophic systems and 'fen' for minerotrophic systems. This is essentially the major division of Du Rietz (1954). Further terms need to be specifically defined when being used. However, care should be taken when interpreting any reference to 'bogs' as the term is not always used in its strictest sense, and even major recent classification schemes refer to 'bogs' which are clearly affected by runoff, although this may be only a minor influence.

1.5 HYDROMORPHOLOGICAL PEATLAND CLASSIFICATION

The classification systems of different countries provide a bewildering array of hierarchies, terms and definitions and it is difficult to provide a summary of ideas that will be applicable to all countries. Inevitably, most work has been done in north-west Europe, Scandinavia and North America and the systems discussed here are predominantly from these regions. Although the characteristics described above are used in many ways in different systems, a common idea is to use the overall shape of the peat deposit and the underlying ground, together with a basic idea of the site hydrology (where the water comes from and flows to), to provide a set of basic site types, which are then further subdivided on the basis of a more detailed understanding of their

vegetation, water chemistry and peat stratigraphy (Figure 1.1). This is known as a hydromorphological classification and the different site types discussed below are based on this idea. Although it is largely derived from north-west Europe and North America, many of the ideas are transferable to other regions of the world and there are several good examples of this. For example, 'raised mires' are recognised in South America (Pisano, 1983; Rabassa et al, 1996) and New Zealand (Campbell, 1983). Although the descriptions given here are simplified and not exhaustive, it should be possible to fit most other peatland types into these broad categories. The following section describes the characteristics of the different mire types in more detail.

1.5.1 Ombrotrophic peatlands

Ombrotrophic peatlands form the simplest group of peatland types and the group where there is probably most unity between national systems of classification.

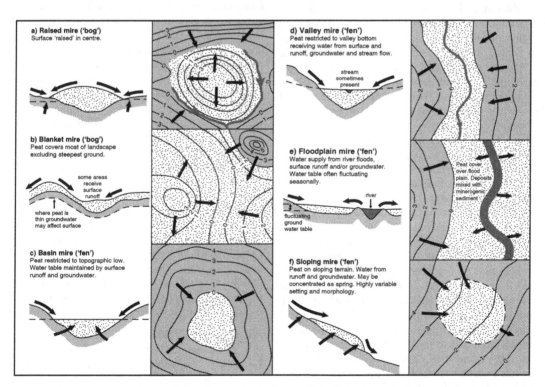

Figure 1.1 Schematic cross-sections and plan views of key hydromorphological mire types. Other variations on each of these basic types exist (see for example the more detailed breakdown of basin fens in Figure 1.6), and regional typologies and classifications vary considerably in the terminology used and the subtypes that are recognised. However, the main types illustrated should fit most peatlands in both temperate and tropical regions, in general terms at least.

The main characteristics that determine different types of ombrotrophic mires are their morphology, stratigraphy and hydrology. The types described below are equivalent to the 'grossformen' of Aario (1932) or the now more commonly used 'mesotopes' of Ivanov (1981) described more fully in Chapter 2.

1.5.1.1 Raised bogs

Raised bogs are peatlands that have a distinctly convex profile (a peat 'dome') and as a result their water supply is derived only from atmospheric inputs. Runoff from streams, springs and floods does not reach the surface of the ombrotrophic area in undamaged raised bogs but it may produce a surrounding ring of fen, often known as the 'lagg'. Lindsay (1995) provides a more extensive summary of the main features of undisturbed raised bogs, but these are the key features. Because of the dependence upon precipitation, such mires can only form where there is a positive water balance and an excess of precipitation over evaporation and runoff from the peatland surface. In north-west Europe, this is around the southern part of Scandinavia, Britain and the plains of the Netherlands, Germany and through into Poland, and the Baltic states. In North America, the Atlantic provinces of Canada, the north-eastern part of the USA and parts of the Pacific coast hold most of the raised bogs. However, there are also raised bogs in the Southern Hemisphere, for example the Kopouatei peat dome on the North Island of New Zealand (Newnham et al, 1995) and the extensive systems in South-east Asia (e.g. Reiley and Page, 1997). There are several different types of raised bog that are distinguished, depending on the precise shape of the peat dome and the underlying substrate.

Concentric raised bogs are approximately symmetrical domes with concentric rings of pools and hollows if they occur (Figure 1.2). Where they do not occur, it may be difficult to distinguish between types of raised mire from a simple inspection. In contrast, *eccentric* raised bogs have the apex of the dome off-centre and again this is most evident in the arrangement of pools and hollows on the surface. The difficulty of distinguishing easily between these two types, which are commonly referred to in Scandinavian literature (e.g. Sjors, 1983) where surface pool features are lacking and there has been damage to the system, perhaps explains the absence of such terms in some other peatland classification schemes (Lindsay, 1995). Raised bogs where the surface is horizontal and flat are often referred to as *plateau raised bogs* (Moen, 1985) and

on such sites surface patterning is not arranged in any particular orientation. Good examples of such mires can be found on the west coast of Newfoundland and New Brunswick, Canada (Wells and Hirvonen, 1988; Figure 1.2), although they are called 'Atlantic plateau bogs' in this case, which illustrates the differences in terminology, even in the recent literature.

1.5.1.2 Blanket bog

Blanket bogs are somewhat of an anomaly in this section since they are really a type of mire complex, made up of various subtypes. However, their main distinguishing feature is that they are ombrotrophic peatlands, which cover virtually all the landscape in a particular area, including slopes and mounds. Although there may be 'raised bogs' hidden within the blanket bog landscape, these are impossible to discern without an extensive peat stratigraphic and topographic survey. In addition, there may be minor minerotrophic elements around springs and seepages, which may occasionally form significant peatland types in their own right (e.g. Charman, 1993), but these do not form an appreciable proportion of the total peatland cover. There will almost always be minor areas of rock outcrops through the peat cover and some steeper slopes that have not been enveloped in peat. It is in such landscapes that the definition of peatland by peat depth (>40 cm) becomes very difficult to apply since deep peats of 6–7 m may adjoin a relatively thin area of sloping peat of c.1 m depth which thins to a variable 25–50 cm peat on a steeper slope. In this case it is practically impossible to separate peatland from non-peatland as defined by soil characteristics without very intensive peat depth surveys. Hence, estimates of blanket bog area for particular regions or countries may be subject to considerable error at the local scale. In order for peat to grow over sloping terrain, extreme climatic conditions are required with continuously cool and wet weather. Absolute amount of rainfall is probably less important than the distribution throughout the year and Lindsay et al (1988) suggest the following conditions are necessary for blanket bog formation:

(i)   Minimum of 1000 mm rainfall.
(ii) Minimum of 160 wet days (>1 mm rain).
(iii) Mean temperature of <15 °C for the warmest month.
(iv) Minor seasonal fluctuation in temperature.

As a result of these rather special conditions, blanket mire is restricted to temperate hyperoceanic areas of the

(a)

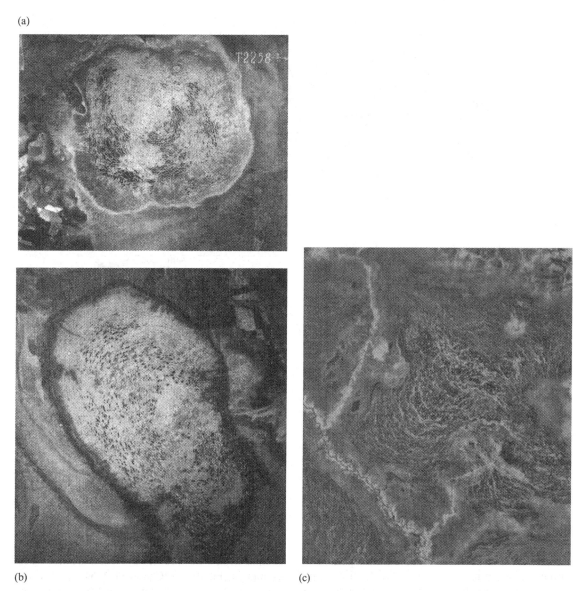

(b)   (c)

Figure 1.2   (a) Laukasoo bog, Estonia, a typical concentric raised mire in Europe. The surface patterning emphasises the morphology of the peatland. The highest part is the unpatterned zone in the centre and the concentric rings of pools and ridges reflect the contours running down to the peripheral fen shown by the light coloured band. (b) Valgeraba bog, Estonia, showing more eccentric raised mire patterning with the high unpatterned area off-centre. Photographs supplied by Edgar Karofeld from Aaviksoo *et al* (1997). (c) Eccentric patterning within a blanket mire complex at Claish Moss in western Scotland. Reproduced by permission of HMSO.

world, apparently with the exception of the Ruwenzori Mountains in Uganda (Figure 1.3).

There had been little attempt to separate different elements of blanket bogs into subtypes until recent major surveys of blanket bog in Britain (Goode and

Lindsay, 1979; Lindsay *et al*, 1988). Although the vegetation of different elements of blanket bog may not always be very different, there are a number of separate hydromorphological subcategories that can be identified (Figure 1.4). This helps enormously with

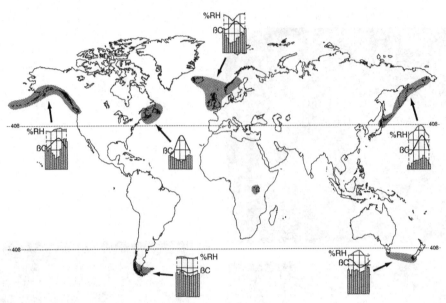

Figure 1.3 Global distribution of climatically suitable areas for blanket mire formation and areas where blanket mire has actually been recorded (darker shading). Redrawn from Lindsay *et al* (1988) by permission of Scottish Natural Heritage.

survey and description of such areas, but is also fundamental to an understanding of the ecological functioning of the system; for example, choice of site type is highly significant in palaeoclimatic studies (Blackford, 1993; Chapter 8). In this case, watershed sites are essential since they are the only areas of blanket bog where it is certain no runoff from adjacent blanket bog areas is received. Thus any proxy measures of past surface wetness should be more clearly related to past climatic change. The full range of variation is described in detail based on peatlands in Britain by Lindsay (1995) but these concepts can be easily extended to other parts of the world where no such detailed schemes exist at present. *Watershed mires* occur on hilltops and no ground is higher than the peatland apart from minor rock outcrops. As the ground falls away on all sides, no water other than precipitation can affect the bog surface. Where pools exist they are often large, deep, permanent water bodies. *Valleyside mires* occur on sloping ground, bounded by a watercourse at the base of the slope which may inundate the base of the bog from time to time and any pools are linear and oriented along the contours of the slope. *Spur mires* are similar to watershed mires but are affected by surface runoff from the steep slope where the spur is joined to the higher hillside. *Saddle mires* occur between two summits and consequently can be affected by flow from the two adjacent steep slopes. In both saddle

and spur mires surface pools will be predominantly linear but more rounded towards any apex that exists.

1.5.1.3 Intermediate blanket-raised bogs

While a subtype of raised bog is usually determinable on the basis of the precise shape of the peat dome, there are problems when an individual mire unit such as the raised bog is linked to another peatland type or extends beyond the usual definition of a raised bog. The concept of mire complexes will be discussed further in Chapter 2, but the problems are well illustrated by the example in Figure 1.5 (Chapman, 1964). In this case, two raised bogs have grown together to form a single expanse of ombrotrophic peat and the normal terminology is difficult to apply. A number of different systems of terminology have been proposed to describe such peatlands. Hulme (1980) suggests 'semi-confined' mire or, where a system merges into a blanket bog landscape, an 'unconfined' mire. Other terms are Atlantic raised bogs (Moen, 1985), ridge-raised bogs and intermediate bogs (Lindsay, 1995), and 'compound' mires (Masing, 1982). While such mire types may be locally significant (e.g. for Britain, Lindsay, 1995), this has not proved to be such a controversial aspect of peatland classification outside of Europe, and in terms of a uni-

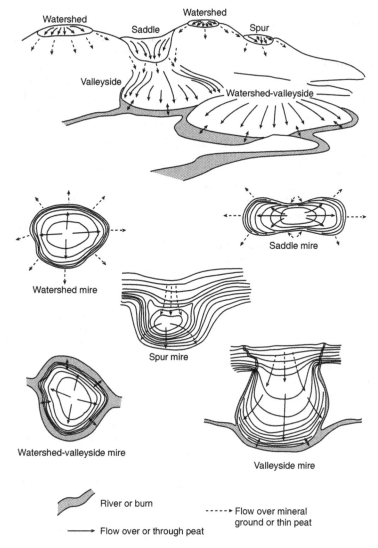

Figure 1.4   Subcategories of blanket bogs according to Lindsay (1995). Redrawn by permission of Scottish Natural Heritage.

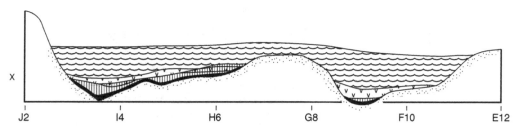

Figure 1.5   Profile of Coom Rigg Moss showing how the peatland has formed from two coalesced peat bodies formed in separate basins. The mire is clearly not a single simple raised mire, nor can it be classed as full blanket mire (see Figure 1.2). Redrawn from Chapman (1964) by permission of Blackwell Science Ltd. Shading represents different peat types reflecting hydroseral succession (section 7.3).

versally recognisable system it may be more sensible to consider them as simply small mire complexes.

1.5.2 Minerotrophic peatlands

There is a vast array of minerotrophic peatland types throughout the world and it is difficult to provide a comprehensive summary of terminology or to describe a system that adequately caters for all of this variation. However, there are certain major hydromorphological types of minerotrophic peatland which can be identified globally and these are described below. A major division exists between topogenous fens and soligenous fens in many cases (e.g. Wheeler, 1984), but some of the permafrost systems in northern Eurasia and North America do not fit easily into this major subdivision. Even if different names are used in other accounts, it should be possible to identify which of the following types it refers to given an adequate description of the site.

1.5.2.1 Topogenous fens

Basin fens/schwingmoors: a variety of peatlands form in basin situations, but their common characteristic is that the topographic depression is totally confined and does not have a major inlet or outlet. The water source is therefore from upwelling groundwater or from surface runoff from the basin edges. Sometimes the basin is completely infilled with peat but there is often an area

of open water, usually associated with the deepest area of the basin. There may often be a floating mat of peat and vegetation over some of the mire surface. This is often termed a *schwingmoor* in Europe but also as a *floating fen* or (confusingly) *floating bog* in North America (Tarnocai *et al*, 1988). Further subdivision of these mires is common (e.g. Damman and French, 1987; Figure 1.6) but these subtypes all come within the broad category of basin fens.

Floodplain fens form in association with a river system and receive much of their water supply in the form of seasonal floods, but the soils are usually waterlogged all year round. Such systems can be very extensive even in intensively used landscapes such as the Norfolk Broads in England. However, due to the input from riverine sources, soils are often transitional between organic and mineral soils and not all could strictly be termed peatlands. There are several other terms which are used in the literature to refer to topogenous peatlands. 'Open water transition fens' (Wheeler, 1984) are developed around the edges of open water, as are 'shore fens' (Tarnocai *et al*, 1988). However, many of these ecosystems with fen vegetation will not be peat forming and are thus outside the scope of this book.

1.5.2.2 Soligenous peatlands

Valley mires are formed along the floors and lower slopes of valleys that have a dispersed flow of water through them. Besides the flow of water down the

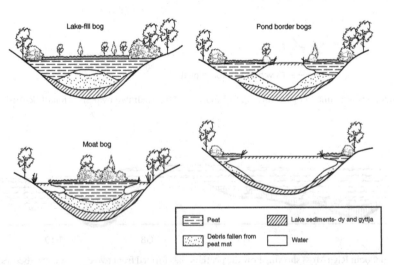

Figure 1.6 Various subcategories of basin mires according to Damman and French (1987). Although these are all termed 'bogs' in the North American literature, this is not used in the strict sense of ombrotrophic peatlands, as the surface vegetation is still influenced by groundwater and runoff. Redrawn from Damman and French (1987).

valley, springs and seepage from the valley sides may also help maintain the water table. These are also referred to as 'channel fens' in the Canadian literature (Tarnocai *et al*, 1988). Spring fens form in areas around an emergence of groundwater and may or may not be peat forming. Although the water supply is a point source, its influence may stretch some distance.

1.5.3 Boreal and subarctic peatlands, patterned fens and aapa mires

At least some of the fen systems described above will be familiar to most people living in relatively urban settings with some experience of the agricultural country-side. However, there is a whole range of fen systems that mainly occur in more remote regions, particularly in the northern areas of Eurasia and North America. Many of these systems are characterised by complex patterning on their surfaces formed by the arrangement of pools, hollows and hummocks in a similar way to several of the bog systems already described. These peatlands do not fall easily into soligenous or topo-genous categories, although both influences may be present in different situations and they are therefore described separately here. The description and classification of these areas have received most attention in Scandinavia, particularly in Finland, and a series of zones of mires can be recognised (Figure 1.7). The

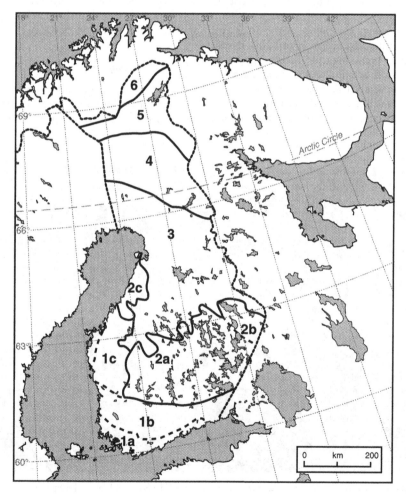

Figure 1.7   Mire zones in Finland, showing division into regions. 1 = concentric raised bogs, 2 = eccentric raised bogs, 3 = southern aapa mires, 4 = northern aapa mires of Peräpohjola, 5 = northern aapa mires of Forest Lapland, 6 = palsa mires. The subdivisions of classes 1 and 2 reflect regional variations of the same main peatland types. Redrawn from Eurola *et al* (1984).

zones represent 'mire complex' types, but are named after the predominant peatland types within each region. The southern part of the region is dominated by bogs of various kinds but north of this is the 'aapa fen complex' (Eurola *et al*, 1984). *Aapa fens* may be flat or sloping, but usually have some expression of surface patterning with ridges and/or hollows. This patterning is most characteristically of very elongated, often very narrow ridges (strings) and hollows or pools running parallel with each other across slopes of very low gradient (*c*.1%). In Canada the equivalent of sloping aapa fens are *ribbed fens* and good examples occur in Newfoundland, Labrador and throughout the boreal region (Zoltai *et al*, 1988b). Seen from the air, these form some of the most impressive peatland systems anywhere in the world (Figure 1.8). Similar patterning may occur on a smaller scale on fen systems within other climatic regions associated with areas of increased runoff from adjacent ombrotrophic mires, and these have been termed *ladder fens* in Newfoundland (Wells and Pollett, 1983) and simply *patterned fens* in Scotland (Charman, 1993).

As one moves north, another distinct mire type is evident and is characterised by mounds raised above the surrounding peatland surface. These are the *palsa mires*, where permafrost occurs within the peat and results in mounds of 1–8 m above the surrounding mire area. The palsa mire itself is simply the mound and immediately surrounding area, but other peatland types are usually also present, such as aapa mires throughout northern Finland (Eurola *et al*, 1984). These palsa mires also occur throughout subarctic Canada (Zoltai *et al*, 1988a). They may also coalesce to form more extensive raised areas or contorted ridges. Smaller mounds are often referred to as *pounikkos* in Scandinavia (Eurola *et al*, 1984). Further north still are the extensive arctic mire systems characterised by *polygon mires* of various forms (Tarnocai and Zoltai, 1988). The ground is divided into polygons with ice wedges formed between the polygons themselves. The polygons may be high or low centred, forming domes or depressions within the polygons.

1.5.3.1 Mire complexes

Many peatlands do not occur only as discrete types that fall easily into the categories described briefly above. Instead, they are complexes containing a variety

Figure 1.8 Kilsiaapa-Ristivuoma, a typical aapa mire in northern Finland, showing elongated ridge and hollow patterns running across the direction of slope. Areas of forest occur on drier localities within the mire complex. Photography by Aarno Torvinen. Reproduced by permission of Aapamire-Life (EU LIFE project on the Protection of Aapa mires in Lapland and Ostrobothnia), Lapland Regional Environment Centre, Finland.

of peat landforms and types. In order to describe these complexes satisfactorily, a clear hierarchy of terms is needed to differentiate individual peatland areas. Russian peatland specialists have appreciated this need for some time and developed such a generic system of description (Ivanov, 1981). Ivanov's concept of different scales of hydromorphological features and peatland areas is described fully in Chapter 2. However, even within mire complexes it should be possible to recognise individual elements of the peatland landscape that conform to the hydromorphological peatland types discussed above.

1.6 MIRE DISTRIBUTION

1.6.1 The global picture

The distribution of the world's peatland resources is not known with any accuracy. The extent and nature of tropical peatlands in particular are poorly known, but even in Europe and North America we only have approximate ideas of the precise extent and distribution and we must make some significant assumptions in calculating peat volumes (e.g. Gorham, 1991). For example, recent evaluation of Russian peatland areas suggests much greater extent than was previously known (Botch et al, 1995). However, there are some excellent regional accounts of mires in Gore (1983b) and attempts at global summaries in Pfadenhauer et al (1993) and Immirzi et al (1992). Lappalainen (1996a) is one of the most comprehensive attempts to assess the extent of the global peatland resource. In all such assessments, it is not simply that data for all regions are incomplete, but that the definitions and terminol-

ogy applied are different and thus any global compilation is problematic and inevitably involves some compromise and ultimately guesswork. Immirzi et al (1992) summarise the difficulties involved the use of existing inventory data as:

(i) Variable detail, accuracy and reliability.
(ii) Errors and variations in classification.
(ii) Uncritical acceptance of previous estimates.
(iv) Omission of data, especially in terms of depth and stratigraphy.

Their estimate of total peatland extent, defined as soils with more than 50% organic matter and greater than 30 cm deep, is between 386 and 409 million ha (Table 1.2). The total extent is probably somewhere in excess of 400 million ha, although estimates of tropical peatland area are likely to rise in future. Armentano and Menges (1986) suggest 393 million ha, and Kivinen and Pakarinen (1981) arrive at a figure of 421 million ha, excluding Africa and parts of South America, which would bring the total to more than 500 million ha. The lowest recent estimate is 320 million ha (Pfadenhauer et al, 1993), but this is almost certainly an underestimate and some of their data are rather confusing with respect to peatland areas in the former Soviet Union. Other broader estimates of 'wetland' area are of 1300 million ha (Bouwmann, 1990) and 526 million ha (Matthews and Fung, 1987), although these figures clearly include a large area of non-peatland ecosystems. Whichever estimates of global peatland distribution are used, it is clear that certain countries and regions are particularly important on a global basis for the sheer amount of peatland they contain. Europe, including the former Soviet Union, and

Table 1.2 Estimates of peatland areas in the world. In the two left-hand columns of figures, data refer to extent of peatlands before extensive anthropogenic disturbance in historical times (before c. AD 1800), and upper and lower estimates are given (after Immirzi et al, 1992). In the right-hand column of figures, data are for all remaining peatlands, although many of these may be damaged by human activities (after Lappalainen, 1996b). Estimates for similarly named regions differ between authors because of different treatment of boundaries but the total estimates for global peatland area are similar. All figures in ha.

Immirzi et al (1990)			Lappalainen (1996b)	
Region	Lower estimate	Upper estimate	Region	
Africa	4 976 500	4 976 500	Africa	5 800 000
Asia	24 031 000	43 610 500	Asia	111 900 000
Central America	2 290 569	2 614 000	C. and S. America	10 200 000
North America	170 967 000	171 127 000	North America	173 500 000
South America	6 174 780	6 174 780		
Europe and former Soviet Union	177 213 510	179 763 849	Europe	95 700 000
Pacific	184 000	275 000	Australia and Oceania	1 400 000
Total	385 837 359	408 541 629		398 500 000

North America are by far the most significant continents in terms of the absolute extent of peatlands. Within this, Canada, the former Soviet Union, the USA and the Nordic countries (especially Finland) contain the largest peatland areas (Table 1.3). There are some considerable discrepancies in the figures for

Table 1.3 Estimates of original peatland areas within individual countries in North America and Europe. Figures from Immirzi *et al* (1992) are upper estimates. All figures in ha. Work is currently under way by the International Mires Conservation Group (IMCG) to update these figures (www.imcg.net), to derive more precise estimates of peatland areas in these and other countries in the world. Other estimates are also given in the individual chapters of Lappalainen (1996a).

Country	Immirzi *et al* (1992)	Pfadenhauer *et al* (1993)
North America		
Canada	111 327 000	111 330 000
USA (Alaska)	49 400 000	49 400 000
USA (S. of 49° N)	10 400 000	10 240 000
Total	171 127 000	170 970 000
Europe		
Austria	30 000	20 000
Belgium	18 000	1 000
Bulgaria	1 000	1 000
Former Czechoslovakia[1]	31 370	33 000
Denmark	120 000	60 000
Finland	11 800 000	10 400 000
France	120 000	120 000
Germany	1 599 000	1 479 000
Greece	5 000	9 000
Hungary	100 000	100 000
Iceland	1 000 000	1 000
Ireland	1 180 000	1 175 600
Italy	120 000	120 000
Luxembourg	200	0
Netherlands	280 000	45 000
Norway	3 200 000	3 000 000
Poland	1 350 000	1 300 000
Romania	7 000	7 000
Spain	6 000	6 000
Sweden	7 000	7 000
Switzerland	55 000	5 500
United Kingdom	1 640 280	1 508 700
Former USSR[1]	150 000 000	71 500 000
Former Yugoslavia[1]	100 000	100 000
TOTAL	179 762 900	98 990 800

[1] Data for recently divided countries are not available from most studies, although Lappalainen (1996a) provides some data.

European mires, mainly because estimates often take into account the condition of the mires. Larger areas will always be estimated where any peatland is included. Smaller areas result from only including peatlands that have a natural vegetation cover. For example, estimates of peat cover for the Netherlands vary widely. A figure of 280 000 ha of potentially exploitable peat deposits is quoted by Immirzi *et al* (1992) based on data from Bord na Móna (1984). This is much higher than the 45 000 ha figure of Pfadenhauer *et al* (1993), which is not clearly referenced but probably refers to less damaged peatland area. Both figures are much lower than the likely maximum area during the Holocene period (Pons, 1992). This illustrates the considerable problems of evaluating the distribution of peatlands on a global basis, especially in areas where there has been significant alteration of natural peatland ecosystems. Despite these problems several estimates now agree that there are probably around 400 million ha of peatland worldwide.

The areas of the world that are most important for peatlands are shown in Figure 1.9. The Northern Hemisphere above 45° N is clearly the most important region for peatlands on a global basis, although within this area the climate is too arid across much of Asia for extensive mires to form. However, tropical peatlands in South America, Africa and particularly South-east Asia are also important and there are smaller areas of temperate peatlands in the Southern Hemisphere.

1.6.2 North America

Excellent summaries of North American wetlands, including peatlands, are available and it is not a function of this book to describe regional peatlands in detail (e.g. National Wetlands Working Group, 1988 and Glooschenko *et al*, 1993, for Canada; Hofstetter, 1983, Wilen and Tiner, 1993 and Mitsch and Gosselink, 2000 for the USA; Olmsted, 1993, for Mexico). The North American systems of habitat classification consider wetlands as a whole (Cowardin *et al*, 1979; Mitsch and Gosselink, 2000; National Wetlands Working Group, 1988; Table 1.4), which makes the clear separation of peatlands sometimes difficult, although in some cases there are categories of wetlands which can be regarded as freshwater peatlands (Table 1.4). Within the Canadian wetland classification system bogs and fens are entirely peatlands, while marshes and swamps contain some peatlands. Shallow-water wetlands are entirely non-peatland. In the USA and Mexico, true peatlands are much less

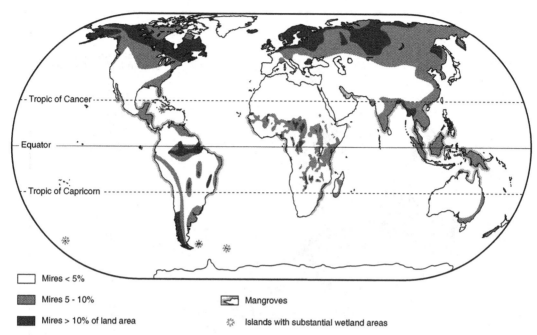

Legend:
- Mires < 5%
- Mires 5 - 10%
- Mires > 10% of land area
- Mangroves
- Islands with substantial wetland areas

Figure 1.9 Peatland distribution expressed as a proportion of the land surface for different parts of the world, based on Gore (1983a) and Lappalainen (1996a). This map provides a general idea of where peatlands are an important part of the landscape and is based on incomplete data. Redrawn from Lappalainen (1996a) by permission of the International Peat Society.

Table 1.4 Summary of the wetland types in different wetland classification systems for Canada and the USA that contain significant amounts of peatland area.

Cowardin *et al* (1979) – Overall wetland classification	
System	*Class*
Palustrine	Moss–lichen wetland
	Emergent wetland
	Scrub–shrub wetland
	Forested wetland

National Wetlands Working Group (1988), Zoltai (1988)	
Wetland class	*Wetland form*
Bog	Atlantic plateau, basin, blanket, domed, flat, floating, lowland polygon, northern plateau, palsa, peat mound, peat plateau, polygonal peat plateau, slope, string
Fen	Atlantic ribbed, basin, channel, collapse scare, feather, floating, horizontal, ladder, lowland polygon, northern ribbed, slope, spring
Swamp, marsh	Various wetland forms are included in these two classes. None are always major peat formers, although some examples will contain significant amounts of peat

abundant than mineral-based wetlands, and there are relatively few categories which include peatland elements in the wetlands and deep-water habitats system – only 4 out of 55 (Cowardin *et al*, 1979). These are the moss–lichen, emergent, scrub–shrub and forested wetland classes.

The occurrence of peatlands generally increases northwards throughout North America until north of about 60° N, where conditions are less optimal for peat formation due to decreased temperatures (Figure 1.10). Peatlands form distinct zones within Canada, principally related to temperature and oceanicity (National Wetlands Working Group, 1988). The Low Arctic and High Arctic wetland regions are found north of the tree line and peatlands are all influenced by permafrost with a variety of polygon, peat mound and palsa peatlands. Peatlands in the subarctic regions are also affected by permafrost, but this is absent as one moves into the Atlantic subarctic region. Here, peatlands are peat plateaux, palsas and polygons, all of which are strongly related to permafrost action, but 'northern ribbed fens' and slope peatlands also occur in more temperate areas (Glooschenko *et al*, 1993). The greatest extent of peat occurs throughout the boreal region and a great variety of peatland types occurs here, including domed bogs,

(a)

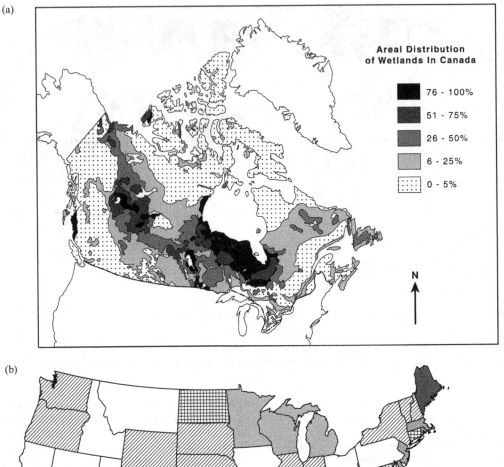

(b)

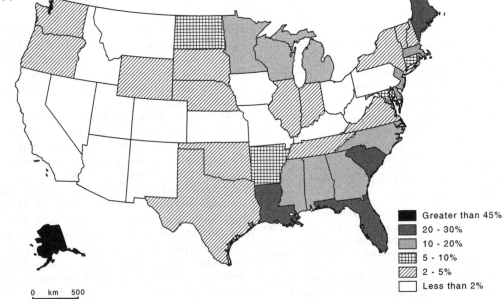

Figure 1.10   Wetland distribution in (a) Canada (National Wetlands Working Group, 1988) and (b) United States of America by states (Dahl, 1990; Wilen and Tiner, 1993). The Canadian wetlands are predominantly peatlands throughout the country, and dominant peatlands in the USA are the northern peatlands in Minnesota and New England, the pocosins in North Carolina and extensive swamps in Florida and Georgia (the Everglades and Okefenokee Swamp) which are only partially peatland. Redrawn from (a) National Wetlands Working Group (1988) by permission of Clayton Rubec and Kluwer Academic Publishers (b) Dahl (1990).

plateau bogs, basin bogs, ribbed and horizontal fens and coniferous swamps. The influence of permafrost is still apparent in some of these peatlands, and palsas can still be found in the north of the region. More oceanic peatlands and especially domed bogs occur in the Atlantic boreal region over much of Labrador, Newfoundland and Nova Scotia. More southerly wetland regions are very different in character. The prairie wetland region is dominated by marshes where little peat forms, and the eastern temperate region is dominated by swamps of various kinds. Extreme oceanic systems are found in the Atlantic oceanic and the Pacific oceanic wetland regions, including blanket bogs in Newfoundland. Finally, the mountain regions in the west carry an assortment of small bogs and fens where the terrain is suitable. Biogeographically, these regions extend into the USA and the 'prairie potholes' are the same as those north of the border. The extensive northern peatlands of Minnesota (Wright *et al*, 1992) are the southward extension of the low boreal region of Canada, and there is some extension of the Atlantic boreal region into the coastal areas of the New England states. Rather different systems are found further south in the USA, including the pocosins. These are mainly found in Carolina and once covered *c*.1 million ha (Wilen and Tiner, 1993). These are 'scrub–shrub wetlands' dominated by evergreen shrubs and are reminiscent of the bogs of more northern areas as they grow on top of hills. They have sometimes been termed bogs although whether they are truly ombrotrophic systems is not clear. Other large wetland areas which are at least partly peatlands are the swamps such as the Everglades, Big Cypress Swamp and Okefenokee Swamp in the southern states of Georgia and Florida (Cohen *et al*, 1984; Duever, 1984; Gunderson and Loftus, 1993). These are complex mosaics of minerotrophic wetland communities on shallow peat and mineral soils with forested and open areas and a wide range of plant communities.

1.6.3 Europe and northern Asia

It is difficult to summarise the distribution and characteristics of the peatlands of this region, which holds a large proportion of the global peatland resource and goes from the intensively studied north-west European systems to the little-known systems of Siberia and the far east of Russia. Several good summaries exist for western Europe (Gore, 1983b; Moore, 1984a and chapters therein; Doyle, 1990; Verhoeven, 1992) and there are numerous individual research papers from

throughout the twentieth century. Much of the extensive Russian and German literature is still relatively inaccessible to those who do not read the languages concerned, but Botch and Masing (1983), Ivanov (1981) and Heathwaite *et al* (1993b) have done much to remedy this. The countries of southern Europe are still rather poorly researched but this is unsurprising in view of the limited extent of peatlands in these countries. Non-European Mediterranean countries should logically be included in this section, but again there is rather little peatland here. Similarly data on China are sparse, although Chai (1980), Pfadenhauer *et al* (1993) and Xuehui and Yan (1996) provide summaries. Japan also contains some extensive peatlands of *c*.200 000 ha (Umeda and Inoue, 1996).

Like North America, distinct zones of peatlands occur throughout this area, although there is no single description that caters for the whole region. Figure 1.11 shows this zonation for the main areas of this region, but the systems do not always coincide at boundaries. However, some general trends and patterns can be identified even at this large scale. In fact the schematic zonation in Figure 1.11(c) (Eurola, 1962; Eurola *et al*, 1984) is probably the most meaningful and simplest presentation of the general trends across the region. Pfadenhauer *et al* (1993) also describe the zonation of Kac (1971), who formulated 12 mire provinces, where similar trends to the other systems can be identified, but the zones are probably less familiar to most peatland specialists than those of Moore and Bellamy (1973) or Eurola *et al* (1984). The north–south differences are similar to those in North America and the oceanic influence is equally obvious. However, the zone where polygon mires are found is much narrower and entirely restricted to the northern part of Asian Russia (Botch and Masing, 1983). It is referred to as the 'unconfined arctic mires' zone by Moore and Bellamy (1973). Palsa mires are likewise a relatively narrow zone in Europe but spread further south in Asian Russia. The aapa mire zone spreads across the main central part of Scandinavia and into Karelia and the White Sea area of Russia. However, it then disappears further east, presumably due to increasing continentality, with much lower precipitation levels. In Europe, south and coastward of this zone, there is a broad band of ombrotrophic peatlands, variously divided by different authors. Eccentric bogs ('raised string bogs' in Russia) are generally dominant in the north and especially to the east of Europe. South of this, concentric raised bogs are more prevalent. The other main group of raised bogs are the 'plateau domed mires' (Moore and Bellamy, 1973) which are flat-topped and often

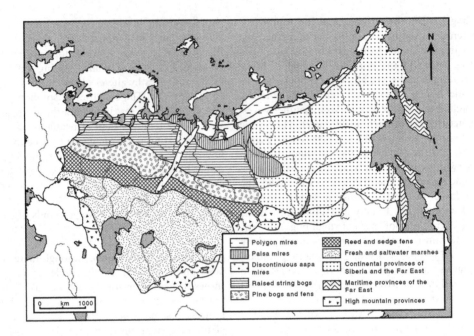

(a)

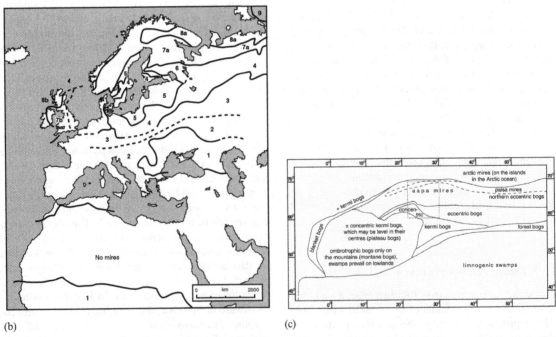

(b)   (c)

Figure 1.11   Zonation of peatland in Europe and northern Asia. Names of zones derive from dominant and characteristic mires within each area and do not imply that these are the only peatland types found within these regions. (a) The former USSR (Botch and Masing, 1983). (b) Europe from Moore and Bellamy (1973). See Figure 8.1 for descriptions of mire types 1–9. (c) An alternative schematic zonation for Europe only outlined without shallow or enclosed seas (Eurola *et al*, 1984). See also Figure 1.7 for mire complex zonation within Finland. Redrawn from (a) Botch and Masing (1983) with permission of Elsevier Science, (b) Moore and Bellamy (1973) by permission of Peter Moore and (c) Eurola *et al* (1984).

have a cover of pine in the east (Botch and Masing, 1983), as a result of the suboptimal conditions for raised mire development in the more continental climates. The British Isles are dominated by extreme oceanic systems, with large areas of blanket bog in the uplands and raised bogs in lowland regions. Moore and Bellamy (1973) suggest an additional zone of 'ridge raised mires' for these oceanic systems which can overflow their immediate confines (see intermediate raised-blanket bogs above). Peatlands in Asian Russia become less and less abundant in the continental provinces of Siberia and the far east and are generally valley fens or topogenous mires of a kind. Likewise peatlands become less abundant to the south with basin mires, valley fens and other minerotrophic peatlands only present. Finally peatlands peter out altogether south of the Mediterranean. However, the Pacific coast area of Russia has extensive peatlands, especially in west Kamchatka where over 80% of the ground is covered with blanket bog (Botch and Masing, 1983), but also in Sakhalin to the south.

1.6.4 Temperate peatlands in the southern hemisphere

Although the peatlands of southern South America, Australia, New Zealand and the islands of the South Atlantic do not appear to be of major significance in terms of their contribution to the global peatland resource, these areas possess some regions where peatlands are locally extensive. In addition, due to their biogeographical differences from northern temperate peatlands, they are very different in terms of their vegetation, animal communities and peat stratigraphy.

The peatlands of the southern part of South America have been referred to as the 'Magellanic tundra complex' (Pisano, 1983) but this is a little misleading, as although the plant communities are similar in terms of plant physiognomy, there is no permafrost and the mire systems have more similarities with other oceanic areas of the world, such as Newfoundland and north-west Europe (Rabassa et al, 1996). Rainfall is between 800 and more than 6000 mm distributed evenly throughout the year with low temperature fluctuations about an annual mean of usually less than 8 °C. Conditions are therefore ideal for peat formation in a variety of settings. Description of these peatlands is largely phytosociological, but they are dominated by ombrotrophic systems, apparently with raised and blanket bogs present (Pisano, 1983), although several of the woodland and heath plant communities occur

on much shallower organic soils. The raised bogs are dominated by *Sphagnum magellanicum*, liverworts (Hepaticae) or other cushion-forming bryophytes. The 'blanket mires' are also described as 'high rainfall bogs' (Rabassa et al, 1996) and have a wider range of plant communities, dominated by cushion-forming plants, dwarf shrubs (e.g. *Empetrum rubrum*) and grasses (e.g. *Poa flabellata*). These blanket peatlands appear similar to those found on the Falkland Islands, where a large percentage of the land area is covered by peat, with vegetation dominated by *Sphagnum* or cushion plants such as *Oreobolus obtusangulus* and *Astelia pumila* (Lindsay et al, 1988).

The cushion 'bogs' occur in many other areas of the Southern Hemisphere and extend particularly to New Zealand and Tasmania. The same genera of plants also occur with species of *Donatia* and *Oreobolus* (Campbell, 1983). These are mainly restricted to the mountainous areas on the South Island of New Zealand and often include patterning similar to that found in Northern Hemisphere systems (Mark et al, 1995). True ombrotrophic bogs mainly occur in the South Island of New Zealand, with blanket mires along the fjords of the west coast and the Chatham Islands, 750 km south-east of Wellington (Pfadenhauer et al, 1993). Raised mires occur on both North and South Islands. Their vegetation and peat are very different from raised mires elsewhere in the world. They are dominated by Restionaceae rather than mosses and these sedge-like plants take the place of *Sphagnum* as the main peat former. 'Pakihi' mires evolved over gleyed soils and are dominated by a more diverse flora including Restionaceae, sedges and *Sphagnum* (Dobson, 1979). A variety of minerotrophic peatlands occurs throughout New Zealand, Tasmania and in the less arid parts of mainland Australia, mainly the northern coastal and east, south-east and south-west coastal areas (Campbell, 1983). However, many of these mires are swamps and other wetlands, which may be marginal for peat formation despite high productivity, due to relatively high temperatures. The peatlands that do occur in the north of Australia are better considered under tropical peatlands.

1.6.5 Tropical peatlands

Tropical peatlands are very different from the temperate systems on which much peatland research has been based. Conventional classifications are difficult to apply strictly, but there is some commonality in terms of overall form, if not vegetation cover and physiognomy. Many of the wetlands can be classified as types of swamp and have predominantly mineral soils. However, they will receive some coverage in this book as the soils are often highly heterogeneous and there will be localised peat accumulation within swamp areas. Thompson and Hamilton (1983) in discussing peatlands and swamps of Africa note the difficulty of separating peatlands from other wetlands in tropical areas, and it is one of the main reasons why estimates of tropical peatland area are so uncertain and inevitably approximate. Pfadenhauer *et al* (1993) provide the best global summary of tropical peatland occurrence and characteristics, but more detailed regional accounts can be found, particularly within the broader wetland literature. The summary of global peat resources by Lappalainen (1996a) contains large sections on Asia, Africa and Central and South America, within which a number of authors provide individual accounts for particular countries. More detailed ecological descriptions are provided by the following: Anderson (1983), Gopal and Krishnamurthy (1993) and Reiley and Page (1997) for South and South-east Asia; Junk (1983, 1993) for tropical South America; Thompson and Hamilton (1983), Denny (1993a,b,c), John *et al* (1993)

and Breen *et al* (1993) for Africa; Osborne (1993) and Finlayson and von Oertzen (1993) for tropical Australia and Papua New Guinea; Hofstetter (1983), Olmsted (1993) and Wilen and Tiner (1993) for the southern United States and Mexico. Many of these accounts focus on the more widespread mineral wetland types but there is also discussion of peat-forming wetlands.

Pfadenhauer *et al* (1993) divide tropical peatlands into lowland mires, forested mires and mountain mires. The lowland mires are wholly minerotrophic and are divided into floating meadows, oligotrophic and eutrophic fens. Floating meadows occur in lakes and slow-moving waters throughout Amazonia, central Africa and also in South-east Asia but rarely form extensive or deep peats. Oligotrophic fens is a broad category and would include any relatively acidic floodplain, lake margin or valley fens throughout the tropics and may form rather deeper organic deposits. Eutrophic fens tend to be of low species diversity and dominated by very tall-growing grasses and Cyperaceae. Forest mires are perhaps the most significant in terms of the amounts of peat which have been accumulated. They can be very thick (up to *c*.16 m) and with a wide range of nutrient status. Well-known

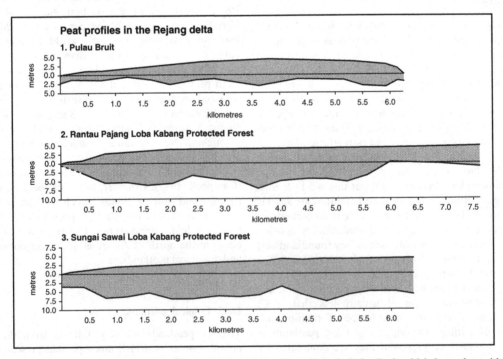

Figure 1.12   Peat profiles in the Rejang Delta, Sarawak. Note the domed profile in the Pulau Bruit, which descends on either side. The other profiles are only partial but display similar profiles on the parts shown. Redrawn from Anderson (1983) by permission of Elsevier Science.

examples are the swamp cypress forests in the south-eastern USA such as the Okefenokee Swamp in Georgia, but they also include the largest ombrotrophic peatlands in the tropics in the lowlands of Indonesia and Malaysia. Although relatively young (*c.*4500 years), the rapid accumulation rates (2.2–2.8 mm yr^{-1}, Anderson, 1983) have led to deep deposits of wood within a loose slurry peat. Despite difficulties of applying temperate peatland classification systems to the tropics, these are undoubtedly raised mires, with domed profiles (Figure 1.12). These cover very large areas – 27 million ha in Indonesia alone (Anderson, 1983). Many of the mountain mires to be found in the tropics are similar to those described for temperate areas in the Southern Hemisphere and are dominated by cushion-forming plants. The 'paramo' are probably the best known of these, occurring at high altitudes in South America, but others can be found in Africa, Sumatra and Hawaii. *Sphagnum* mires also occur in the mountains of South America and Africa, and other sedge and grass-dominated fens also occur here in relatively restricted areas. Occasionally blanket peatlands can form at high altitude in the tropics, as in the Ruwenzori Mountains (Thompson and Hamilton, 1983). There remains much to be discovered regarding the present characteristics, development and classification of tropical mires, but despite their differences,

it should be possible to develop the existing hydro-morphological categories to cater for these systems to provide an all-encompassing typology at the broadest scale.

1.7 SUMMARY

This chapter has presented some key concepts in the classification and definition of peatland systems. It will be apparent that a large number of terms have to be grasped in order to appreciate the complexity and range of habitats that are grouped together under the title 'peatlands'. It will be equally apparent that some of these terms do not mean the same things to different people in different parts of the world and that strict definitions of peatlands are often not easily applicable to some groups of wetlands such as marshes and swamps. Depending on our interest in peatlands, we see them in different ways – perhaps as a resource to be harvested, an ecosystem with biodiversity value, or an archive of past environmental change. These viewpoints determine the ways in which we define and classify peatlands. There are no easy solutions to these difficulties but it is vital that an attempt is made to understand the terminology used by individual authors if their meaning is to be correctly understood.

2

Peat Landforms and Structure

2.1 INTRODUCTION: PEAT LANDFORMS

It is already apparent from the introduction to peatland classification and distribution in Chapter 1 that the build-up of peat gives rise to distinctive peat landforms of various shapes, sizes and morphologies. Peatlands are like other landforms composed of rock and unconsolidated mineral deposits, except that they differ by being composed of biologically derived materials. The key difference is that they are shaped principally by biological processes rather than physical processes. Physical processes can work with biological processes to determine peat landform structure but not all peatlands are affected in this way.

Good examples of the influence of physical processes can be found in permafrost-dominated landscape climates where freeze–thaw processes result in major changes such as the development of palsas or distinctive polygon patterns (see Chapter 8). On the whole, however, biological processes governed by productivity and decay determine peatland shape, although they may be strongly influenced by other factors such as hydrogeochemistry and climate. In turn, once a peatland begins to take shape, it will also start to influence the factors that led to its initial development. For example, the growth and accumulation of peat may ultimately be sufficient to isolate the peatland surface from the influence of groundwater, leading to the development of a raised peatland system, with a much different hydrology and chemistry compared with that which existed when the peatland began to form (Chapter 7). Thus, peatland structure is an essential determinant of the nature of the entire ecosystem and is intimately entwined with the key ecological and developmental controls. This chapter describes the main controls of development of the peat landform, the different scales of landform expression and the techniques for survey and description of peatlands at large scales.

2.2 LANDFORM DEVELOPMENT: FORM, PROCESS AND TIME

In considering peatlands as landforms, we can view the entire peat mass almost as a 'living landform' which, once initiated, grows laterally and vertically, changing shape and character through time as it ages. Considering peatlands in this way is important to understanding the ecology and functioning of the system and is one of the factors that distinguishes peatlands from other ecosystems and landforms. Peatlands are a unique combination of both ecosystem and landform. While all ecosystems evolve spatially, there are few parallels to the vertical changes that take place in peatlands. Many ecosystems are influenced by their earlier phases of development but only in peatlands does this ultimately influence the actual surface morphology of the landform. In peatlands, the physical mass of earlier growth phases becomes part of the system itself, with strong influences on processes such as hydrology, carbon cycling and nutrient balance. This 'organismic' concept of peatlands is also required in management as there are strong functional links between one part of the peatland and another. This is particularly crucial to hydrological management, where effects of forestry, drainage and peat extraction on one part of the system can have serious consequences for adjacent parts. Many of the aspects of peatlands discussed throughout this book depend on an understanding and visualisation of peat landforms, so that particular parts of the system can be placed in a broader context.

There are several factors that ultimately determine the morphology of any particular peatland but Figure 2.1 attempts to identify the key parameters in a general sense. These ideas are loosely derived from Moore and Bellamy's (1973) 'templates' of peat formation, hydrological, climatic and geochemical, but I have

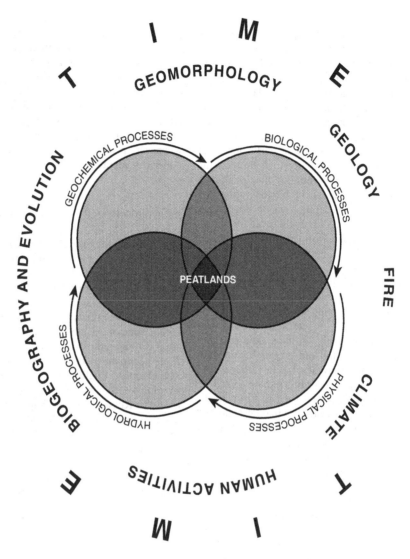

Figure 2.1 Conceptual model of the main influences on peat landform development. Factors external to the peatland are positioned around the outer rim of the model with key processes internal or immediately impinging on the peatland represented by the four interlocking circles. Both are influenced by, and change over, time, which surrounds the entire model. The peatland is situated in the centre of the model as a product of the combined influence of all these factors. Modified from Warner (1996).

attempted to show the interaction of these with time and with the processes which affect the growing peatland. The basis for peatland formation is that organic productivity must exceed decay. Both productivity and decay are initially determined by factors external to the peatland system and it is these factors that enable or prevent the initiation of peat. The climate must be wet enough to provide adequate water for plant growth and to provide sufficient waterlogging for at least part of the year so that decay of dead material is inhibited. The temperature must also be adequate for plant growth but also low enough so that evaporation is limited and a waterlogged substrate is maintained. The underlying geology must be relatively impermeable to ensure sufficient water retention on the land surface and it will also determine the nutrient status of the groundwater. The geology and geomorphology together also determine the topography, which must be suitable for the collection of water or at least flat enough to restrict runoff. Exceptions exist in hyperoceanic regions where precipitation is in constant supply so that peatlands can exist on steep-sloping topography or at high altitudes where clouds and condensation may also support peat-forming vegetation and peatland growth. Human activities may also have a role to play in some situations, for example in clearing forest and altering the hydrological balance (e.g. Moore, 1993) and natural fire can often play a role in early phases of peatland growth by allowing water to collect on the surface. An additional differentiating factor that is important on a global scale is the biogeography and evolutionary history of a region, as it determines which plants are present to begin the peat formation process. Since plants differ both in productivity and decay rates, this can be an important determinant of peat landform development. Of course it is ultimately the interaction of these factors that make any particular location suitable or unsuitable for peat formation. For example, peat can develop to great depths in the tropics (Anderson, 1983), where temperatures are warm so decay rates are high, yet more than compensated for by very high rainfall and very high plant productivity. Likewise peat may develop in rather warm, dry climates if geology and geomorphology yield an impermeable basin for water to collect in. At the opposite extreme, peatlands may also form in desert and arid grasslands where groundwater discharge supplies water necessary to support peatland systems.

The external factors determine the nature of the processes that take place on the embryonic peatland. The interlocking circles on Figure 2.1 represent these processes. As has been hinted at, hydrological processes are critical in peatland functioning and ecology but the geochemistry is also crucial. The concepts of fen and bog (Chapter 1) derive from a consideration of these processes, and the primary influence on peat landform development is through the amount of nutrients and minerals that are supplied to the peatland surface. Physical processes become increasingly important as the peat landform evolves and begins to develop its influence on other processes such as the rate of surface runoff. Finally, biological processes of productivity and decay are central to peat formation and accumulation. Initially all these processes will be determined by the external factors, but once the peat landform begins to expand then a microcosm of internal factors becomes established which can ultimately completely control some aspects of the peat landform. These 'autogenic' factors and their relative influence on landform development are discussed in some detail later (Chapter 7).

Both the factors external to the peatland and the processes on the peatland change over time, which encircles the whole of Figure 2.1. Some influences and processes will change gradually and others may alter in a stepwise fashion, while there may be long periods of constancy in others, but time is fundamental to many of the ideas and arguments presented in this book. These issues are the framework for peat landform development and functioning and need to be borne in mind whenever peatland ecology and management are under consideration.

2.3 DESCRIPTION OF PEAT LANDFORMS

A systematic description of landforms is crucial to the understanding of fundamental processes in peatlands. The classifications based on hydromorphological criteria presented in Chapter 1 depend heavily on a consideration of size and shape, and the basis for these ideas is enshrined in the Russian school of peatland science most closely associated with, and accessible through, the work of K.E. Ivanov. The book *Water Movement in Mirelands* (Ivanov, 1981), made available in English by Thompson and Ingram, summarises the approaches to peatland hydrology in post-war Russia and is perhaps best known for its influential presentation of broad concepts of peat landform units. The ideas presented by Ivanov are based on earlier work (principally that of E.A. Galkina) and emphasise the importance of scale in consideration of peat landforms through the use of the terms microtope, mesotope and macrotope, explained below.

2.3.1 Microtope

This is 'part of the mire where the plant cover and all other physical components of the environment connected with it are uniform' (Ivanov, 1981, p. 16). This is an area of peatland that has no clear major boundaries within it, and is also referred to as a mire facies (borrowed from the geological term). In practice, and in most of the examples given by Ivanov, this is defined principally by vegetation, so that typical microtopes in Russian peatlands are birch with black alder, birch–sedge–reed and sedge–*Sphagnum* (of homogenous structure). These microtopes refer to areas of uniform, homogenous vegetation over moderately large areas of perhaps tens to hundreds of metres across or even larger. However, many peatlands contain smaller features such as the pools, hollows, ridges and hummocks referred to in Chapter 1. As a result, within the definition of microtopes, a 'complex', or vegetation that is heterogeneous on a small scale, is possible. An example from Ivanov is a complex consisting of *Eriophorum* (cotton grass) and *Sphagnum* mosses on ridges with *Scheuchzeria palustris* (Rannoch rush) in adjacent pools. Each individual pool or ridge may be no more than 1–2 m across and a continuous area of such patterning would comprise the microtope. The individual pools or hummocks are also sometimes referred to as 'microforms'.

2.3.2 Mesotope

These are defined as 'isolated mire massifs formed from one original centre, and possessing at each stage of their development a pattern of microtope distribution that conforms to clearly defined principles' (Ivanov, 1981, p. 24). Although the translated description is a little clumsy, this is perhaps the most useful unit of classification for survey and classification, and the idea is the

underlying basis on which many hydromorphological classifications and descriptions are founded. In the simplest examples these are areas which are separate peatlands such as raised mires or basin mires which are surrounded by mineral ground, having developed from a single focus of peat initiation. However, they can also be parts of much larger peatlands, in which case their boundaries are not so clearly defined. The individual elements of a blanket mire (watershed, spur, saddle mires) described in Chapter 1 are good examples of this. Figure 2.2 shows how an area of blanket mire can be subdivided into separate mesotopes on the basis of topographic position and related surface patterns. It should be noted that within most mesotopes such as the watershed mire, there are at least two microtopes. In this case there is a pool and hummock complex in the centre surrounded by a margin where no patterning exists. With no data on vegetation it is not possible to say whether there should be further divisions into separate microtopes (for example, the margin mesotope may be divided into a *Calluna vulgaris* dominated area on steeper ground and a

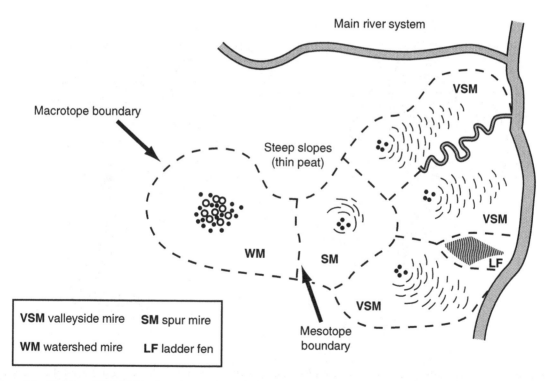

Figure 2.2 Division of Scottish blanket mire macrotope into separate mesotopes on the basis of topography. The shaded areas represent pools and the macrotope boundary is placed at the edge of the peatland extent. Redrawn from Lindsay (1995) by permission of Scottish Natural Heritage.

Sphagnum–Eriophorum area on flatter ground. Other mesotopes (for example the ladder fen shown in Figure 2.2) may comprise only a single mesotope if there is no internal variation in vegetation types.

2.3.3 Macrotope

This is 'a geotope that has been formed by the fusion of isolated mire mesotopes' (Ivanov, 1981, p. 26). Macrotopes are complexes of a number of mesotopes joined together. They occur principally in areas where peat growth is almost ubiquitous due to high rainfall, low temperatures and flat or undulating ground. Northern Eurasia and North America are the principal regions for the development of large macrotopes, but they also exist in hyperoceanic areas in the Southern Hemisphere. Figure 2.3 shows an example of a macrotope consisting of two mesotopes which are coalescing and have no dividing mineral ground between them. In this example, Ivanov also gives detailed boundaries for the principal microtopes. This figure also emphasises the close relationship between landform and hydrol-

ogy, which was one of the principal reasons for the development of the system. The shape of the peatland surface is one of the principal determinants of the direction and rate of water flow. The distinction between separate mesotopes within a macrotope is not always easy once an advanced stage of development is reached, as the outlines of the original fused mesotopes may not be clear. Ultimately, as acknowledged by Ivanov, it is really only possible to fully resolve this by stratigraphic study using techniques described in Chapter 6, although the kinds of divisions employed in surveys of Scottish blanket mires provide a reasonable working methodology where this is not possible (Lindsay *et al*, 1988).

2.4 LANDFORM SURVEY TECHNIQUES

The majority of peat landforms are difficult to perceive accurately from ground level. This is mainly due to their size and the fact that most peatlands occur on relatively flat ground, but even within small peatland

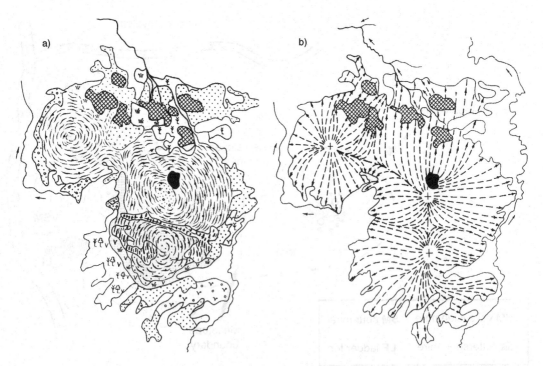

Figure 2.3 An extensive mire macrotope composed of three raised mire mesotopes fused together, with their individual component microtopes identified (a). The concentric patterning of pools and ridges within each of the raised mire centres shows that these are domes of peat. As a result of the topography, the main direction of water flow can be inferred and drawn as a flow net (b). Redrawn from Ivanov (1981). See text for further discussion.

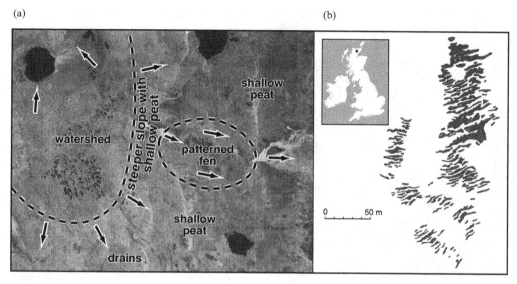

Figure 2.4   (a) Aerial photograph of a blanket mire landscape in the Flow Country, northern Scotland, near Strath Halladale. The photograph covers an area of 1.2 km across. The main mesotopes are outlined and arrows indicate main direction of water flow. Air photograph courtesy of the University of Cambridge. (b) Map of main microforms based on an enlargement of the patterned fen area highlighted in (a), redrawn from Charman (1998).

areas, views can be obscured by vegetation, especially in forested peatlands where plant cover is particularly dense. As a result, comprehensive views of large peat landforms can only be gained from the air. In many cases, the development of aerial photography and survey is strongly linked to the discovery and description of large areas of peatlands, particularly in the post-Second World War era. The nature of the landforms of the peatlands in Minnesota, USA, was only discovered in the late 1940s with the first set of comprehensive air photographs (Glaser, 1992a), about the same time as aerial photography was being used to develop peatland survey in Russia (Ivanov, 1981). Aerial photographs are still one of the key sources of information in surveys of the nature and extent of peatlands (e.g. Parkyn and Stoneman, 1997; Reid *et al*, 1997) and in the monitoring of change (Grünig, 1997), but there are now also other kinds of photography and images which provide additional data from airborne and satellite-based equipment (e.g. Stove and Hulme, 1980; Stove, 1983; Fan, 1987; Glaser, 1989; Gorozhankina, 1993; Markon and Derksen, 1994; Reid *et al*, 1994, 1997).

2.4.1 Aerial photography

Aerial photographs remain one of the best techniques for mapping peat landforms, and are often used simply

to describe main landform features on which to base further ground-level studies. Figure 2.4 shows an aerial photograph of a blanket peatland area in the Flow Country, northern Scotland. Part of the photograph was enlarged and used to plot a detailed map of the main microforms on the surface of the patterned fen highlighted on the photograph. Some of the most extensive work involving aerial photography and satellite imagery has been carried out in Minnesota (see Glaser, 1992a for a review) and other areas of North America (e.g. Glaser and Janssens, 1986; Glaser, 1987a). Almendinger *et al* (1986) and Glaser *et al* (1990) used aerial photography to delineate landforms in the Lost River peatland and based hydrological and stratigraphic studies on this to determine the influence of groundwater on peatland development and ecology. Glaser (1987a) used aerial photographs to identify streamlined bog islands in continental bogs in North America (e.g. Figure 2.5). Bog islands occur within many of the large peatland areas in North America as complexes of teardrop-shaped ombrogenous islands separated by a network of poor fen water tracks. Aerial photographs were used as a basis for a morphometric analysis of this particular peat landform. The outlines of 40 individual bog islands were traced and the maximum lengths, widths and areas calculated. Strong relationships between these dimensions were found and an analysis of shape showed that the

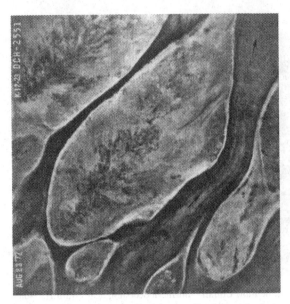

Figure 2.5 Aerial photograph of streamlined bog islands in the Red Lake peatland, northern Minnesota, USA. The dark area in the centre of the island is the forested crest of the bog, and the dark areas surrounding the island are the fen water tracks which flow from bottom left to top right in the photograph. Reproduced from Glaser (1987a) by permission of the author and the Regents of the University of Colorado.

island outlines corresponded to a streamlined airfoil-type shape that minimises resistance to a flowing liquid. From these spatial characteristics, Glaser inferred two modes of development for these landforms. Firstly, they could develop behind obstructions in catchments with irregular relief, where slow flows of water would be induced. Secondly, they could develop in large bog areas by fragmentation due to water track development (Figure 2.6). This study shows how aerial photographs can be used for more than simply the subjective description of peat landforms, including the derivation and analysis of data that would be otherwise impossible to collect.

Aerial photographs taken on colour infrared film can be even more informative, principally because the image is temperature sensitive as well as using the visible wavelengths. Glaser (1992a) demonstrates this most clearly in photographs of peatlands in Minnesota. In these dramatic images, more densely forested areas are picked out in a deep purplish hue with more lightly wooded *Sphagnum* raised bogs showing up in a brighter red. Open ground with no trees is paler and zones where water is concentrated into tracks are a contrasting bluish green. Surface textures in these

images also help characterise the peatlands, especially in the water tracks where the faintly striated surface is very suggestive of the water movement that takes place there. The insights these images give into large-scale peat landforms have been used to help develop various theories of ecological development of the peatlands (e.g. Glaser, 1992b).

2.4.2 Remote sensing

While aerial photographs have been used successfully and extensively for peatland and landform mapping and interpretation, they have a number of limitations. Firstly, the spatial coverage is limited, usually to areas that have been photographed for other purposes. Aerial photography is also expensive and is not often carried out purely for peatland survey and research. While national coverage exists for some countries, it may be fragmentary in other regions, especially more remote areas where peatlands often occur! Secondly, it is rare for any particular area to be photographed regularly. Examination of changes over time (e.g. Aaviskoo, 1993) requires a series of photographs from the same region at different, preferably regularly spaced, intervals. Decadal changes may be documented in this way but annual or seasonal change is impossible to monitor. Thirdly, coverage of very large areas depends on examination and amalgamation of a large number of images, which may be difficult to perform accurately. Finally, aerial photographs require careful manual interpretation, a process that depends considerably on experience, time and ground knowledge of the terrain to be explored. Partly as a result of these limitations, a number of studies have attempted to use satellite data to study large-scale peat landforms. Satellite data also have the advantage of being able to use reflected radiation in the non-visible wavelengths so they provide additional information on landscape changes. They are more sensitive to some changes in environment such as surface wetness, which has a strong effect on reflected infrared radiation.

The images are derived from sensors on board satellites orbiting above the earth, which systematically scan the entire surface at regular intervals. The exact interval depends upon satellite height and area scanned, but repeat scans of any area are typically every two to three weeks. Digital data are in the form of reflected radiation at particular wavelengths and the area scanned is divided into a series of small squares or pixels. While satellite data have a number of advantages over aerial photography, their main deficiency is

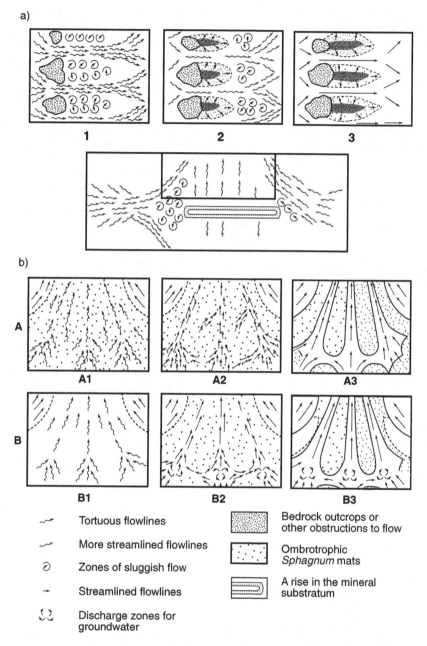

Figure 2.6 Suggested models for development of streamlined bog islands (Glaser, 1987b). (a) Primary bog island development behind obstructions to surface water flow. The relatively stagnant conditions behind the obstruction (1) reduce alkalinity and allow *Sphagnum* to colonise and expand to form a bog community (2), its lateral expansion being limited by the zones of more rapid flow between obstructions (3). (b) Development of secondary bog islands in a large bog with an underlying mineral ridge. In this case, the existing uniform bog area is dissected due to the development of flow lines from either surface flow (A) or groundwater discharge (B). In both cases, increased flow of water is concentrated into water tracks where more enriched conditions occur due to the increase in the amount of water, supplemented in B by groundwater input. The water tracks develop sedge vegetation and subsequent channelling of water in the tracks sustains and develops the new landforms. Redrawn from Glaser (1987b) by permission of the Regents of the University of Colorado.

resolution, or the size of each pixel. Typical satellite data which have been used for peatland surveys are based on Landsat thematic mapper (TM) or SPOT imagery with pixel sizes of 30 and 10–20 m respectively, depending on the particular sensors and wavelengths used (Lillesand and Kiefer, 2000). This clearly limits the usefulness of the images at small scales as individual microforms and even microtopes may be smaller than this. However, satellite data have been used with success for detecting large-scale features and for understanding processes operating at this scale.

2.4.2.1 Landscape mapping

Mapping of tundra land cover has been carried out by Markon and Derksen (1994) using SPOT imagery. Peatlands were included within this survey which was carried out to assess the distribution abundance of waterfowl habitats. Other studies using satellite data for mapping include Gorozhankina (1993), Varjo (1996), Cruickshank and Tomlinson (1996), Matthews (1991) and Stove and Hulme (1980). Landscape mapping from satellite data has the advantage of being comparable between studies carried out by different researchers at different times. The CORINE land-cover methodology was devised by the European Commission to standardise description from satellite data between European countries. Peatlands are one of the categories within this and have been mapped in Northern Ireland by Cruickshank and Tomlinson (1996). This study highlights the problems of resolving small-scale features such as damage from hand cutting of peatlands and the possible confusion with other vegetation types at certain times of the year.

2.4.2.2 Landscape processes

Studies using satellite imagery can go well beyond basic landform and land-cover description. Glaser (1989, 1992c) has demonstrated this particularly well in studies of the major peat basins in North America and especially the large peatland landscapes in Minnesota, USA. Large areas of 50–100 km across are well suited to even the relatively coarse pixel size of Landsat TM and multispectral scanning (MSS) images. These studies show the value of images taken at different times of the year. Images taken in spring clearly show the areas of active water movement as darker areas within a generally highly reflective snow-covered landscape. The mineral ground and raised bogs are frozen and snow-covered, but water tracks are open and flow-

ing with warmer water that can only have come from upwelling groundwater. Flow patterns can also be detected and discerned in some detail at other times of the year. Areas of peatlands that appear nearly uniform on standard aerial photographs are clearly highlighted by different coloration and by banded vegetation aligned with the flow lines. Major peat landforms are picked out in colourful detail by the classification of the imagery and the major components of the vegetation can also be established over very large areas. For example, lichen and *Sphagnum*-dominated peatlands are clearly separated (Glaser, 1989). These studies are just some of the evidence for groundwater control of peat landforms and ecology in the Minnesota peatlands, discussed in greater detail in Chapters 7 and 8.

2.4.2.3 Hydrological characteristics

Other work has attempted to use remotely sensed images to determine hydrological status of peatland surfaces. Vogelmann and Moss (1993) suggest that this may be easier in *Sphagnum*-dominated vegetation than it has been in vegetation dominated by vascular plants, where the variability in moisture content is too small to be reliably determined. However, Gilvear and Watson (1995) have shown that depth to water table can be mapped in some detail using airborne thematic mapper (ATM) images. These images have the advantage of being taken at low altitude by aircraft and are therefore of high resolution with a pixel size of 2 m. The data can be transformed to water table depth values using equations developed from ground truth data of actual water table depths in selected locations. However, there are still considerable problems in the use of satellite-based remotely sensed data for more sophisticated interpretation of peatland landscapes, including the more limited range of bandwidths carried by Landsat and SPOT and the larger pixel size of these platforms.

2.5 PEAT LANDFORM SURVEY: AN EXAMPLE FROM SCOTLAND

The use of landform terminology and survey techniques is perhaps best illustrated using a detailed example. The blanket peatlands of Scotland extend to approximately 1 million ha (Lindsay, 1995) and are of high conservation value as a threatened habitat within Europe. They are also under threat of change from changing land-use, including afforestation. As a result a large amount of effort has been spent in attempting to fully describe their character and extent. In the 1980s a major survey was undertaken of the peatlands in Sutherland and Caithness, sometimes known as 'The Flow Country', using aerial photograph and field survey techniques (Lindsay *et al*, 1988). The only available aerial photograph survey for the entire region dated from the late 1940s and early 1950s and these were used as the basis for the identification of sites requiring field visits. Sites that had been damaged extensively by forestry, agriculture or major erosion were not visited, but as far as possible all sites with some remaining intact peatland were visited. Field survey was based on identifying major landforms according to Ivanov's (1981) terminology with mesotopes forming the main unit for survey. Typical blanket mire mesotopes were identified, and each site visited was assessed for site damage and vegetation. Also described were the main microtopes and the nature of the microforms within each of these. Macrotope areas were subsequently defined by amalgamating adjacent

mesotopes and using major hydrological boundaries such as lakes and rivers or the edges of deep peat areas as boundaries. This hierarchy of survey scales made the description of a complex peatland landscape relatively straightforward, although some subjective decisions had to be made on site boundaries. An analysis of vegetation and microform characteristics was used to characterise a series of mesotope types (Table 2.1) and these were used as a basis for mapping and conservation assessment of individual areas. Figure 2.7 shows the extent of the peatland macrotopes described and identifies locations where a particular mesotope can be found. A typical watershed blanket mire mesotope is shown in Figure 2.8.

Other survey work on Scottish peatlands has followed this detailed work on one particular area, but it has not been possible to replicate the full field survey techniques employed over the whole of Scotland's 1 million ha of blanket mire. Instead a methodology based on satellite imagery has been developed as air photo interpretation cannot give any detailed

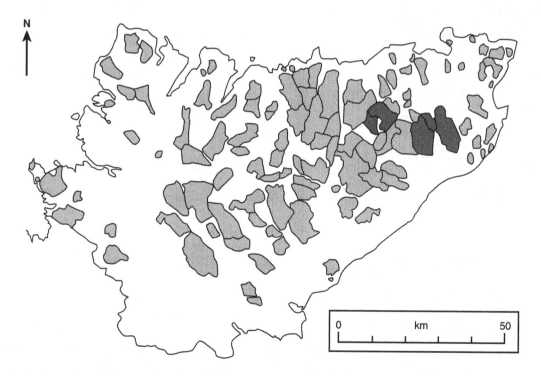

Figure 2.7   Map of Sutherland and Caithness, Scotland, showing the outlines of the major peatland macrotopes in the 'Flow Country'. The shaded macrotopes all contain one or more 'Eastern watershed blanket bog' mesotopes (Type 9). This particular mesotope is characterised by extensive deep round pools with hummock–hollow complexes within the surrounding peatland expanse. See Table 2.1 and text for further details. An example of this mesotope is also shown in Figure 2.8. Redrawn from Lindsay *et al* (1988) by permission of Scottish Natural Heritage.

Table 2.1 Synoptic table showing the constancy of different microforms and vegetation types for mesotope types in the peatlands of Sutherland and Caithness. Type 9 is the eastern watershed blanket bog for which the distribution is shown in Figure 2.8. The constancy gives an indication of the percentage of sites in which each microform or vegetation type is present. For clarity, only vegetation types with >40% constancy are shown here. Based on Lindsay et al (1988, Table 4). Constancy values, V = 81–100%, IV = 61–80%, III = 41–60%, II = 21–40%, I = 1–20%. Constancy figures in bold indicate microforms are particularly abundant (as well as constant) within a site type. The three watershed blanket bog site types (8, 9, 10) are distinguished by having very high constancy of deep drought-sensitive (A3) and permanent (A4) pools. The terminology for microforms (A and T zones) was developed by Lindsay et al (1985) in an attempt to formalise the description of microtopographic features on mires.

Site type (mesotope)

	1	2	3	4	5	6	7	8	9	10	11	12	13	14	15
Microform															
T5 – Peat mounds						I	I			I					I
T4 – Erosion hags		I	I	I	II	I	I	I		II	III		V	IV	V
T3 – Hummocks	I	V	IV	V	V	V	V	V	V	V	V		V	IV	V
T2 – High ridge	IV	V	V	V	V	V	V	V	IV	V	V	III	V	V	V
T1 – Low ridge	V	V	V	V	V	V	V	V	II	V	V	V	IV	V	IV
A1 – *Sphagnum* hollow	I	IV	V	V	V	V	V	V	V	V	III	III	III	III	IV
A2 – Mud-bottom hollow	V	V	V	V	IV	V	V	V	III	IV	V		V	V	V
TA2 – Erosion channel		II			I	I		I		II	V		V	IV	IV
A3 – Drought-sensitive pools	III	III	III	III	IV	III		V	V	V	III		III	II	II
A4 – Permanent pools	I	I	I	I	II			V	V	V	II				I
Plant community type															
Empetrum–Hylocomnium splendens mounds/hummocks									III						
Racomitrium–Cladonia hummocks and hags												III	III	III	III
Racomitrium–Molinia hummocks and hags								III							
Racomitrium–Pleurozia wet low ridge		III	IV					III							
Sphagnum fuscum hummocks								III	III						
Sphagnum imbricatum hummocks		III		IV	IV	III	III	IV	III	IV	III	III			
Sphagnum rubellum–Odontoschisma dry ridges					III										
Mixed *Sphagna* (*S. rubellum* hummocks)			III				III	III		III	III				
Sphagnum papillosum–Molinia ridge					IV	III									
Sphagnum–Eriophorum vaginatum ridge					IV			III	III		IV				
Sphagnum compactum ridge											III				
Sphagnum magellanicum–S. subnitens	III														
Sphagnum magellanicum–S. rubellum ridge							III	IV	III		IV				
Sphagnum–Arctostaphyllos–Betula nana mire								III		V					
Rhyncopspora alba–Sphagnum low ridge				III											
Carex panicea damaged mire														III	
Microbroken mire												V			III
Pure *Sphagnum cuspidatum* carpets						III			III						
Typical *Sphagnum cuspidatum* carpets					IV	IV	III	IV	IV		IV		III		
Sphagnum cuspidatum–Carex limosa				IV											
Eleocharis multicaulis mud-bottom hollows				III											
Rhynchospora alba mud-bottom hollows				III											
Sphagnum auriculatum bog pools							IV	IV	IV	III		III			
Deep pools					I		III	IV	III	IV	V	IV	III	III	
Molinia–Myrica ridges (ladder fens)	IV	III													
Carex rostrata–C. lasiocarpa (ladder fen 'flarks')	V	III													

information on vegetation changes within most blanket peat areas (Reid *et al*, 1997). Briefly the method involves preliminary identification of areas with peat soils based on a digitised version of drift maps from surveys undertaken by the British Geological Survey at 1 : 50 000 scale. Landsat TM data are then extracted to match these particular areas and analysed in a series of appropriate bandwidths. A classification of peatland areas is then generated from these data where each pixel is assigned to a particular class. Although some

Figure 2.8   An extensive area of watershed blanket bog in the Flow Country, northern Scotland. The large, round, deep pools are typical of these sites, which are divided into three separate types (8, 9 and 10; see Table 2.1) by Lindsay *et al* (1988) by permission of Scottish Natural Heritage.

manipulation of the classification, based on ecological knowledge of the areas, is used to adjust this initial classification, the actual meaning of the classes has to be established by field survey of a sample area of each class. This allows the image classification to be refined and then by combining the satellite classes with the ground truth data, maps of particular vegetation types according to standard terminologies (Rodwell, 1991) and complexes can be produced with reasonable confidence. While these techniques will never replace field survey completely, the organisation responsible for landscape conservation, Scottish Natural Heritage, is able to identify the main areas of potential conservation value and use these to re-examine habitat designations and to target ground survey more efficiently (Reid *et al*, 1997).

2.6 HYDROLOGY AND PEAT LANDFORMS: THE GROUNDWATER MOUND HYPOTHESIS

Ivanov's terminology for peat landforms is intimately connected with the understanding of hydrology and particularly with the determination of surface water movement through the establishment of 'flow-nets' indicating the direction of water flow as a series of lines over the mapped area. The work in the patterned

peatlands of Minnesota (Glaser 1992a,b) already discussed has demonstrated the important role groundwater can play in controlling the nature and distribution of peat landforms. Indeed hydrology drives many of the processes of peatland ecology and development as we shall see throughout the remaining chapters. While water and water flows are crucial on very large scale suited to study by remotely sensed images, they are also crucial at smaller scales. One particular demonstration of this is what has become known as the 'groundwater mound hypothesis' (Ingram, 1982, 1983, 1987). This hypothesis seeks to explain the size and shape of raised mires and it has subsequently been used to inform management and restoration of raised mires damaged by peat extraction (Ingram, 1992; see Chapter 10). Although we have already described raised mires as domed peat deposits in Chapter 1, when the facts of raised mire morphology and structure are plainly stated, the issue of how and why they can exist at all becomes an obvious problem. Ingram (1992, p. 85) states this most clearly:

'We are accustomed to think of a raised mire as an accumulation of peat whose centre may be several metres higher than its margin; and the concept of natural dung-heap raises no conceptual problem. But when we realise that water accounts for nearly all the volume of this heap it becomes clear that we are actually dealing with a kind of lake: a lake whose surface turns downward toward the shore and which continually drains away in that direction, getting rid of its surplus water through lagg streams. Regarded in this way the raised mire becomes a vastly more interesting and puzzling phenomenon.'

Until relatively recently it was thought that capillary action was responsible for holding the water table above the surrounding land surface and that the growing peat surface draws up the water table through time (Gosselink and Turner, 1978), but that process is now thought to be unimportant as it does not explain the dome shape and implies unrealistically small pore sizes to exert sufficient capillary rise (Ingram, 1982). Instead, it is now understood that the domed profile is primarily a result of the low hydraulic conductivity of the main mass of peat ('catotelm' – see Chapter 3) underneath the looser surface vegetation and incompletely decayed peat ('acrotelm'). The details of the hydrological equations and the Dupuit–Forchheimer theory used to derive the model are described in full elsewhere (Ingram, 1982, 1983, 1992), but the model basically states that the size and shape of the raised mire will depend on the recharge and hydraulic conductivity of the catotelm. Ingram was able to demonstrate that a real raised bog, Dun Moss in Perthshire, Scotland, actually had approximately the same profile as the model predicted, despite some likely errors in the estimation of the main parameters and in some of the assumptions of uniform topography being violated (Figure 2.9). Similar results have been obtained from Ellergower Moss, Wigtownshire, Scotland, with a remarkably good fit between actual shape of the mire surface and the half-ellipse predicted from the formula (Ingram, 1987).

The groundwater mound model for predicting the form of raised mires makes a series of assumptions that may not always be applicable in real mires. As discussed above, some raised mires do actually fit the model very well despite these assumptions, but there are ways of accommodating more complex variations in hydraulic conductivity (rate of water flow through the peat) and topographical influence (Armstrong, 1995; Bromley and Robinson, 1995). Armstrong (1995) has suggested a method for incorporating variable hydraulic conductivity with depth. This is one attempt to make the models more realistic and results in predictions of higher mire centres with less pronounced steep edges to the peat mound. However, this method is based on a decreasing conductivity from the surface to the base of the peat. While it may in general be true that conductivity at the surface is higher than at depth, it may not always be the case that this is unidirectional throughout the profile and there may be layers of high or low conductivity within the deeper peat. Armstrong's assumption is that con-

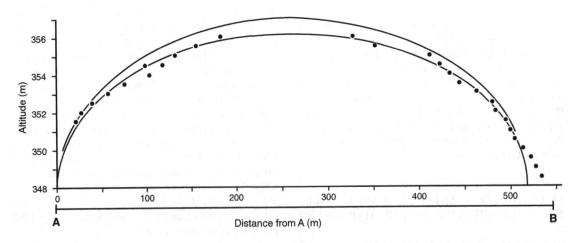

Figure 2.9 Profiles of groundwater mound predicted by the groundwater mound hypothesis and actual altitudes of surface water on Dun Moss, Perthshire. The thick line represents the predicted ellipse based on a value of 9.95×10^{-3} for U/K where U is net recharge and K is hydraulic conductivity. This is derived from morphological data from the mire and the model. The upper line is a predicted ellipse based on an estimate for the driest year for which full hydrological data exist and $U/K = 1.2 \times 10^{-3}$. The difference between measured and estimated U/K may be attributable to the difficulty of accurately measuring U and K and to topographic variability from the assumed uniform flat surface. Redrawn from Ingram (1982) by permission of *Nature*.

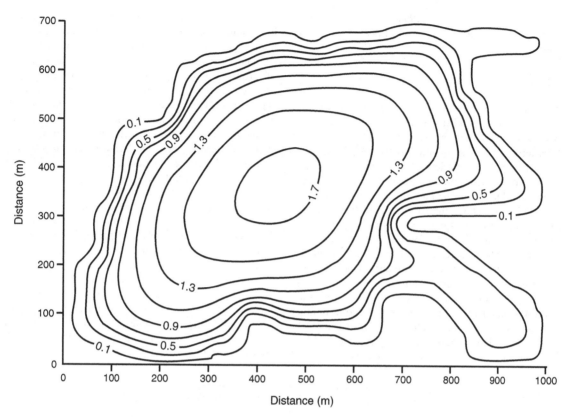

Figure 2.10 Numerical water table predictions for water table contours (in m AOD) for an irregularly shaped raised mire. In a real peatland, these values would correspond to the water levels in pools as measured by Ingram (1982) and shown in Figure 2.9. The actual surface of the peatland would be similar to this but would be modified by microtopographic variability. Redrawn from Bromley and Robinson (1995) with permission. © John Wiley & Sons Limited.

ductivity shows an exponential decline with depth but this has been challenged by Baird and Gaffney (1996), who also point out that methane gas bubbles may block pores and reduce hydraulic conductivity (Baird and Gaffney, 1995). Both Baird and Gaffney (1996) and Armstrong (1996) agree that more field data are necessary to really test the models based on variable hydraulic conductivity. Meanwhile the existing field data (Ingram, 1982, 1987) still seem to fit the original model for these particular sites!

Another study which attempts to provide a more realistic model for raised mire mound forms is discussed by Bromley and Robinson (1995). The application of a numerical groundwater model (MODFLOW) to model peat landforms perhaps has greatest future potential as it operates in time steps and divides the region of interest (in this case the peat body) into a series of subregions which can have different values

for conductivity. Such models can also incorporate mires of irregular form, which is one of the principal limiting assumptions of Ingram's (1982) original model. This kind of approach has principally been used to demonstrate the prediction of water table change in mires of known form (Figure 2.10) but also to assess the impact of peat cutting on water table depths (Bromley and Robinson, 1995). Clearly these more sophisticated models will ultimately be a way of modelling peatland groundwater forms with sufficient reliability to be able to predict the impact of changes in peatland form on water tables. They also provide a more flexible tool which is likely to be applicable to a greater range of peatland types than have hitherto been considered in this way, including ombrogenous mires in the tropics (Bragg, 1997). Applications of models for peat mound shape based on Ingram's (1982) ideas and incorporating models of decay have been presented by Winston (1994, 1996) and shown to reasonably represent actual tropical peat landforms in Sarawak.

2.7 SUMMARY

This chapter has emphasised the importance of a consideration of peatlands as landforms, which need to be considered in their entirety, in order to fully understand long-term processes and functioning. The landforms are a product of a variety of external factors which unite to produce unique conditions for each individual site, and which form the basis for the large variety of peatlands throughout the world. These factors determine the peatland-forming processes that are then modified and altered by the formation of the peat landform itself. The hierarchical system of peat landforms from the Russian school of peatland science (Ivanov, 1981) seems to be the most universally applicable and useful system of description. It has been successfully employed implicitly and explicitly in a large number of peatland surveys and studies. Remote-sensing techniques including aerial photography and satellite imagery are most appropriate for the description and study of large-scale peat landforms and they are also increasingly used for the study of developmental and hydrological processes. More detailed prediction of peat landform morphology is possible using models of groundwater shape which have now developed beyond the idealised models initially presented by Ingram (1982). The use of these predictive models will be discussed further in consideration of impacts of peatland exploitation and techniques for restoration later in Chapters 10 and 11. The discussion has mainly been limited to larger peatland areas and especially the extensive northern nutrient-poor mire systems, because the landforms are easier to observe at these scales. However, many of the concepts can be applied at much smaller scales as well. Many more nutrient-rich fen systems are less dependent for their morphology and landform development than bog or poor fens, as they are defined more closely by the topography of the underlying mineral ground. Despite this influence, peat growth plays some role in most of these systems. The concept of peatlands as landforms is important when considering the various aspects of processes and functioning of all peat-forming ecosystems, issues which are the subject of the next major section of this book.

PART 2

PEATLAND PROCESSES

Viewing the whole peatland as a landform is a crucial part of understanding the processes and functioning of the system. However, the overall development of the system is dependent upon a huge range of processes operating at a smaller scale. The next part of the book considers some of the fundamental processes operating in peatlands that determine how the peatland functions and responds to change. Understanding of these processes is critical in studying the long- and short-term changes discussed in Part 3 and in managing the peatland resource, as explored in Part 4. The first and more lengthy chapter here (Chapter 3) examines the fundamental aspects of hydrology and ecology of peatlands. Chapter 4 tackles the question of how and why peat begins to form in the first place, whether as a result of wetland development through basin infill or by forming directly over previously dry mineral soil. Finally, Chapter 5 looks at the way in which the continued accumulation of peat is ensured through processes of production and decay. More emphasis and space are given to these last two chapters as it is the formation of peat that defines the very existence of peatlands. Other wetlands have many common ecological and hydrological characteristics, but peat formation is a process unique to peatlands.

PART 2

PEATLAND PROCESSES

3

Peatland Hydrology and Ecology

3.1 INTRODUCTION

'The most characteristic feature of wetlands is that they are wet' (Wheeler, 1999, p. 127). This is not so much stating the obvious as emphasising a key factor in wetland ecology and development and it could equally be applied to peatlands in their natural state. It will already have become apparent that hydrology is one of the central factors determining the type of peatland which can develop, but the very existence of any peatland ultimately depends on there being an excess of water which prevents complete decay of the biomass produced by vegetation growth. It is not easy to appreciate the sheer quantity of water held in a peatland, even when standing on the surface. A solid or only slightly soft surface conceals the fact that in many cases peatlands consist of well over 95% water by weight – some peatland ecologists have commented that this means that peat has less solids in it than milk! In effect, when standing on a peatland surface, one is supported by a giant bubble of water, held together by a mass of living and dead plant material. Given the importance of water to the existence and functioning of peatlands, this section of the book begins with an examination of peatland hydrology. The main aim is to explain the water budget and the ways in which water affects peatland functioning, as a basis for further discussion in later chapters. The second part of the chapter aims to explain principal aspects of peatland ecology. Since plants and animals that live on peatlands are strongly affected by the quality, quantity and distribution of water, this is a logical progression from an understanding of basic peatland hydrology.

3.2 HYDROLOGY AND WATER BALANCE

The hydrology of peatlands has become a large area of study over the past few decades and we will not provide a comprehensive coverage of the details here. Several

excellent reviews of the principles already exist. Ingram (1983) is still the most detailed and comprehensive source of information and incorporates many of the important ideas generated in the Russian literature, particularly those of Romanov (1968) and Ivanov (1981). Eggelsmann et al (1993) is also a good summary and Hughes and Heathwaite (1995b) provide a collection of hydrologically related papers from Britain. Price and Waddington (2000) is a review of recent advances in wetland hydrology from Canada. Here we are concerned with the fundamental hydrological relationships on peatlands and the way in which these affect peatland functioning overall. Many aspects of peatland hydrology are discussed in relation to these other aspects of peatland development and change elsewhere in the book.

3.2.1 Water balance

In its simplest form, the water balance in a peatland consists of influx, efflux and storage of water, and the changes in these three components must balance (Eggelsmann et al, 1993), so that

$$\text{Influx} - \text{Efflux} - \text{Changes in storage} = 0$$

Influx of water is sometimes described as *recharge* to the peatland and efflux as *discharge* from the peatland. Within the overall balance, there is a large amount of variability between specific mire types and individual sites. A slightly more complex representation of water balance is given in Figure 3.1, which shows the main routes by which water enters, flows through and leaves the peatland system. Principal inputs to the mire are from precipitation, surface runoff from surrounding slopes or from groundwater upwelling, such as springs. Outflows from the system are from runoff, seepage to groundwater or from interception and evapotranspiration. Not all of the processes will operate in all peatlands. For example, in the case of simple ombrotrophic mires, the only input is precipitation. In contrast, for

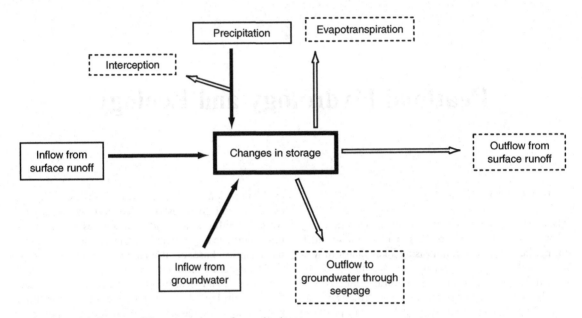

Figure 3.1   Main features of the water balance of a peatland system.

many fens direct precipitation may be much less important than inflows from surface runoff or groundwater.

Many texts and specific studies refer to the idea of an overall water balance for peatlands, but there are relatively few undamaged sites where a total budget has been measured. For an ombrotrophic bog, Hemond (1980) is an excellent example of the estimation of a total hydrological budget and the resultant geochemical flows associated with this. Gilvear *et al* (1992, 1993) calculate overall water balance for fens in East Anglia, England, and Richardson (1983) provides an example for pocosins in the USA. In a study of Thoreau's Bog, Massachusetts, USA, Hemond (1980) determined the direction of flow of groundwater using piezometers installed within and outside of the peatland area. He found that water flow was consistently out from the peatland, demonstrating that water supply to the bog is solely from precipitation. Precipitation was measured as 1448 mm in the 15-month study period with 1019 mm lost as evapotranspiration and only 246 mm as surface and subsurface runoff, with the remaining 183 mm as positive net storage (Table 3.1). In this case, the main exchanges of water are with the atmosphere; there is no input from other sources and only a relatively small proportion of water leaves the peatland as runoff. Seasonal variability is strong, with most recharge of the peatland taking place during the winter (December to March) period with occasional further

recharge during months when precipitation is sufficient to overcome evapotranspiration (August, October).

Estimation of the overall budget of peatlands where groundwater plays a more important role is more difficult, mainly because measurement and calculation of groundwater inflow are not straightforward (Lloyd and Tellam, 1995). Gilvear *et al* (1992, 1993) used piezometric head data and hydraulic conductivity measurements on the different underlying geological deposits, together with groundwater flow modelling to estimate groundwater flow for a series of fen sites in East Anglia, England. Combined with surface flow measurements and precipitation data, total budgets were calculated for three sites in contrasting settings. Of these sites, only one (Catfield fen) had a significant peat cover, but it is clear that both groundwater and surface inflow are a significant addition to precipitation as a source of water. Drexler *et al* (1999a) provide a further example of a water budget for a groundwater-dominated peatland, showing that over two years, between 84 and 88% of the total water supply was from this source (Figure 3.2). Other non-peat wetlands, such as swamps and marshes, may be even more dominated by surface or groundwater inputs, including extreme cases such as the alluvial cypress swamp which received 50 times the annual precipitation input in a single flood event (Mitsch, 1979, cited by Mitsch and Gosselink, 2000).

Table 3.1 Estimated annual water balance for Thoreau's Bog, Massachusetts, USA for March 1976 to June 1977. All data in mm. March to June data are averages of 1976 and 1977 and the coldest winter period is given as a single value. Raw monthly data from Hemond (1980).

Month	Precipitation	Runoff	Evapotranspiration	Net change in storage
16–31 March	90	27	23	40
April	110	45	66	−1
May	73	47	55	−29
June	62	12	151	−101
July	260	0	281	−21
August	189	0	152	37
September	86	0	125	−39
October	191	3	97	91
November	28	1	69	−42
1–15 December March	359	111	0	248
Total	1448	246	1019	183

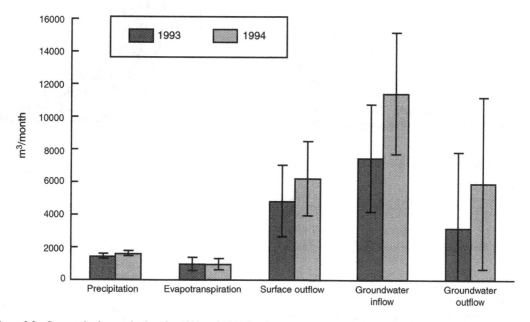

Figure 3.2 Summarised water budget for 1993 and 1994 for the McLean Preserve Fen, New York, USA, showing mean water flux of each component with 95% confidence intervals. Groundwater inflow was clearly the dominant input to the system. Redrawn from Drexler *et al* (1999a) by permission of the American Water Resources Association.

3.3 WATER MOVEMENT WITHIN PEATLANDS

Once water has entered the peatland system, it moves through the peat or is stored over a period of time. Diagrams such as Figure 3.1, showing changing storage as a simple 'black box', belie the complexity of pro-cesses that operate within the peatland and the contro-versy that surrounds these. Peat does not appear to behave in exactly the same way as mineral soils and it may not always be possible to directly apply techniques and theory derived from them to peatland hydrology. The rate at which water will move within a peatland is a function of the pressure of water and the resistance to it. This is often formally expressed as an equation known as Darcy's law:

$$Q = -k(\delta h/\delta l)$$

where Q is the rate of flow, k is the hydraulic conductivity of the peat and $\delta h/\delta l$ is the hydraulic gradient (Eggelsmann *et al*, 1993).

Dense peats, with a fine peat matrix and small pore spaces, generally have very low hydraulic conductivity and water movement is therefore impeded. Less compact peats with larger plant fragments tend to have larger pore spaces and therefore higher hydraulic conductivity. Although there are considerable problems in making reliable measurements of hydraulic conductivity in peats (e.g. Baird, 1995), comparisons between different peats of different botanical composition and amount of decay show considerable variation (Figure 3.3). Rycroft *et al* (1975a), in reviewing the literature available at the time, suggest the range of variation in catotelm peat is from 6×10^{-6} to $5 \times 10^{-3}\,\mathrm{cm\,s^{-1}}$. Chason and Siegel (1986) suggest a slightly wider range of 10^{-1}–$10^{-7}\,\mathrm{cm\,s^{-1}}$ for field measurements compared with 10^{-1}–$10^{-6}\,\mathrm{cm\,s^{-1}}$ for laboratory studies. In general *Sphagnum* peats are some of the least permeable, whereas peats composed of higher plants such as *Carex* and especially larger taxa such as *Phragmites*, are most permeable (Ingram, 1983).

3.3.1 Soil layers and water movement

It is obvious to anyone who has dug or cored into a peatland that the structure of peat changes with depth. The surface is covered by loose, living vegetation, which gradually changes with depth into brown or black peat much denser in structure. Clearly water is not likely to be able to move through the deeper peat as quickly as it can through the surface matrix of living and partially decayed plant material. However, the change from surface vegetation to the deeper solid peat is not gentle and there is a relatively sharp transition between the upper and lower layers. Ingram (1978) gave the name *acrotelm* to the upper layer and *catotelm* to the lower layer and suggested contrasting characteristics for these two zones (Table 3.2). The practical definition of the boundary between acrotelm and catotelm is not always easy, although it can be estimated by a rapid change in bulk density or by the deepest point to which the water table descends in an annual cycle. Visually, it is apparent in some profiles as a change from loose, pale peat to less structured solid, saturated peat (Figure 3.4). However, there is no precise definition that can be applied in all situations. The acrotelm

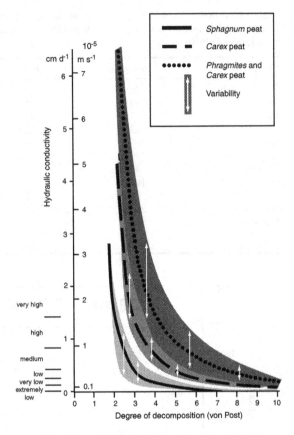

Figure 3.3 Changes in hydraulic conductivity with different peat types and degree of decay (after Eggelsmann *et al*, 1993; Baden and Eggelsmann, 1963). © John Wiley & Sons Limited. Reproduced with permission.

is sometimes known as the 'active' layer, and indeed it is the zone in which most growth and decay occur, and in which living organisms mostly exist. It may be misleading to think of the catotelm as the 'inactive' layer as processes of decay and change still proceed, albeit slowly (see Chapter 5), and water movement is still important (Baird and Heathwaite, 1997).

3.3.2 Groundwater flow within peatland systems

Given the very high hydraulic conductivity of the acrotelm compared to that of the catotelm, it has often been assumed that water flow in the catotelm is unimportant. Most of the studies measuring hydraulic conductivity have been on peats less than 1 m deep and intuitively one would expect deeper peats to have even less potential for water movement. As a result, on deep peats,

Table 3.2 Main characteristics of the acrotelm (upper layer) and catotelm (lower layer) in peatlands. Based on Ingram (1978, 1983).

Character	Acrotelm (upper layer)	Catotelm (lower layer)
Water table	Fluctuating	Absent
Water content	Variable	Constant
Aeration	Periodically aerobic	Anaerobic
Microbial activity	High with aerobic and anaerobic activity present	Low with only anaerobic activity present
Water movement	Relatively fast. Variable from surface to base of acrotelm	Very slow, constant
Exchange of energy and matter	Rapid	Slow

movement of water would be expected to be dominated by surface seepage through the acrotelm. However, in some peatlands, it can be shown that hydraulic conductivity varies considerably with depth and deeper peat does not necessarily have lower values. In addition, the potential for horizontal water movement at depth may be much greater than that of vertical movement (Chason and Siegel, 1986). Figure 3.5 shows some of the results of the study carried out on the Lost River Peatland in Minnesota by Chason and Siegel (1986). Three different areas of peatland were tested, representing different peat types, and both laboratory and field measurements of hydraulic conductivity were made. In all three areas, there was considerable vertical variability in average hydraulic conductivity. The raised bog profile shows a zone of particularly low values at about 1 m depth with higher values below this. The 'external fen' site has highest hydraulic conductivity in the shallowest and deepest parts of the profile. A comparison between the horizontal and the vertical hydraulic conductivities (Figure 3.5b) suggests that in general there is greater potential for horizontal movement of water than for vertical transport. Partly as a result of these kinds of measurements, there has been a growing realisation that water movement in deep peats may be more important to the functioning of the peatland, and much greater in magnitude, than was initially thought (e.g. Siegel, 1983; Siegel and Glaser, 1987). However, the peat structure is likely to be an important consideration, and the more woody peats of continental North America may be much more prone to such variability than the oceanic peats of western Europe where much of the earlier hydrological work was carried out. The oceanic peats are composed of *Sphagnum* and smaller graminoid and ericaceous taxa, which are likely to have a much more homoge-

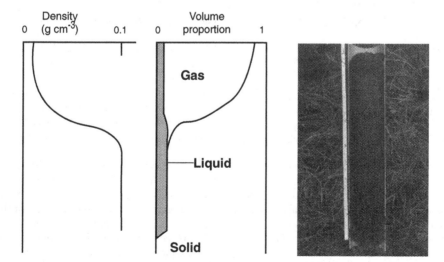

Figure 3.4 Schematic representation of the changes with depth in density, proportion of space occupied by gas, liquid and solid and the water table changes through an annual cycle. The photograph shows typical visual changes through the surface layers of peat. Adapted from Clymo (1992a).

a)

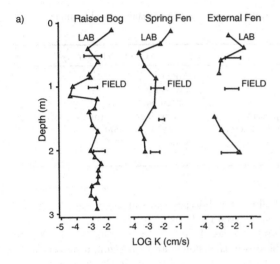

b)

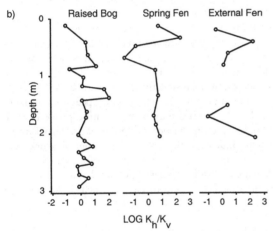

Figure 3.5 Changes in hydraulic conductivity in three peatland types with peat depth. (a) Average values of horizontal and vertical measurements made in the laboratory (continuous curves – LAB) and in the field. Field measurements are represented as the average of eight measurements at different times with 95% confidence intervals. (b) Ratio between horizontal and vertical hydraulic conductivity measured in the laboratory. In both (a) and (b), the x-axis values are shown on a log scale. Redrawn from Chason and Siegel (1986) by permission of Lippincott Williams and Wilkins.

nous structure with less potential for the development of large pores and channels for water to move through. Despite this, one study on an open *Sphagnum* bog in Sweden shows periodic reversals in flow (Figure 3.6; Devito *et al*, 1997).

Further studies on the continental North American sites, especially those in the extensive peatlands of Minnesota (Wright *et al*, 1992), have confirmed that

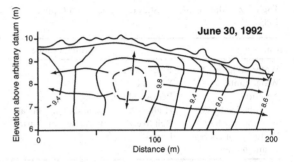

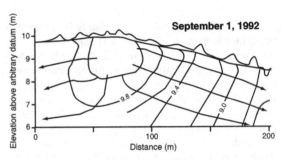

Figure 3.6 'Flow nets' for transects of piezometers on an eccentric raised bog near Umeå, Sweden for two days in 1992. The measurements for June were preceded by an extended dry period with low surface water tables, which led to upward flow of water from mid-depths in the peat. By 1 September, increased precipitation had recharged the bog and a more normal downward flow was resumed. Redrawn from Devito *et al* (1997) with permission. © John Wiley & Sons Limited.

complex patterns of water movements can take place within both bog and fen sites (Siegel and Glaser, 1987; Siegel, 1992). Figure 3.7 shows how groundwater can move up through raised bog peats as well as in spring fens and water tracks in Minnesota peatlands. Prevailing climatic conditions may determine the level that groundwater can penetrate in the peatland. During very dry periods, the groundwater can move higher in the peat profile than during wet periods, an effect which has important implications for carbon cycling (Glaser *et al*, 1997; see also Chapter 9). Thus regional groundwater plays a major role in controlling the hydrology, chemistry and vegetation patterns found in these systems. Figure 3.8 illustrates the relationship between raised bogs and groundwater flow in the glacial Lake Agassiz region in Minnesota. In theory, there is inadequate atmospheric moisture for raised bogs to form in this area and it is clear that groundwater flow is critical to the existence and distribution of peatland landforms, including ombrotrophic raised bogs (Glaser *et al*,

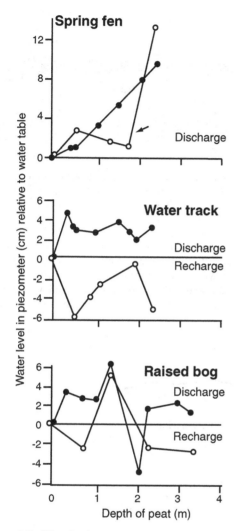

Figure 3.7 Water level measurements in piezometers for three peatland areas in the Lost River Peatland, Minnesota, USA in summer (July, closed circles) and autumn (October, open circles) 1983 (Siegel and Glaser, 1987). When water levels in the piezometers are higher than the water table, this indicates discharge from the groundwater to the bog. Where levels are below the water table, recharge to groundwater is taking place. Redrawn from Siegel and Glaser (1987) with permission of Blackwell Science Ltd.

1997). On the basis of findings such as these, Moore (1997) has suggested that some of the traditional assumptions on the linkage between raised bogs and ombrotrophic conditions may need a rethink. However, the groundwater flow on the Minnesota bogs may be heavily influenced by a unique combination of regional geology and peat composition (see

above). While large-scale differences in peatland systems and landforms are clearly strongly affected by the influence of groundwater, fine-scale differences are also likely to play a significant role in peatland functioning (Drexler *et al*, 1999b).

3.3.3 Changes in the surface and water table elevation

The hydrological status of a peatland surface is most conveniently expressed by changes in the water table (see for example Chapter 8). It should be noted, however, that this is not necessarily an accurate measure of the amount of water held within the peatland as the surface of the peatland itself may rise and fall in relation to the surrounding mineral ground (Ingram, 1983). Although this phenomenon (sometimes known as mooratmung – 'mire breathing') is most often associated with raised bogs, it may be just as significant or more so in fen systems. For example, in schwingmoors, where a mat of vegetation and peat floats over a water lens, the peatland surface actually follows the changes in water level in the system precisely. Water table measurements from the surface are clearly of little use in this situation for hydrological research, although they will still be useful for ecological studies relating the water table depth to the occurrence or activity of organisms living on the surface. Indeed water table depth is one of the most important ecological factors in most mire systems (see below). An example of surface elevation fluctuations measured over a four-year period is shown in Figure 3.9. West Sedgemoor is a drained fen in Somerset, England, which retains considerable value as a wildlife resource. The measurements taken between August 1986 and October 1990 at six-monthly intervals clearly show a seasonal variation in the elevation of the peat surface as well as a long-term decline, possibly as a result of a series of dry summers (Gilman, 1994).

Water table is relatively easily measured with simple dipwells or more complex mechanical or electronic devices (see Ingram, 1983; Bragg *et al*, 1994; Anderson, R., 1997). Belyea (1999) provides what is probably the simplest of all techniques devised to date. Fluctuations in water table occur over a variety of timescales and in response to regular cyclic factors as well as particular hydrological events. At the finest scale there are diurnal changes in response to evapotranspiration superimposed on seasonal cycles. Added to these are responses to external changes in inflow and outflow such as rainfall events, groundwater fluctuations and human activities such as drainage (see

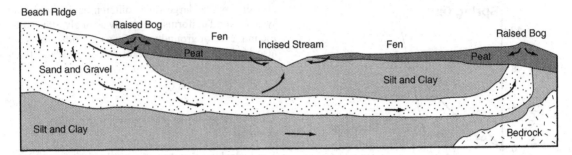

Figure 3.8 Schematic representation of the relationship between groundwater flow and the occurrence of raised bogs and fens in the glacial Lake Agassiz region, Minnesota, USA. Groundwater flows readily through the sands and gravels and provides a water supply for the development of bog landforms where it emerges at the surface through sand lenses. Redrawn from Glaser *et al* (1997) by permission of Blackwell Science Ltd.

Chapter 8). The position of the water table can be characterised in a variety of ways. The simplest is to describe water table changes over time. Figure 3.10 shows a typical schematic pattern of changes over a short period of time in an ombrotrophic peatland. A stepwise decline in water table levels is a typical diurnal pattern, reflecting increased evaporative losses during the day with little change overnight. Occasional precipitation events recharge the peatland, and water tables rise rapidly in response to these. Annual cycles are

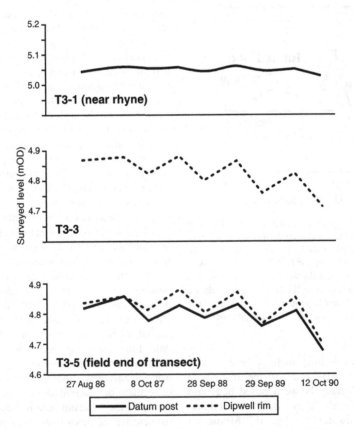

Figure 3.9 Changes in the elevation of the peat surface at West Sedgemoor, Somerset, England. Redrawn from Gilman (1994) by permission. © John Wiley & Sons Limited.

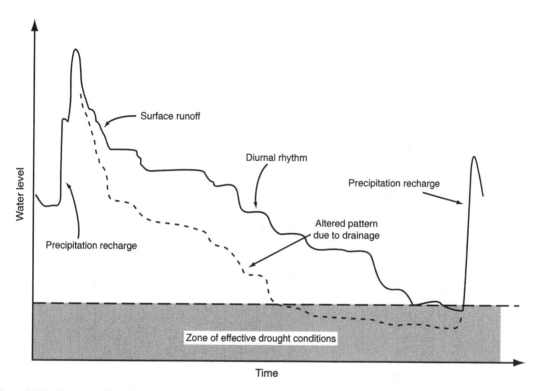

Figure 3.10 Short-term changes in water table for a peatland in Argyll showing typical diurnal fluctuations and response to rainfall events. Solid and dashed lines refer to separate locations. Redrawn and modified from Lindsay *et al* (1988) by permission of Scottish Natural Heritage.

evident in records over longer periods of time such as those shown in Figure 3.9 with summer water table drawdown followed by recovery over the autumn–winter–spring period (e.g. Gilman, 1994, p. 26). These time series data can be converted to mean or modal values over a particular period, but an alternative method of characterising the water table position is to plot 'residence time'. Such plots summarise the detail of particular water table behaviour accurately and help to explain differences between locations and peat layers. Water table behaviour is often strongly skewed, with values within a small range near to its maximum value for a large proportion of the year, but descending to greater depths for shorter periods of time (Ingram, 1987).

3.4 OUTFLOWS

3.4.1 Runoff

Water can leave the peatland by a variety of pathways including surface runoff (either as diffuse or channel flows), seepage and pipe flow. Subsurface losses are usually through seepage to underlying and surrounding mineral soils and bedrock. In many peatlands, surface runoff is the predominant outflow from the system and it is often strongly linked to fluctuations in the water table. Because the upper layers of peat have a high hydraulic conductivity (see above), which decreases further towards the surface, as the water table rises, the runoff tends to accelerate. The upper limit of the water table is thus self-limiting to an extent. This relationship is clearly expressed in measurements of water table depth and discharge in associated streams. Figure 3.11 shows that rising water tables on a blanket mire in northern England lead to rapidly increasing discharge in the stream that the bog feeds. The highest discharges are associated with water tables within 5 cm of the surface, suggesting that significant overland flow cannot occur until the peat is saturated to this point.

The nature of runoff may also vary with the hydrological state of the peatland and with changing inputs to the system. Branfireun and Roulet (1998) determined flow paths within and between a peatland and the

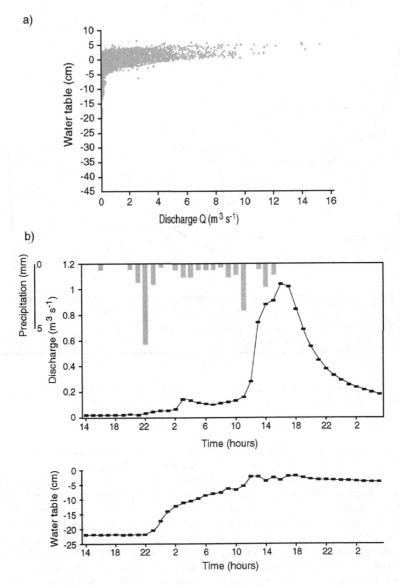

Figure 3.11   (a) Relationship between water table on Moor House blanket peat and discharge in the associated Troutbeck catchment over a three-year period (hourly data). (b) Data showing the rise in water table (bottom) associated with increased rainfall (top vertical bars) and the delayed increase in discharge from the catchment. A large increase in runoff does not occur until the water table has risen to 5 cm below the surface. Redrawn from Evans *et al* (1999) by permission of Elsevier Science.

adjacent forested upland during storm flow and 'normal' baseflow conditions. They describe complex changes in pathways of water movement with changes in the condition of the peatland and in the rainfall events occurring. During baseflow conditions, seepage from the hillslopes and groundwater maintained a relatively wet peatland surface and streamflow in an associated stream. The response of the system to major rainfall events depended on whether dry or wet conditions prevailed in the preceding days and weeks ('antecedent' moisture storage). If the antecedent conditions were dry, rainfall resulted in rapidly raised water tables on the mire and only a small but rapid increase in the stream discharge. In wet conditions, a larger but slower response was recorded with much increased runoff from both peatland and hillslopes. In wet conditions, the hydrological pathways were much more strongly connected between peatland and surrounding hillslopes than during dry conditions. In the context of the overall landscape, the contribution of a peatland to runoff may only be partial, with other flows from mineral soils at least as important (Burt *et al*, 2001). Smit *et al* (1999) describe a method for separating peatland and non-peatland contributions to the hydrologic budget of a catchment. The response of the peatland-derived runoff is usually delayed by comparison with that from mineral soils, especially where there has been a prolonged dry period before the rainfall event. In the case of Dun Moss, Scotland, Smit *et al* (1999) show this can be more than 26 hours after the rainfall begins and up to 22 hours after runoff has significantly increased from the mineral soils.

3.4.2 Evapotranspiration

Evapotranspiration is the combination of water that has vaporised directly from water, soil or plant surfaces (evaporation), together with that which has passed through plants to the atmosphere (transpiration). The loss of water from living plant surfaces is described as interception, as distinct from loss of water from the non-living peat surface or areas of open water. However, it is hard to differentiate empirically between these processes and in quantifying losses in mire water budgets, they are usually considered as a single term. Evapotranspiration from mires is extensively reviewed by Ingram (1983) and by Eggelsmann *et al* (1993). It is a function of a variety of factors including meteorological conditions of temperature, relative humidity and wind speed, together with factors dependent on the peat or plants in a particular situation (moisture status, physiology, architecture, etc.).These come together to produce a pressure differential between the water, peat or plant surface and the surrounding air. The greater the differential in pressure, the greater the evaporative loss. Thus the measurement and prediction of evapotranspiration can be difficult and a variety of techniques has been used to estimate it (Ingram, 1983; Mitsch and Gosselink, 2000).

Different vegetation types will have different evaporative limits and there have been many measurements made to estimate these. Eggelsmann *et al* (1993) give a range of values for raised mires in Germany but all are close to 500 mm pa, unless they are forested, in which case the value rises to over 900

mm pa. Ingram (1983, quoting Neuhäusl, 1975) gives four categories of raised mire vegetation and their related evaporative and transpiration losses:

1. Low transpiration (< 2 mm day^{-1}) and low evaporation where evaporation of the vegetation stand (E) is less than that from an open water surface (E_0). Examples of these are *Sphagnum magellanicum* communities and *S. recurvum* and *Eriophorum vaginatum* successional vegetation in old peat cuttings.
2. Negligible transpiration (around 0 mm day^{-1}) but high evaporation ($E = E_0$). These are very wet *Sphagnum* communities (*S. cuspidatum*, *S. recurvum*) in old peat cuttings.
3. Moderate transpiration ($2–4$ mm day^{-1}) and high evaporation ($E > E_0$). Sedge fens and some lightly forested peatlands come into this category.
4. High transpiration (> 4 mm day^{-1}) and high evaporation ($E \gg E_0$). These are communities such as sedge fens and meadows and plantations of *Picea abies* (Norway spruce).

These are all values for a single peatland type in Europe and there may be much larger variability elsewhere. For example, Campbell and Williamson (1997) assess total evaporation from a restiad-dominated raised bog in New Zealand and find values much lower than other Northern Hemisphere peatlands with low-growing vegetation (Table 3.3). The ratio of evaporation to the open water potential evaporation is also considerably lower than that found on other sites when the canopy is dry but similar to the northern peatlands when the canopy is wet. The low evaporation rates are found despite the fact that the water table is at or very close to the surface of the bog. Campbell and Williamson (1997) describe the system as a 'wet desert' on the basis of its hydrological characteristics. It appears that the canopy of the low-growing restiad species *Empodisma minus* acts as a mulch, preventing heat reaching the surface of the mire and restricting the movement of water vapour from the surface.

3.5 HYDROCHEMISTRY

The sources and throughflow of water in a peatland exert a strong control on water and peat chemistry and therefore on the ecology of the mire system. Some of these influences have been mentioned in connection with peatland classification in Chapter 1. In this section we briefly examine some of the important chemical processes that are related to hydrology and the sources and sinks of important elements in the peatland

Table 3.3 Comparison of mean daily evaporation rates (E) and the ratio between evaporation and potential open water evaporation (E_0). Data from Campbell and Williamson (1997).

Peatland type and location	Evaporation (mm day^{-1})	E/E_0	Source
Raised bog, New Zealand			Campbell and Williamson (1997)
– Dry canopy	1.54	0.34	
– Wet canopy	2.29	0.77	
Subarctic coastal wetland, Canada			Lafleur (1990)
– Dry site	2.6	0.74	
– Wet site	3.1	0.9	
Quaking fen, Netherlands	2.5	0.77	Koerselman and Beltman (1988)
Sedge and *Sphagnum* fens, Hudson Bay lowlands	2.5	0.75	Lafleur and Roulet (1992)
Blanket bog, Newfoundland	2.5	–	Price (1991)

system. The chemistry of water and peat within a mire system is determined by two principal factors: the quality of the water coming into the system and the chemical transformations within the system itself, including processes of decay. Ross (1995) provides an overview of hydrochemistry and solute processes in British wetlands, but many of the principles apply equally to wetlands elsewhere (including peatlands).

3.5.1 Chemical inputs

External inputs to the system are mainly associated with water flows, as discussed above. Precipitation, surface runoff and groundwater inputs therefore normally account for most of these inputs. Dry deposition (aerosols, wind-borne particles from natural and anthropogenic sources, etc.) may also be a significant input in some situations, especially in coastal settings (salt spray) and close to industrial areas (heavy metal pollution and acid rain).

3.5.1.1 Precipitation

Precipitation is relatively dilute in solutes compared to groundwater, and peatlands that are exclusively dependent on this as a source of water are typically nutrient-poor and acid (Chapter 1). However, even ombrotrophic bogs are not uniform in their chemical characteristics for two reasons. Firstly, precipitation may vary considerably in its composition, even over relatively short distances. Secondly, the amount of precipitation can have a significant influence on the annual flux of solutes to the system. Proctor (1992) showed that ombrotrophic mires in Britain and Ireland varied particularly strongly in sodium (Na^+), magnesium (Mg^{2+}) and chlorine (Cl^-) ions with distance from the coast,

reflecting the influence of sea-spray components. The relative amounts of marine and non-marine calcium ions (Ca^{2+}) confirmed this idea, with more than 50% of Ca^{2+} derived from marine sources in western Ireland and less than 20% in inland Britain. Atmospheric pollution and terrestrial inputs also influence the ionic composition of rainwater and dry deposition of Ca^{2+} from terrestrial sources such as agriculture and quarrying may be locally very important (Proctor, 1992). The nature of precipitation is also an important consideration. For example, fog deposition can be a significant component of the hydrological budget in extreme oceanic settings (Price, 1992) and it also carries much greater concentrations of major ions than rainwater (Reynolds and Pomeroy, 1988; Price, 1994). Conversely, the supply of hydrogen ions (H^+) is greater in rain than in fog (Price, 1994). Other factors such as vegetation structure may alter the efficiency with which ions are captured from the atmosphere. Schauffler *et al* (1996) show that forested sites receive much greater input of salt and nutrients from atmospheric sources than open shrub, sedge and *Sphagnum* sites in similar areas.

3.5.1.2 Surface runoff and groundwater

Surface water runoff in many cases reflects the rainwater chemistry, particular in storm conditions (Cresser *et al*, 1997), but it will clearly vary much more than precipitation. Groundwater inputs vary widely in their chemistry, with factors such as geological strata, status of surrounding soils and rate of throughput. They are also much affected by pollution, especially from agricultural sources in many areas. Sites overlying calcareous bedrock are particularly rich in carbonates and have a major influence on the water chemistry of the site (e.g. Gilvear *et al*, 1993), but glacial deposits can also contribute considerable carbonate input (Almendinger and Leete, 1998a, b). These site-specific conditions can be very important in determining the conditions for plant growth and in maintaining the biodiversity and conservation interest of such sites (Almendinger and Leete, 1998b). Mixed sources of water are the norm in all fen sites (since they all receive precipitation plus an additional source). In the absence of a measured water budget, mixing models may help elucidate the relative contributions of different sources (Glaser *et al*, 1990).

3.6 CHEMICAL PROCESSES WITHIN PEATLANDS

The supply of organic and inorganic chemicals to peatlands is only part of the hydrochemical system. Once within the peat, water or living parts of the peatland, there are a large number of chemical and biochemical changes that take place and which critically affect the availability and storage of elements in the system. Throughflow interacts with interstitial water mediated by various factors such as vegetation, detritus, fauna, microorganisms and sediments (Howard-Williams, 1985; Ross, 1995). Ross (1995) highlights three types of hydrochemical processes within wetlands: (i) solute chemical and biochemical transformation, (ii) solute storage and retention processes and (iii) solute transport processes. Reviews of chemistry and hydrochemical processes in peat are given by Clymo (1983), Sikora and Keeney (1983), Gorham *et al* (1985), Shotyk (1988), Ross (1995) and Mitsch and Gosselink (2000), for wetlands in general.

3.6.1 Redox conditions

A key factor in determining chemical transformations in peatlands is the degree of aeration. In saturated peat, the pore spaces are filled with water and oxygen can diffuse only slowly through the peat. Conditions are therefore anaerobic and any oxygen present is rapidly consumed. Redox potential (or oxidation–reduction potential, usually expressed as E_h) is a measure of the electron availability in a solution, or the tendency of the peat/water to oxidise or reduce substances. With increasingly anaerobic conditions the redox potential decreases and a series of chemical transformations can take place as a result of bacterial activity (Figure 3.12; Ross, 1995). In moderately reduced conditions,

after oxygen depletion has occurred, nitrate, manganese and iron reduction are the first processes to occur, carried out by facultative anaerobic bacteria. The facultative bacteria are able to use both oxygen and other chemical species in their metabolic processes, which allows them to continue functioning even when redox potentials drop. As redox potentials continue to decline, sulphate reduction by true anaerobic bacteria in the complete absence of oxygen follows. Finally carbon dioxide reduction occurs at very low redox potentials in extreme anaerobic conditions. While the relative sequence of change is the same under any conditions, the absolute redox potential at each of these stages is variable with pH levels. At higher pH levels, lower redox values are associated with each stage. For example at pH 7, sulphate reduction occurs at around -220 mV, whereas at pH 5 the same process occurs at approximately -70 mV (Ross, 1995).

The effects of these transformations are important for plants and animals for several reasons. First, the reduction of nitrate releases the gas nitrous oxide (N_2O) to the atmosphere, which means that nitrogen is lost, a factor which is especially important as the supply of this important nutrient is often severely limited. Second, the availability of many elements changes with reduction from one chemical species to another. For example, iron and manganese can reach toxic levels as a result of this process and plants may need adaptations to cope with this (see below). Iron may precipitate to such an extent that it forms 'bog-iron' deposits within the peat (Naucke *et al*, 1993).

Redox potentials are often assumed to be a function of water level and as redox potential can be difficult to measure reliably, water table is often used as a proxy. However, in an extensive series of measurements, de Mars and Wassen (1999) showed that redox potential is also strongly affected by mire type, explained by differences in peat composition. Moderately rich fen peats have lower redox potentials than poor fens and bogs due to the less decomposable organic matter in poor fens and bogs. Haraguchi (1995) has successfully modelled changes in redox potentials in a Japanese mire using water table depth, temperature and microbial activity as predictor variables.

3.6.2 Nitrogen and phosphorus transformations

Major nutrients (N, P, K) are in limited supply in many peatlands and therefore sometimes subtle changes in nutrient cycling can have significant impacts on the growth of plants. Nitrogen and phosphorus are most

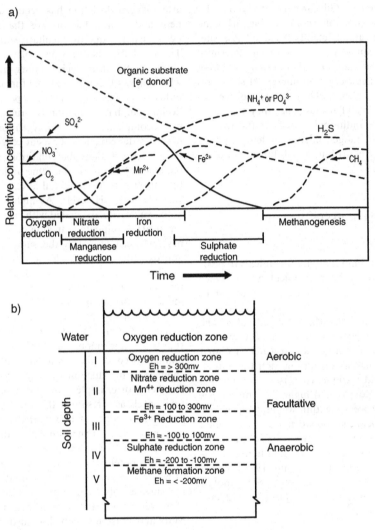

Figure 3.12 Sequence in (a) time and (b) depth of chemical transformations with increasingly reduced conditions or increasing depth in peat. Redrawn from Reddy and D'Angelo (1994) by permission of Elsevier Science.

often the limiting factors to plant growth in many wetlands, including peatlands, and potassium (K) is the least likely to restrict plant growth (Wheeler, 1999). Mineralisation of organically bound nitrogen and fixation of atmospheric nitrogen are important additional sources to the external sources mentioned above. The release of nitrogen from the peat mass (mineralisation) is usually to ammonium via a series of pathways collectively known as mineralisation (Mitsch and Gosselink, 2000). Ammonium can be used directly by plants or it may be oxidised to nitrate through nitrification by *Nitrosomonas* and *Nitrobacter* species in the upper

aerobic layers of peat or in the oxidised rhizosphere of plants. In highly waterlogged conditions nitrification is likely to be most limited, although many peatlands have significant rates recorded with higher levels closer to the surface, presumably associated with greater oxygen availability, even in flooded peats (Verhoeven *et al*, 1990). Denitrification is a dominant process in many peatlands due to low redox potentials, converting nitrates to the gases nitrous oxide and nitrogen as explained above.

Phosphorus is the other major nutrient that is of critical importance in peatlands. It, too, may be a limiting factor for plant growth. For example, experimental fertilisation of sawgrass communities (*Cladium jamaicense*) in the Everglades of Florida shows that they are phosphorus limited, as additional applications of nitrogen had no effect on biomass production (Craft *et al*, 1995). The dynamics of phosphorus in wetland soils are complex (Sikora and Keeney, 1983) and in many aspects they are poorly understood with few estimates of overall budgets (Ross, 1995). Although phosphorus exists in many organic and inorganic forms in peatlands, that which is available to plants is often measured as soluble-reactive phosphorus (SRP). Phosphorus in other forms may be present in some abundance but unavailable for plant growth. Redox potential also has a major indirect influence on phosphorus. Some phosphorus is unavailable in oxidised conditions as it is bound to ferric iron (Fe^{3+}). Under anaerobic conditions such as those in flooded peatlands, this is reduced to ferrous (Fe^{2+}) iron and the phosphorus is released into solution and is available for uptake by plants. Thus the production of SRP in peatlands during wetter periods of the year may exceed the uptake, especially as this often coincides with the winter non-growing season period (Ross, 1995).

3.6.3 Sulphur

Sulphur is an abundant element in peats and is another element strongly affected by the reducing effect of waterlogged conditions. The behaviour of sulphate differs considerably between waterlogged undrained peat and drier conditions, particularly where these are brought about by drainage (Ross, 1995). Under waterlogged, reducing conditions, sulphates and other oxidised forms of sulphur can be reduced by *Desulphovibrio* bacteria to the gas hydrogen sulphide (H_2S) or to pyrite (iron sulphide, FeS), depending on the availability of iron. Hydrogen sulphide is the most noticeable product of peat hydrochemistry as it is what produces the memorable smell of rotten eggs when sediments are disturbed! It can also be oxidised in aerobic conditions by *Thiobacillus* bacteria, creating iron sulphates and sulphuric acid. The production of sulphuric acid contributes to the acidity of many sites, especially after drainage of sulphide-rich peatlands. Processes of sulphur transformation are generally more important in fen systems than in bogs.

3.6.4 pH and ionic processes

Many peatlands are strongly acid. This is initially principally an effect of the relative importance of different sources in the hydrologic budget. In ombrotrophic bogs, where only rainwater feeds the peatland, acid conditions are most likely. As more groundwater influences the water supply, then pH levels tend to be higher, although the factors such as underlying geology make this a highly variable effect. Even relatively small proportions of groundwater can raise the pH status of a peatland considerably; 10% of the total may be adequate to raise the pH from 3.6 to 6.8 (Glaser *et al*, 1990). However, while inflows go some way to explaining pH levels on peatlands, there are several internal processes that are also involved, including exchange of cations in the water and the release of organic acids through decay (Gorham *et al*, 1984). Both peat and some peatland plants have a high cation exchange capacity (CEC). CEC is a loosely used term (Clymo, 1983) but generally means the removal of cations from solution and their replacement with hydrogen ions, reducing pH. Cation exchange with peat and *Sphagnum* provides the principal sources of hydrogen ions in many peatland systems, although factors such as sulphide oxidation, ultimately to form sulphuric acid, may be important in some settings. Organic acids may also be a significant contribution to lowering pH (Gorham *et al*, 1984).

Sphagnum is particularly efficient at removing cations from solution on oligotrophic and ombrotrophic mires where it grows in abundance. Carboxyl groups (COO^-) in uronic acids are the exchange sites and these constitute between 10 and 30% of the dry mass of *Sphagnum* (Clymo and Hayward, 1982). Different species of *Sphagnum* show considerable differences in their CEC, with hummock species generally having higher values than pool species (Clymo, 1983; Clymo and Hayward, 1982). The pH level in a *Sphagnum* community is a balance between supply of the cations and the rate at which they can be replaced with hydrogen ions. Field measurements of growth rates of *Sphagnum*, and changes in pH in different microhabitats compared with precipitation and potential evaporation (Figure 3.13) show that pH rapidly declines as growth rate increases and the supply of water declines. This is equally rapidly reversed in the late autumn and early winter, when pH levels in all microhabitats are close to that of the rainfall. Although the role of *Sphagnum* is usually emphasised in the creation of acid conditions, other taxa may be as efficient in the role. In New Zealand peatlands, the root layer of *Empodisma minus* has a CEC at least as high as that of the most common *Sphagnum* species on raised mires in the region (Agnew *et al*, 1993a).

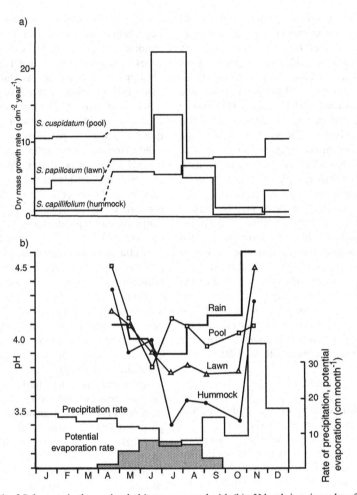

Figure 3.13   (a) Growth of *Sphagnum* in three microhabitats compared with (b) pH levels in rain and surface waters of the same locations and precipitation and potential evaporation throughout a year. The median value of five measurements is plotted for pH levels. Ranges are not shown for clarity, but see Clymo and Hayward (1982). Differences between pH levels in the microhabitats are not explained by differences in growth rates but by differences in CEC in the *Sphagnum* species in the order *S. capillifolium* > *S. papillosum* > *S. cuspidatum*. Redrawn from Clymo and Hayward (1982) with kind permission of Kluwer Academic Publishers.

While many peatlands are generally assumed to be effective 'traps' for cations because of the effectiveness of the cation exchange process within peats and some plants such as *Sphagnum*, there have been relatively few attempts to quantify total fluxes to and from peatlands. Long-term studies have suggested that both inputs and outputs of major cations vary considerably between cations and over time (Urban *et al*, 1995).

3.6.5 Seasonal variability

The different combinations of sources, rates of flow and internal hydrochemistry dynamics do not result in temporally uniform hydrochemistry conditions. There is considerable variation over time in chemical conditions, which is brought about by variability in the factors discussed above. Vitt *et al* (1995) describe the variation in major nutrients, cations and metal ions over the course of a year on continental peatlands varying from a rich fen to an ombrotrophic bog. In general, the variability of bog chemistry was much less than that in fens, and some characteristics (pH, alkalinity, conductivity, ammonium, sulphur and SRP) remained relatively constant when compared to base cations, metal ions and total phosphorus. The rich fen to bog gradient was very clear in pH values and relative values remained constant throughout the year. Calcium concentrations also reflected the overall differences between sites, but fluctuations were much greater in magnitude in the moderate and rich fens than at the bog and poor fen sites. Changes in zinc concentrations showed no clear relationships with the site type but varied a lot throughout the year, apparently inversely related to precipitation variability (Vitt et al, 1995). Weather conditions are also thought to be important in controlling metal ion (Zn^{2+}) concentrations in raised and blanket bogs in England, with short-term changes in cations such as Na^+ and Ca^{2+} also showing a strong response to rainfall totals (Proctor, 1994). Chloride also reflects changing precipitation but is less strongly affected than Na^+.

Interactions with biological factors have strong seasonal effects on nutrient water chemistry, with generally increasing mineralisation releasing N, P and K during warmer, drier conditions. Conversely, plant growth often leads to increased consumption of nutrients during the growing season. The net effect of these processes is evidenced in major fluctuations in concentration of these elements in the annual cycle. In Proctor's (1994) work on English blanket bogs, nitrates were only evident in significant quantities during the coldest winter period, and at other times of the year any inputs appear to be rapidly consumed by biological activity. Variability in other nutrients is also strongly controlled by plant uptake, with potassium (K^+) often showing low values in surface waters during the growing season (e.g. Bragazza et al, 1998). Limited microbial activity in Florida Everglades peatlands has been shown to result in suboptimal rates of mineralisation and release of phosphorus during periods of high temperatures and low water tables (Koch-Rose et al, 1994). Small-scale variability in topographic and biological patterns may also dampen or amplify seasonal effects. Bragazza et al (1998) show that the pH decline over the spring to autumn period is enhanced in hummocks and dampened in hollows.

The influence of changes in the amount of water from particular sources also becomes evident over longer periods. For example in times of flooding, excess runoff dilutes ionic concentrations on floodplain mires (e.g. Giller and Wheeler, 1986b).

3.7 ECOLOGY AND ECOHYDROLOGY

3.7.1 Ecohydrology

The link between ecology and hydrology in mire systems is so close that the term 'ecohydrology' was coined

for peatlands (Göttlich, 1977 in Eggelsmann et al, 1993), although the term is now used much more widely (Baird and Wilby, 1999). I began this chapter with a consideration of key hydrological features of peatlands for the very reason that many of the most striking characteristics of plant, animal and microbial populations are ultimately related to hydrological controls, although plants also have some influence on the hydrochemical conditions in which they grow. An appreciation of these controls is needed to be able to understand the community dynamics and adaptations of plants and animals in peatlands. This section is not intended to be anywhere near comprehensive in its coverage. There is a vast literature on the biology and ecology of peatlands, especially concerning plant communities and with a strong geographical bias towards north-west Europe. For particular peatland types and areas of the world, the regional accounts and summaries referred to in Chapter 1 are probably the best immediate source of more detailed information. The aim here is to highlight the chief environmental factors that affect the ecological functioning of peatlands and to enumerate the limiting factors for the existence and growth of plants and animals, together with a summary of the ways in which they cope with such an apparently stressful environment.

3.8 LIMITING FACTORS FOR PLANTS AND ANIMALS

Peatlands are potentially hostile places for plant life, being waterlogged and often poor in nutrients and high in available toxins such as reduced forms of iron, manganese and sulphur. The direct effects on animals are perhaps not so important although many peatlands have characteristic faunas as well as specialist floras.

3.8.1 Adaptations in plants

The principal factors directly affecting plant growth are:

(i) Low oxygen availability. Oxygen diffusion rates are about 10 000 times slower in water than in the atmosphere so that oxygen availability to the root system is very limited in waterlogged conditions.

(ii) Mobilisation of toxic elements (Fe, Mn, S).

(iii) Low nutrient availability. Both nitrogen and phosphorus are often in very low supply (see above).

(iv) Acidity. The low pH levels of many peatland waters (3.5–4.0 is not unusual in ombrotrophic systems) are below those of many mineral soils.

In addition Wheeler (1999) points out that water stress may actually be a problem for many plants. Although this sounds illogical, many plants that are used to growing in wet conditions have higher rates of water loss from their leaves or inefficient stomatal control of transpiration. Seedlings can be particularly badly affected. Bryophytes have little ability to limit water loss. As a result, during occasional drought conditions, plants may suffer even though conditions are still relatively wet compared to mineral soils. Given slow growth rates in less productive peatlands, prevention of herbivory may be important. *Sphagnum* is known to be a plant that few animals consume; this may be due to allelopathic effects of excreted organic acids (Verhoeven and Liefveld, 1997).

3.8.2 Oxygen availability

Adaptations to the lack of oxygen (anoxia) include a number of structural and morphological characteristics as well as specialist physiological processes to enable plants to obtain enough oxygen for growth. These are mainly adaptations in the root system in vascular plants. Many plants obtain a greater oxygen supply by developing air spaces (aerenchyma or lacunae) in both roots and stems so that oxygen can diffuse from above-ground parts to the roots. Up to 60% of the root volume may be pore space and the greater the root porosity the more flood tolerant the plant is (Mitsch and Gosselink, 2000). Some plants (e.g. willow – *Salix,* and alder – *Alnus*) produce 'adventitious roots' from above-ground parts, usually just above the anaerobic zone, to take in more oxygen. Adaptations in other wetland plants include the development of special aerial parts such as lenticels and pneumatophores, and pressurised gas flow from the surface to the rhizosphere (Mitsch and Gosselink, 2000), although it is not clear to what extent these are important in peatland plants. Whatever mechanism is used to get oxygen to the roots of the plant, it may also provide additional oxygen to the rhizosphere in anaerobic layers of peat. This can produce nitrification promoting plant growth, as well as conditions where toxins oxidise and are therefore not taken up by the plant (Wheeler, 1999).

A number of plants attempt to avoid waterlogging by developing very shallow root systems that are above the water table for most of the time. Coniferous tree taxa are well known for this, examples being Sitka spruce (*Picea sitchensis*) growing as plantation forestry and black spruce (*P. mariana*) in natural boreal forests, although this can cause stability problems in windy areas. Another avoidance strategy is to only occupy drier microsites such as the tops of hummocks formed by mosses or other plants. Reproductive avoidance strategies are only useful in peatlands with strong seasonal fluctuations in the water table and include production of tubers, seeds and other perenating organs. The development of an anaerobic metabolism in some plants is a further strategy but it may only be important in surviving short periods of intense waterlogging rather than long-term tolerance (Wheeler, 1999).

3.8.3 Mobilisation of toxic elements (Fe, Mn, S)

As discussed above, waterlogged conditions lead to more available forms of some metallic ions. Although in small quantities these are nutrients, concentrations can easily rise to toxic levels. Strategies are therefore needed to find ways of either preventing these toxins entering the plant or of isolating them once taken up. The mechanisms used to cope with anoxia are largely responsible for dealing with this problem by oxidation of toxins in the rhizosphere or within the roots themselves.

3.8.4 Low nutrient availability

A variety of mechanisms exists in peatland plants for making the most of what is very often an extremely limited supply of the major nutrients for plant growth. Many peatland plants on nutritionally impoverished peatlands have low growth rates and therefore require only relatively low nutrient inputs, but plants that have higher growth rates and can compete for the available nutrients may be at a competitive advantage. Some taxa can take up different forms of nitrogen, for example. Chapin *et al* (1993) found that *Eriophorum vaginatum* (Hare's tail cotton grass) can take up amino acids instead of inorganic nitrogen, and that it may do so preferentially. Thus it does not have to wait until the breakdown of larger molecules has occurred to obtain nitrogen, and of course it reduces the nitrogen supply to the rest of the plants.

Other plants may exploit mycorrhiza to increase their supply of nutrients. Mycorrhiza are mutualistic associations between the roots of higher plants and fungi. They are important to promoting nutrient availability in many plants from a wide range of habitats but relatively little is known of their occurrence in association with peatland plants. Thormann *et al* (1999) surveyed a range of taxa in boreal peatlands in Canada and found that many of the higher plants had mycorrhiza of various kinds. The occurrence of these organisms in peatlands suggests that many plants obtain additional nutrients from the different mechanisms known to operate in similar mycorrhiza in other habitats. Phosphorus availability is enhanced and nitrogen may also be affected.

Carnivory is probably the best known of the methods by which peatland plants obtain additional nutrients, especially nitrogen. Trapping and digestion of animal life take a number of forms. *Drosera* (sundew) and *Pinguicula* (butterwort) species have sticky leaves to trap insects and the pitcher plants (*Sarracenia* species) entice insects into containers from which it is impossible to escape. Other plants have rapidly reacting 'trigger' mechanisms that physically ensnare the prey. The Venus fly-trap is the best known of these and has become a popular houseplant, but *Utricularia* species (bladderworts) also have an ingenious trap which operates under water to catch small water creatures such as *Daphnia*.

3.8.5 Acidity

Recent reviews have suggested that acidity itself is not the limiting factor on plant growth, although it may often appear to be strongly associated with plant community gradients in many peatlands (Bridgham *et al*, 1996; see below). Many plants tolerate low pH levels in peatlands but the limiting factor on their growth is nutrient availability rather than basic cations. Both Mg^{2+} and Ca^{2+} occur in high enough concentrations for plant growth even in ombrotrophic bogs (Malmer, 1986) and artificially raising Ca^{2+} concentration can reduce *Sphagnum* growth, for example (Clymo and Hayward, 1982). Only occasionally do additions of Ca^{2+} result in increased growth of a few species (Kooijman and Bakker, 1994; see also Bridgham *et al*, 1996 for a review). Therefore, although tolerance of low pH levels may be a necessary prerequisite for survival in some conditions, the response of plants may actually be in relation to nutrient levels rather than pH *per se*. In addition to *Sphagnum*, which has particular tolerance to acidity and actively increases acidity levels in surrounding waters, ericaceous species

are well known as plants which can tolerate very low pH. This tolerance is partly due to effective use of NH_4^+ rather than NO_3^- since the latter are all but absent in low pH, waterlogged conditions, but is also due to resistant roots (Kinzel, 1983 and Runge, 1983, cited in Bridgham *et al*, 1996).

3.8.6 Adaptations in animals

Animal life on peatlands covers a wide range of organisms and it is not my intention to provide comprehensive coverage of this area of peatland ecology. This is partly due to the fact that many animals, especially mammals but also birds, amphibians and reptiles, do not live their entire life cycles on peatlands alone. Other organisms are dependent on peatlands for their existence and have developed specialised life cycles and physiological mechanisms. Unfortunately rather less attention has been given to zoological than to botanical studies on peatlands and even less to palaeoenvironmental studies involving animal remains (Chapter 6). However, we know that many organisms respond to similar environmental gradients to those of plants (see below) and zoological arguments are an important component of many conservation issues (see Chapters 10 and 11).

Besides the availability of water, which differentiates between obligate aquatic animals and those which tolerate or require wet but not fully aquatic conditions, the main factors affecting animal life on peatlands are summarised by Heathwaite *et al* (1993b) as:

(i)   Humic acids and acidity of the mire water.
(ii) Large fluctuations in temperature.
(iii) Low nutrient concentrations.

These factors are principally limitations on raised and blanket bogs but also affect some fens to a lesser extent. For example, most mollusca and crustaceans are absent from acid peatlands because of the lack of calcium carbonate.

Most of the adaptations to these conditions occur in the invertebrates, aside from some behavioural adaptations such as selection of nesting locations in birds. Most of the occurrence of specific peatland fauna in birds and mammals arises out of secondary dependence on invertebrate food sources or other habitat characteristics. For example, Gaither (1994) found that the understory avifauna of South-east Asian peat swamps was much richer than had previously been supposed; most of the specialised peatland species were dependent on insect food sources or fruit at particular times of the year. Mammal species may use peatlands as refuges in areas modified by people. For example, before it became locally extinct, woodland caribou (*Rangifer tarandus*) was restricted to peatland habitats in Minnesota (Glaser, 1987b).

Batzer and Wissinger (1996) review the ecology of insects in non-tidal wetlands and many of their comments apply to peatlands. Finnamore and Marshall (1994) provide a detailed review of arthropod species and communities on peatlands. Hydrology is a key factor in invertebrate survival. There is a strong distinction between insect communities in ephemeral pools and those in permanent pools, and temporary extreme conditions can have a damaging effect. For example, insects that are dependent on a drought-resistant stage in their life cycle suffer badly if the habitat remains flooded (Neckles *et al*, 1990). Smaller organisms such as protozoa are also adapted to temporary drying out of the substrate by an ability to encyst and survive through atypical conditions (Sleigh, 1989).

Water acidity does not appear to be as important as the availability of water for insects, although reduced predation from fish in acid conditions is advantageous (Batzer and Wissinger, 1996). In exceptional cases, however, even fish can tolerate very low pH levels and can result in a specialised fauna in tropical peat swamps, for example (Ng *et al*, 1994). Invertebrates are also extraordinarily adaptable to very specific conditions and in inhabiting particular niches in the peatland environment. The symbiosis between purple pitcher plants (*Sarracenia purpurea*) and three species of Diptera are well known (Damman and French, 1987; Hardwick and Giberson, 1996). In this case the insect larval stages occur within the pitcher which catches other insects. Although the exact relationship is not clear, it appears that the insect larvae help to break down the food for easier absorption by the plant. Variability in adaptation between closely related species is also apparent. For example, the emergence of pupating tipulid species in British peatlands is linked to temperature, but different responses are recorded in different species (Coulson *et al*, 1998).

3.9 ENVIRONMENTAL GRADIENTS

Patterns of vegetation variation are everywhere in peatlands; the more subtle ones are perhaps only detectable by those familiar with mire ecology, but many are easily observed by anyone who can differentiate between a tree, a shrub and a moss. The change from floating

mats of *Sphagnum* or sedges at the edge of a water body through to taller vascular plants away from the water, perhaps to a shrub community and a forested margin at the edge of the peatland, is obvious. Many of these patterns are brought about by gradual spatial variations in environmental conditions, known as environmental gradients. Bridgham *et al* (1996) and Wheeler and Proctor (2000) review environmental gradients on peatlands. Although Wheeler and Proctor (2000) concern themselves primarily with north-west European mires, their summary of the main directions of variability which have been identified over the years is applicable to all peatland systems:

(i) The minerotrophic–ombrotrophic gradient, i.e. the water source gradient (atmospheric–telluric water).
(ii) The acid and/or base richness gradient. Related to pH, calcium and other base cation variability.
(iii) The fertility gradient, normally related to the availability of nitrogen and phosphorus.
(iv) The water table gradient.
(v) The lithotrophic–thalassotrophic gradient. Related to the influence of the ocean (thalassotrophic) as opposed to the influence of rocks (lithotrophic).
(vi) The mire expanse–mire margin gradient, related to distance from the centre of the peatland.
(vii) The deep peat–mineral soil gradient related to the depth of the peat.
(viii) The spring–flush–fen gradient, related to proximity to the emergence of groundwater and the rate of flow of the water.

The acceptability of these gradients as principal determinants of vegetation variability has been recently debated and a new paradigm is becoming evident in mire ecology and classification (Bridgham *et al*, 1996; Wheeler and Proctor, 2000). Any view of the relative importance of different factors will also depend on the scale under consideration. For example, water table gradients are highly significant over 10–100 m in individual mires (see below) but are much less recognisable in comparisons between different sites (Wheeler, 1999; Wheeler and Proctor, 2000). Another gradient not identified so far but active on a much larger scale is the effect of climate, especially temperature (Gignac, 1994). Traditionally there has been a view that the minerotrophic–ombrotrophic, acid and base richness gradients are the most important controls on plant community variation and an implicit assumption that these two gradients are essentially the same. Bridgham *et al* (1996) and Wheeler and Proctor (2000) have challenged this view and suggest that the primary gradients which should be used for description of mires are those of acidity and fertility (Figure 3.14). They advocate rejection of the ombrotrophic–minerotrophic gradient in studies of vegetation variability as it is not possible to consistently identify the relative influences of these factors in either water chemistry or vegetation. Of course, this does not invalidate the consideration of this important concept when considering peatland hydrology, landforms and development. In the discussion below we will attempt to summarise the key gradients from the list above which affect the ecology of the mire and especially the plant communities.

3.9.1 The acidity gradient

The confusion over the difference between the acidity/calcium gradient and that of ombrotrophy-minerotrophy arose from the use of plants as indicators of the situation in one particular part of the world to identify the 'mineral–soil–water limit' (Du Rietz, 1954). Since the pH level is a function of water sources

and internal processes (see above), quite similar pH and calcium concentrations can be found in bog and poor fen sites. Over larger regions the position becomes more confused since rainfall water chemistry and amount of rainfall vary. The influence of oceanic inputs becomes important in determining hydrochemistry and vegetation (see below). For example, vegetation on raised bogs in oceanic areas such as Britain has strong similarities with sloping fen communities in parts of Scandinavia and continental Europe. Recent work has attempted to resolve this issue by suggesting that the major change along the acidity gradient is between bogs/poor fens (low pH, low Ca) and rich fens (high pH, high Ca) (Figure 3.15; Bridgham *et al*, 1996; Wheeler and Proctor, 2000), rather than between bogs and fens.

There are vast numbers of studies that have related vegetation variability to the acidity gradient from many areas of the world, and despite its inadequacy as a reflection of the actual limiting factors on plant growth, it has been used as a proxy for the numerous related variables of water chemistry for a long time (Figure 3.16). Effects appear to be on all groups of plants, including the less abundant lower plants such as lichens and liverworts (Anderson *et al*, 1995; Albinsson, 1997). Acidity clearly is important in its own right but the relationships are most evident in many of the northern peatlands. Further work is still needed to see whether it is as important in other areas of the world, given that at least some systems at lower latitudes do not appear to respond to pH as a primary variable (e.g. Bridgham and Richardson, 1993). Vitt and Chee (1990) found that vascular plant communities appeared to respond mainly to changing nutrient concentrations, whereas bryophytes changed with pH/alkalinity levels so that the response may be partly taxonomically determined. Other characteristics of plant communities may also change with the acidity gradient. For example, plant diversity is often positively correlated with pH level in northern mire systems (e.g. Gunnarsson *et al*, 2000). On peatland areas where there is a lot of structural diversity in vegetation, major changes in plant physiognomy may be associated with pH. Walbridge (1994) found that the major groups of vegetation on fen peatlands in West Virginia, USA, were characterised by different pH ranges. Forest and tall shrub communities occurred with more alkaline surface waters (pH 4.6–5.0) than low shrub and bryophyte communities (pH 4.0–4.4).

Other organisms also show an association with pH gradients, although given the doubts about its overriding importance in peatlands, it would be surprising if these relationships were not examined in greater

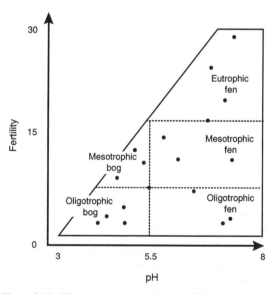

Figure 3.14 The approximate position of different major wetland vegetation types in north-west Europe on gradients of pH and fertility. The points represent different plant community types in Britain as examples, but the concept of the importance of these two gradients should be applicable anywhere, although absolute values will differ. The y axis values are on a phytometrically assessed scale. Redrawn from Wheeler and Proctor (2000) by permission of Blackwell Science Ltd.

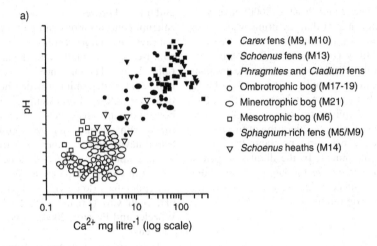

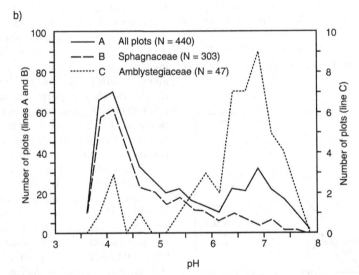

Figure 3.15 Acidity gradients as demonstrated by (a) peatland waters in Britain and Ireland plotted as Ca^{2+} against pH. Note both are on a logarithmic scale to demonstrate the separation into two main groups of data points. Redrawn from Wheeler and Proctor (2000) with permission of Blackwell Science Ltd. (b) Changes in bryophyte flora of North American peatlands along the pH gradient, showing two main groups of site type, characterised by peaks in either *Sphagnum* or 'brown mosses' (Amblystegiaceae). Redrawn after Gorham and Janssens (1992) and Bridgham *et al* (1996) by permission of the Society of Wetland Scientists. Both plots show the division of peatlands into two different groups along the pH gradient. These are not equivalent to 'ombrotrophic bogs' and 'minerotrophic fens' as determined by water sources, but more likely represent 'bogs and poor fens' as against 'moderate and rich fens'. See text for discussion.

detail in the future. Testate amoebae (rhizopods) are a group of the protozoa that are abundant in almost all peatland environments. Their primary response is to the water table gradient, but there is a strong secondary gradient with pH, although as in most studies there is also strong autocorrelation between pH and other hydrochemistry variables (Tolonen *et al*, 1994; Charman and Warner, 1992, 1997). Coleoptera (beetles) also show a general change with acidity, although this is again strongly affected by other natural

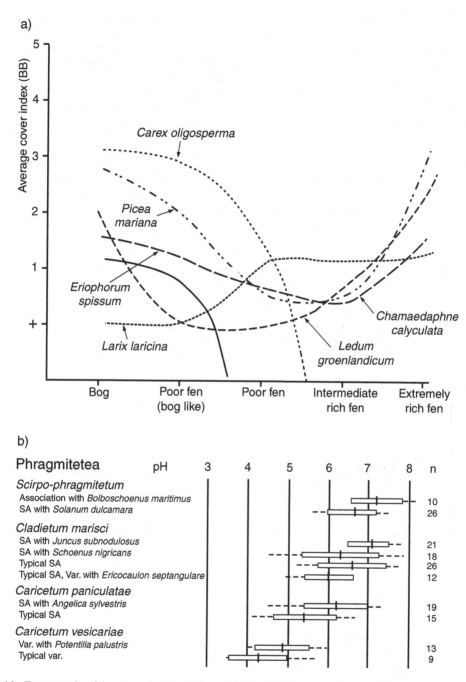

Figure 3.16   Two examples of the pH gradient in relation to (a) changes in plant abundance and (b) plant community occurrence. (a) Changes in individual plant species abundance in Minnesota peatlands. Although the x-axis, representing a peatland type, is not a linear gradient, the change in plant species is clear. The y-axis indicates abundance scores on the Braun–Blanquet scale. Redrawn from Glaser (1987b). (b) Plant communities in relation to pH levels within the Phragmitetea communities described by Dierssen (1982) for north-west European mires. The vertical line indicates the median value with the boxes giving upper and lower quartiles and dashed lines indicate maximum ranges. Extracted from Dierssen (1982) by permission of the Conservatoire et Jardin Botanique de la Ville Genève.

and anthropogenic factors (Holmes *et al*, 1993a, b). Larger animals can also be related to this gradient. Some general associations between bird populations and the acidity levels of open water areas within peatlands have been demonstrated (Fox and Bell, 1994). However, other characteristics of the habitats, such as availability of nesting islands for black-throated divers (*Gavia arctica*), complicate the picture.

3.9.2 The fertility gradient

As discussed above, the importance of fertility (or availability of major plant nutrients) has only recently come to the fore as an independent factor affecting the ecology of peatlands, especially the occurrence of plants and plant communities. Consequently it is now regarded as the second major axis of variation in mire status (Wheeler and Proctor, 2000). The underpinning for this idea comes from numerous studies of limiting nutrients on peatlands (see Bridgham *et al*, 1996, for a review) as well as extensive work on fen systems (e.g. Wheeler and Shaw, 1995a). Figure 3.17 shows the importance of fertility in relation to other environmental factors for British fens. As discussed above, the

fertility gradient may be secondary to that of acidity in many cases, but an important point to note is that it is largely independent of it, as indicated by the orthogonal relationship in the ordination of vegetation data (Figure 3.17).

Perhaps one of the reasons that the importance of fertility has been underplayed in the past is because it has been difficult to measure adequately. Nitrogen and phosphorus are present in different available and unavailable forms and chemical assay may not truly reflect the growing conditions for the plants. Wheeler *et al* (1992) provide a solution to this problem in using so-called 'phytometric' assessments of substrate fertility, using the growth of standard phytometer species in laboratory conditions to indicate the soil fertility. Of course for the detail of which nutrients are the most important in particular settings, measurements of N, P and K must be made separately, even if phytometric data are collected. For example, Wassen *et al* (1990) show that fen vegetation on irregularly flooded fens in Poland is more closely related to P and K than to N content of groundwater and peat. Beltman *et al* (1996a) found that both N and P limited plant growth in Irish blanket mires, but that P was often more impor-

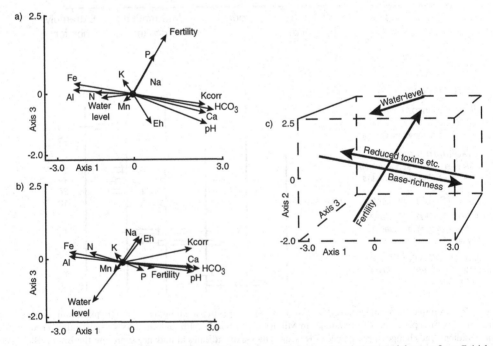

Figure 3.17 Canonical correspondence analysis ordination of a large floristic and environmental data set from British fens. The axes are in decreasing importance from axis 1 to axis 3. (a) Axes 1 and 2 show the orthogonal nature of the pH and fertility gradients. (b) Axes 1 and 3 show separation of water level gradient from hydrochemistry gradient. (c) Schematic three-dimensional ordination summarising main gradients. Redrawn from Wheeler and Shaw (1995a) by permission. © John Wiley & Sons Ltd.

tant than N. On ombrotrophic mires, N may be limiting only where atmospheric inputs are low. In other situations, such as those receiving increased N from atmospheric pollution, P may be more important (Aerts *et al*, 1992). Nutrient gradients may also be artificially produced, typically by pollution from agriculture or urban runoff. Craft and Richardson (1997) studied a P enrichment gradient in a peatland area of the Everglades in Florida and found that changes in macrophytes were highly correlated with this. The dominant species of cattail (*Typha domingensis*) and sawgrass (*Cladium jamaicense*) were particularly affected, with the former increasing and the latter decreasing with P enrichment (Figure 3.18). The pollution from runoff also alters the microbiology of sediments in this region (Drake *et al*, 1996).

3.9.3 The water table gradient

The water table gradient is perhaps one of the more complex gradients to understand in relation to plant and animal ecology. First, it is a complex variable, with significant temporal and spatial variability. Moreover, the temporal variability can be very different in character for different peatland types. For example, ombrotrophic peatlands in oceanic areas tend to have more predictable water table variations with the seasons, and apart from very exceptional severe droughts (e.g. Evans *et al*, 1999), water tables are relatively stable. In contrast, a valley fen may be subject to very major fluctuations, with open water during winter flood events and deep water tables at the height of the summer. As a result, average conditions may not give a reliable comparison between different site types. Second, water table is only one way to measure and express the amount of water. Soil moisture is the chief alternative and may be a more adequate reflection of growing conditions for plants, especially where water tables are deep (Wheeler, 1999). However, temporal variability in moisture conditions is difficult to measure and most studies rely on water table depth to assess the 'surface wetness' conditions of a peatland.

For vegetation, water table gradients are particularly important in determining small-scale patterns of variability. The classic examples of this are in the transitions between hummocks and pools in many mire systems. Although often most strongly associated with northern boreal peatlands, they are also present in the Southern Hemisphere (Mark *et al.*, 1995) and in some subtropical areas (Backéus, 1989). In these situations, plant communities and individual species show distinct associations with particular water table

a)

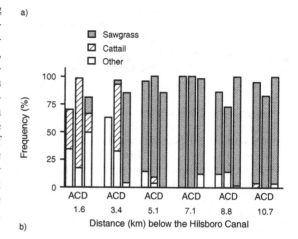

b)

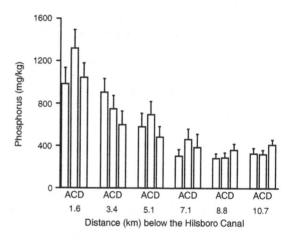

Figure 3.18 Changes in plant species composition and phosphorus with distance from the Hillsboro canal, a source of nutrient-enriched drainage water in the Everglades, Florida, USA. Redrawn from Craft and Richardson (1997) by permission of the American Society of Agronomy.

conditions, creating zones of vegetation which are either concentrically arranged around open water pools or often form linear patterns running across the slope of a peatland surface. Figure 3.19 shows a variety of the relationships that have been found on patterned mires in two different peatland systems. Larger-scale variations in vegetation are also associated with different water regimes within particular sites. Major vegetation types on a kettle hole peatland in southern Finland are related to different average water levels but also to the variability of the water table throughout the year (Reinikainen *et al*, 1984).

The response of individual plants to the water table gradient presumably varies mostly in relation to their ability to cope with waterlogging, but also with competition. Studies on individual plants can demonstrate the degree and likely mechanisms of sensitivity to waterlogging, but may not be adequate to explain the detailed spatial patterns observed in the field. A series of classic studies was published on the tolerance of common European peatland vascular plants to waterlogging in the 1960s (Bannister, 1964a, b; Gore and Urquart, 1966), exploring responses and mechanisms involved in waterlogging. Bell and Tallis (1974) showed how one peatland dwarf shrub, *Empetrum nigrum* (crowberry), responds to changes in water tables with a combination of field and laboratory data. A decline and revival in *E. nigrum* over a three-year period at Wybunbury Moss, Cheshire, England, was attributed to rising and falling water tables. The principal effects of waterlogging on *E. nigrum* were through high hydrogen sulphide and/or carbon dioxide levels with oxygen deficiency, added to elevated aluminium concentrations (Figure 3.20).

Competitive effects are much harder to estimate. Competition in *Sphagnum* is variable. Some species occupy very distinct niches and are relatively stable, but other water table niches can be occupied by a number of possible species, creating strong competition (Nordbakken, 1996). *Sphagnum* species are often effective at excluding higher plants due to their efficiency at removing and locking up nutrients (Svensson, 1995). Other taxa such as hepatics are opportunistic and although they show relationships with water tables, may be influenced to a greater degree by other factors such as occurrence of bare peat patches (Nordbakken, 1996). Competitive interactions are also affected by the persistence of individual species (Nordbakken, 2000).

The water table gradient is also very important to animal communities, and is arguably a more significant determinant of community composition than water chemistry. Certainly aquatic and terrestrial areas of peatlands can have very different insect faunas and the permanence of the water body is often critical (Batzer and Wissinger, 1996). Within particular trophic categories of peatlands, water table is undoubtedly the chief determinant of invertebrate assemblages. Reynolds (1990) suggests three main habitats for invertebrates on Irish ombrotrophic bogs: aquatic, the surface of standing water (the 'neuston') and terrestrial. In specific groups of invertebrates, there is also evidence that water table status is a significant factor.

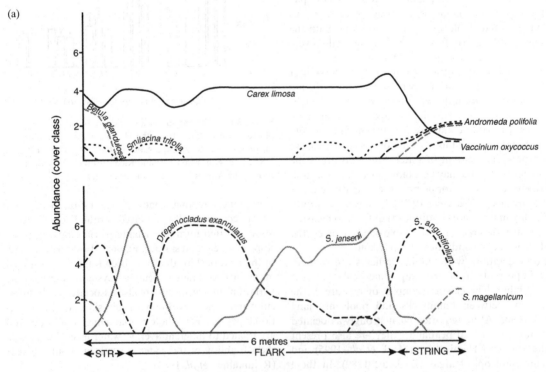

Figure 3.19 Relationships between water table changes and vegetation. (a) Species turnover along a string–flark transition in a patterned fen in central Alberta, Canada. Redrawn from Vitt *et al* (1975) by permission of NRC Research Press.

(b)

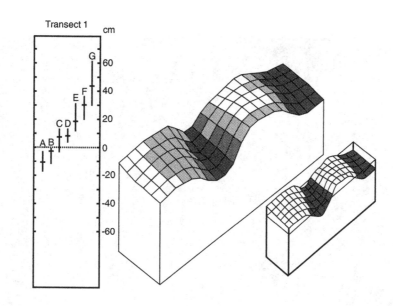

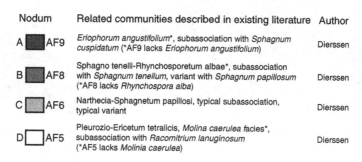

Nodum	Related communities described in existing literature	Author
A ▓ AF9	*Eriophorum angustifolium**, subassociation with *Sphagnum cuspidatum* (*AF9 lacks *Eriophorum angustifolium*)	Dierssen
B ▓ AF8	Sphagno tenelli-Rhynchosporetum albae*, subassociation with *Sphagnum tenellum*, variant with *Sphagnum papillosum* (*AF8 lacks *Rhynchospora alba*)	Dierssen
C ▒ AF6	Narthecia-Sphagnetum papillosi, typical subassociation, typical variant	Dierssen
D ☐ AF5	Pleurozio-Ericetum tetralicis, *Molina caerulea* facies*, subassociation with *Racomitrium lanuginosum* (*AF5 lacks *Molinia caerulea*)	Dierssen

Figure 3.19 Relationships between water table changes and vegetation. (b) Changes in plant communities with surface microtopography over a small (2 × 0.5 m) area in a blanket mire in Sutherland, northern Scotland. The left-hand block diagram indicates the location of each community within the sample area in relation to elevation, and the right-hand block shows the water table position. The box plots show the median and ranges for each vegetation type recorded here and at another location. Redrawn from Lindsay *et al* (1988) with permission of Scottish Natural Heritage.

For example, Holmes *et al* (1993a) note water table depth as one of the key variables associated with communities of ground beetles (Coleoptera: Carabidae). In protozoa, water table depth is undoubtedly the major factor affecting assemblages (Charman and Warner, 1992). Although the mechanisms behind this gradient are not clear for these microscopic organisms, factors involved are likely to be related to efficiency of the desiccation survival mechanisms and size and shape of the outer tests (Heal, 1964; Bobrov *et al*, 1999).

3.9.4 The lithotrophic–thalassotrophic gradient (or oceanic–continental gradient)

This gradient represents the influence of salinity, primarily due to oceanic influences. In fen systems, freshwater fens can intergrade to coastal marshes with a tidal influence (Wheeler and Proctor, 2000), although saltmarshes are outside the scope of this book. However, the oceanic influence is apparent in many freshwater peatlands, through spatial variability in the water chemistry of precipitation with distance

(c)

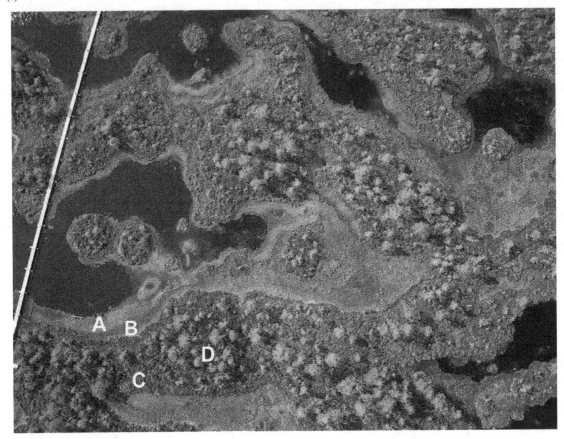

Figure 3.19 Relationships between water table changes and vegetation. (c) Low level aerial photograph of vegetation zonation within pool and hummock system in Männikjärve Bog, Estonia. Plant communities change in association with depth to water table from the centre of a pool to the highest point of a hummock or ridge. A = *Sphagnum cuspidatum* floating in water, B = *Sphagnum cuspidatum* around pool shore, C = *Sphagnum rubellum* above water, D = pine trees on hummocks along with dwarf shrubs – *Empetrum nigrum, Chamaedaphne calyculata, Andromeda polifolia and Calluna vulgaris*. This photograph was taken from a camera fixed to a kite, taken by James Aber (for colour version see http://www. emporia.edu/kite/estonia/color/color.htm). Reproduced with kind permission of James Aber.

from the ocean and direction of prevailing winds. The oceanic–continental gradient is particularly marked in ombrotrophic mires where subtle changes in precipitation composition have a large proportional influence on surface water chemistry. Sodium, chloride and magnesium tend to be enhanced in oceanic peatlands and the total ion concentration is higher too (Comeau and Bellamy, 1986).

Because they tend to be cooler and wetter than continental areas, oceanic regions often have a greater extent of ombrotrophic peatland than continental regions and there have been a number of studies reflecting this influence at different spatial scales (e.g.

Johnson, 1977a, b; Sjörs, 1985; Comeau and Bellamy, 1986). Ionic concentrations can vary significantly even over very short distances. For example, Slack and Hallingbäck (1992) found major changes in Na^+ in peatland surface waters over a range of only 25 km in suboceanic mires in southern Sweden. While the principal change on the oceanic–continental gradient is often assumed to be in water chemistry, the depth and duration of snow cover may also be important in some cases (Johnson, 1977a).

Occasionally, inland peatlands may be affected by a saline influence, not related to proximity to the ocean. For example, the vegetation communities on the

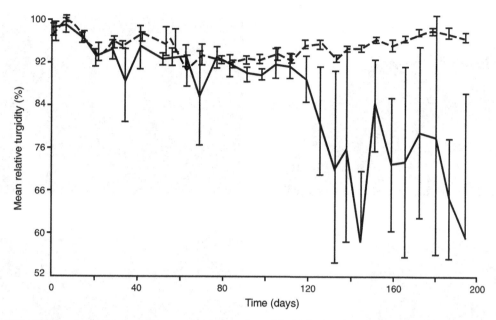

Figure 3.20 Changes in relative turgidity over time in waterlogged (solid line) and control plants (dashed line) of *Empetrum nigrum* (crowberry). Waterlogged plants were grown with a water table at 0 cm and control plants with water table at 9 cm depth. There were three plants in each treatment until day 165 when one waterlogged plant died. Relative turgidity is a measure of the water stress in plants. Bars show ± standard errors. Redrawn from Bell and Tallis (1974) by permission of Blackwell Science Ltd.

extreme rich fens in the Southern Rocky Mountain, Colorado, are found to be related to changes in salinity as well as surface wetness (Cooper, 1996). Halophyte plants such as *Triglochin maritima* and *Glaux maritima* occur in 'salt flat' communities, a result of conditions caused by very low precipitation–evaporation ratios and the formation of sodic peats. Other factors which may induce some salinity in peatlands are penetration of brackish groundwater and the contribution from terrestrialised estuarine sediments (Wheeler, 1999).

3.9.5 The mire expanse–mire margin gradient

The mire expanse–mire margin gradient has always been identified as an important factor in changes in the vegetation of European mire systems (e.g. du Rietz, 1954; Malmer, 1986; Eurola *et al*, 1984). Wheeler and Proctor (2000) reject the gradient as a useful independent ecological factor as it will reflect a number of other changes such as peat depth and the effect of water flow from centre to margins in ombrotrophic mires. As it is associated with the gradient of the surface itself, it also tends to be related to water table depth. This is clearly indicated by the way in

which pool–hummock topography is more pronounced in the centre of raised and blanket mires (van der Molen *et al*, 1994; Lindsay, 1995). However, there is no doubt that on ombrotrophic mires in particular, the gradient is strongly associated with vegetation change in many peatlands. Recent studies show that this is a strong gradient in the tropical peat swamp forests (Phillips, 1998; Page *et al*, 1999). Major forest transitions are strongly associated with the change from shallow peat and mineral soils at the edge of the swamp to the deep peats at the centre (Figure 3.21).

3.9.6 The deep peat–mineral soil gradient

The depth of peat is another factor that is closely related to other factors, which are more likely to be the proximal causes of ecological variability. Deeper peats tend to occur towards the centre of peatlands, so peat depth and the mire centre–margin gradient are often associated with each other (e.g. Charman and Smith, 1992). Equally they tend to be wetter, especially in the centre of ombrotrophic mires and in valley mires. Perhaps the only meaningful direct effect of peat depth is the effect on rooting zone of plants.

Figure 3.21   (a) Principal forest types at Setia Alam, a tropical peat swamp forest in central Kalimantan, Indonesia. The zones of forest types are closely related to the mire centre–mire margin gradient, which reflects a combination of peat depth and surface slope and hydrology. Redrawn from Page *et al* (1999) by permission of the Royal Society and the authors. The photographs show the vegetation in some of the main vegetation zones: (b) marginal mixed swamp forest; (c) low pole forest, located up to 7 km on to the peat dome on peat up to 10 m thick. The trees have a lower canopy height and diameter than the mixed swamp forest and there is abundant *Pandanus* on the forest floor; (d) low canopy forest on the highest section of the peat dome. Here very low growing forest is located on the thickest peat (>10 m). The stunted trees grow on peat 'islands' emerging from shallow blackwater pools. Photographs © Sue Page and Jack Reiley.

Where peat is sufficiently shallow, some plants may be able to root into underlying mineral soils to obtain additional nutrients to those present in the peat.

3.9.7 The spring–flush–fen gradient

This gradient is a combination of two factors. First, the 'spring' gradient which arises as a result of the emer-

gence of groundwater at particular locations in some peatlands. It thus provides a point source of additional water and nutrients, which affect plant growth, diminishing away from the spring head. Second, the 'flush' gradient, which arises as a result of lateral water movement from any source. This can occur even in ombrotrophic mires on sloping areas that receive runoff from higher parts of the peatland or in concentrated areas of surface water flow, which form ephemeral streams on the surface of the peat. Chapman (1964) suggests that the relative cover of *Sphagnum magellanicum* and *S. papillosum* on Coom Rigg Moss, Northumberland, England is related to a flushing effect. Damman (1986) finds that nutrient-demanding species such as *Nymphaea odorata* and *Utricularia vulgaris* increase in response to greater water flow on the lower slopes of ombrotrophic bogs in Newfoundland, Canada. The degree of lateral water movement is important, particularly in relation to the degree of oxygenation and solute supply (Wheeler, 1999). The influence of this factor by comparison with the effect of different sources of water may be difficult to quantify, however. Classifications that invoke this as a significant factor (Kulczynski, 1949; Bellamy, 1968) with the term 'rheophilous' (flow-loving) do not distinguish between the effects of flow *per se* and the influence of groundwater (Wheeler, 1999).

3.9.8 Climatic gradients and other factors

In addition to the gradients discussed above, there are a number of other factors that play important ecological roles. Perhaps the most significant of these is climate, something that only becomes apparent when large spatial scales are considered. In addition, habitat-specific factors such as shade (Belland and Vitt, 1995) and microclimate (van der Molen and Wijmstra, 1994) may be important when particular circumstances are examined at different scales. Likewise, the effects of management and overall habitat characteristics can be important in relation to invertebrate populations (Coulson and Butterfield, 1978; Holmes *et al*, 1993b). Some of the impacts of these external factors will be explored later in the book.

'Climate' encompasses many factors and some of these are related to the oceanic–continental gradient discussed above (e.g. snow cover) but others operate independently of these or almost so. Glaser (1992c) provides an in-depth examination of regional variability in species richness and plant communities for raised bogs in eastern North America, covering an east–west distance of some 4000 km. As with many other areas, there is a continental–oceanic gradient identified but this is not so much related to changes in ionic composition of bog waters, but rather to precipitation and temperature regimes. Species richness was particularly well correlated with mean annual precipitation and annual freezing degree-days (cumulative temperature below 0 °C). Sites from the floristically rich maritime region

have more than 1000 mm annual precipitation and less than 1000 freezing degree-days; sites from impoverished bogs have less precipitation and lower temperatures. This is partly explained by the fact that elements of the northern arctic-alpine flora and of the southern temperate flora can survive in the intermediate maritime sites. In contrast, temperate species are excluded from the northern continental sites and arctic-alpine

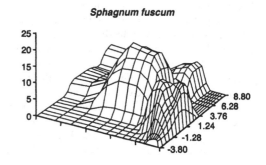

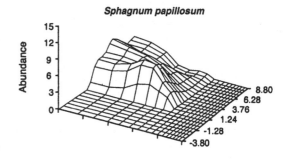

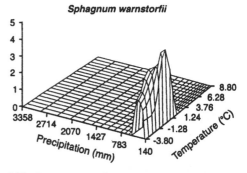

Figure 3.22 Response surfaces showing the changing abundance of selected *Sphagnum* species with major climatic factors in western Canada. *Sphagnum fuscum* occurs over a wide climatic range, whereas *S. papillosum* is restricted to areas of higher precipitation and higher temperatures and *S. warnstorfii* occurs in low-precipitation, low-temperature areas. Redrawn from Gignac (1994) by permission of the Society of Wetland Scientsts.

species are absent from the southern continental sites (Glaser, 1992c).

The distribution of individual species can also be characterised in terms of climatic variables. Gignac (1994) summarises some of the key findings from earlier work (Gignac *et al*, 1991a, b; Gignac, 1992), showing that *Sphagnum* species in western Canada vary considerably in their relationship with mean annual temperature and mean annual precipitation (Figure 3.22). These relationships only become apparent over wide geographical areas related to significant changes in climatic regime. At the individual site scale, summer moisture levels (humidity and precipitation), including those of the previous year, have been found to be important controls on *Sphagnum* growth (Backéus, 1988) so a direct climate control seems to exist.

For animal life, climatic conditions may be more important than for many plants, especially in invertebrates where temperature is a critical control on physiology and life cycles. Over smaller regions such as those across northern Britain, the relationship with climate manifests itself as a relationship with altitude (Coulson *et al*, 1995).

3.10 SUMMARY

In this chapter I have attempted to bring together some of the critical aspects of the hydrology and ecology of peatlands as a basis to further understanding of the dynamic nature of peatland functioning in the rest of the book. There is much more to read about this area of peatland science, especially for those who are interested in the ecology of particular regions or mire types or in tracking through time the changes in ideas and received truths on the subject. The next chapter of the book turns to the beginnings of peat growth and the reasons for the initiation of peatland development. Why and how does peat form in the first place?

4

Origins and Peat Initiation

4.1 INTRODUCTION: TIME AND PEAT GROWTH

So far we have discussed general definitions and classifications of peatlands, the landforms they have created and their contemporary ecology and hydrology. For the next five chapters we will turn to aspects of *time*, to look at the ways in which peatlands have developed and their changes in response to both external and internal processes. Time is of central importance to peatland functioning and also to our own understanding of peatland science for two reasons. First, as we have emphasised earlier, peatlands evolve over the course of millennia and their ecology today depends to a large extent on this long-term development. A raised bog would not be 'raised' if it were not for the differential accumulation of peat forming the dome over the previous 5000–10 000 years! However, many ecosystems have some interdependence on past events, but there are few ecosystems (perhaps with the exception of lakes) which actually record these past events faithfully over thousands of years. Thus, the second reason for the importance of time for peatland study is that the preserved 'archive' of environmental change can be used to unravel long-term processes and functioning. Palaeoenvironmental studies have become a major part of peatland science. Despite the presence of this record, which is often well preserved and provides a multi-faceted outlook on the past, the evidence is not always unequivocal, as we shall see. To begin with, we will tackle a fundamental question that could be asked of any ecosystem – how did peatlands begin their growth?

4.2 FRAMEWORKS FOR PEAT GROWTH

To begin to address this question, we need to go back to a consideration of the fundamental factors that affect peatlands throughout their existence. These were discussed in section 2.2 with reference to Figure 2.1, which shows how a range of internal and external influences affects peatland growth (Warner, 1996). We can adapt these general ideas to consider what it is that determines when and where the initiation of a peat deposit takes place. In the simplest terms, the critical threshold to be crossed is for production of organic matter to exceed decay. Taking this as the fundamental control, then theoretically production could increase to overcome decay. However, in practice any increase in production is often accompanied by increasing respiration in soils and although the organic content of the soil may increase, it is rarely sufficient to form peat. The transition from open to forested landscapes would be a good example of this kind of change. In peatland initiation, it is the decay rate of the biomass produced that controls the production–decay balance sufficiently to cause peat formation. In turn, decay rate is primarily a function of moisture status, with secondary influences from temperature and water chemistry, including the degree of oxygenation in the water. So, as we might expect intuitively, the amount of water is critical. However, the apparent simplicity of this idea belies the complexity of interacting processes that cause peat to begin growth. The lengthy history of debate in the literature on causes of blanket peat initiation in the British Isles (summarised by Moore, 1993) is a good example of the difficulty of interpreting evidence in a particular situation and we will discuss this later.

4.2.1 Factors influencing the hydrological balance at peat initiation

Taking the model in Figure 2.1, if we simply consider the amount of water at the surface to begin with, there are five main factors that are important:

(i) *Climate*. Clearly climate must be of central importance in determining whether there is likely to be a surplus of water available for peat initiation. The precipitation–evaporation balance is critical here and a surplus can result from low precipitation–low temperature or high precipitation–high temperature regimes as well as high precipitation–low temperature conditions.

(ii) *Geomorphology*. Topography creates spatial diversity in the hydrological characteristics of the landscape. Some areas will be natural foci for water collection (basins) and others will tend to shed water rapidly (hilltops). Slope gradient may also be important.

(iii) *Geology and soils*. Permeable substrates cause more rapid water loss and geological features may create particular conditions at specific sites as a result of faulting or hydrogeological features such as spring-lines.

(iv) *Biogeography*. Particular areas of the world may be more susceptible to peat development partly as a result of the presence of particular plants or groups of plants.

(v) *Human activities*. Most development of extant peatlands has taken place in the last 10 000–15 000 years in a period of major human advancement. Human modifications to the hydrological balance are thought to have been critical to the extent of peat development in some areas of the world.

In any particular instance, some or all of these factors will be important and inevitably interactions between them are normally involved. A critical aspect of understanding the causes of peat initiation at a specific location is to identify the conditions that are limiting peat development. In some circumstances some of the conditions may be adequate for peat formation, but peat growth can only occur with a change in the limiting conditions. For example, climatic moisture surplus may be adequate for peat growth over soil with open vegetation but inadequate in forested areas with a higher evapotranspiration rate. Deforestation through human activity or natural fire may therefore be required to tip the scales towards peat growth. The cause of peat growth could be seen as deforestation but climate would be also implicated as a 'background' condition. In this chapter we will attempt to show some of the variety of ways in which peatlands have formed throughout the world and to explore some of the processes and pathways that this involves.

4.3 PATHWAYS TO PEAT GROWTH: TERRESTRIALISATION AND PALUDIFICATION

Excess of water can result in two main pathways of peat formation, termed terrestrialisation and paludification. Terrestrialisation is the process by which a shallow water body is gradually infilled with accumulated debris from organic and inorganic sources. This continues to a point where the water table is at or below the surface for at least some of the year, and peat accumulates over the previously deposited limnic (lake) sediments. Paludification describes peat formation directly over a mineral substrate with no fully aquatic phase and no limnic sediments laid down. In practice, the two processes are often combined in the growth of any particular peatland area, although the relative importance of the two processes is highly variable.

4.3.1 Hydroseral succession and peat initiation

Hydroseral succession is dealt with more extensively in relation to autogenic processes in Chapter 7 but the transition from limnic sediment to terrestrial peat is relevant to any consideration of peat initiation. Peat growth only commences at the point when the aquatic phase has ceased, and the process is most often related to infill and stabilisation of the substrate to allow peat-forming terrestrial plants to colonise what was once open water. The progression from an aquatic to a peat-forming ecosystem can be influenced by a number of factors including climate, topography (particularly as it affects the catchment area of the site), and biogeographical factors such as soils and vegetation around the site, which may significantly alter the water balance.

Hu and Davis (1995) describe a typical hydroseral succession for a raised mire in Maine, USA. An early lake phase lasted from deglaciation through until 8500–8000 radiocarbon years BP.[1] This was followed by accumulation of sedge fen peat and then sedge fen with *Larix laricina* (larch or tamarack) forest before the growth of *Sphagnum* bog from 5500 radiocarbon years BP to the present day. Peat initiation (as distinct from limnic sediments) therefore occurred by around

---

[1] Two different timescales are referred to throughout the book – radiocarbon years and calendar years (cal.). These are slightly different from one another for reasons explained in Chapter 6. Unless otherwise stated, ages are quoted in radiocarbon years. BP – before present.

8000 radiocarbon years ago, although the exact timing varied in different locations on the mire. The causes of initiation in a case such as this are a combination of external conditions (primarily climate) and local hydrological controls. Clearly the climate was wet enough to ensure a positive moisture balance on the mire surface, but the proximal cause of peat initiation was lake infill and establishment of a stable substrate for the peat-forming vegetation to grow. The observed differences between locations within the mire also support this idea, with slightly elevated areas becoming terrestrialised first. Other types of peatland initiation in hydroseral succession show a similar combination of internal and regional climatic factors at work, although the clear separation of any of these effects is often difficult (e.g. Bunting *et al*, 1996).

Hydroseral successions leading to peat initiation can be accelerated or delayed by several factors. In glaciated areas, the process of re-establishing biological activity in a lake may take some time (Hu and Davis, 1995) and physical processes can also influence early development. Warner *et al* (1991a) suggest that the slow melting of a dead-ice block in a kettle hole resulted in a delay to sedimentation in the basin following deglaciation. Renewed glacial activity possibly with increased aridity delayed the onset of peat initiation around a cirque lake in Nepal by almost 10 000 years (Yonebayashi and Minaki, 1997). Climate may also delay the development of the water body needed for hydroseral succession to a peatland. A dry climate during the early Holocene delayed the flooding of a shallow basin near Beauval, Saskatchewan, western Canada, so that eventual peat initiation through hydroseral succession was delayed until approximately 3000 radiocarbon years BP (Kuhry, 1997).

In low-lying coastal areas, sea-level changes can have a significant effect on both the initiation and sequence of hydroseral change. Areas of coastal emergence clearly cannot develop peat until the ground is exposed above the tidal limit (see section on tropical peatlands below). In other areas, marine transgressions can interrupt a succession. The sequence recorded by Smith and Morgan (1989) shows a superb example of the way in which a succession was reversed by sea-level rise. A *Phragmites* reedswamp peat began to accumulate just after 7000 cal. BP, followed by colonisation by a shrub community. A minor marine incursion reverted the succession to reedswamp then through a drier fen stage to ombrotrophic bog at 5750 cal. BP. This relatively mature stage in peat development was cut short by a major marine incursion at 3400 cal. BP.

4.3.2 Paludification

As Sjörs (1983) has pointed out, paludification is a much more important process than terrestrialisation, in terms of the proportion of peatland area that has developed this way. Typically, peats formed through terrestrialisation cover only relatively small areas, although of course they may subsequently influence the rate of spread by paludification of surrounding dry ground. Given an adequate supply of water, terrestrialisation proceeds naturally over time in any suitable water-collecting basin – it does not necessarily require a change in external conditions for peat growth, simply the elapse of enough time to fill in the basin with sediment. On the other hand, for paludification of dry ground to occur there usually has to be a shift in local hydrological conditions to produce waterlogging in the soil sufficient to reduce the decay rate significantly below that of production. It is often assumed that this requires an external factor such as climate to change, but it may also be a result of natural successional change (Davis, 1988; Smith and Taylor, 1989). It has been shown that in some cases, paludification can take place very rapidly on exposure of ground under suitable climatic conditions. In the Gulf of Bothnia, rapid isostatic uplift of up to 9 mm yr^{-1} is exposing land rapidly, which has been colonised by *Pinus sylvestris* (Scots pine) forest (Hulme, 1994). Incipient peat formation is also taking place within the pine forest with invasion of *Sphagnum* and development of a thin peat layer. The plant macrofossils and pollen content of the soil layers show the sequence of change that has taken place over the past 500 years, and especially in the last 50 years since the pine invaded the area (Figure 4.1). Soil waterlogging here results from a low hydraulic head and a shallow hydraulic gradient. Paludification is proceeding as a natural successional process under existing climate, soil and topographic conditions; no shift in hydrological conditions has been necessary. The potential causes of paludification have been debated mostly in relation to blanket peat growth in the British Isles (see discussion below) but the processes of the change are poorly understood. Davis (1988) commented that, 'We have, perhaps, focused too much on consequences and too little on dynamics' and little has changed since then.

Successional change may be a result of soil-forming (pedogenic) processes which progress over centuries or millennia, creating slow alteration of conditions within the soil profile which influence the soil hydrology, nutrient status, acidification and consequently the vegetation (Taylor and Smith, 1980). Leaching and

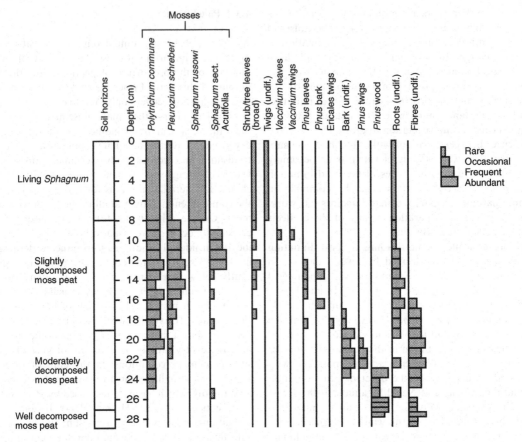

Figure 4.1   Changes in a soil profile and selected plant macrofossils from Hailuoto in the Gulf of Bothnia, Finland (Hulme, 1994). The plant macrofossils show that the mosses have colonised the pine forest (represented by the *Pinus* wood and bark) in the first stages of paludification of the ground recently exposed through isostatic uplift. See text for further discussion. Redrawn with permission of the New Phytologist Trust.

podsolisation are usually implicated in these changes, producing a thin organic soil horizon which can be the precursor of peat formation, termed 'pedogenic peat' by Taylor and Smith (1980). There is good evidence from various parts of boreal North America that this is an important process in paludification (e.g. Alaska – Ugolini and Mann, 1979; Newfoundland – Hilbert *et al*, 2000; Quebec – Payette, 1984). In this process, leaching of iron and aluminium oxides may lead to the formation of a less permeable 'iron-pan' type horizon within the soil which increases water retention in the upper layers of the soil. This is also associated with acidification as base cations are removed and acidification also inhibits decay of organic matter in the surface horizons. Increasing acidity may also encourage the invasion of plants which have further potential for

acidification and water retention, such as *Sphagnum* (Foster, 1984; Noble *et al*, 1984). The formation of a thin organic horizon at the surface can also inhibit drainage of water through the profile, further increasing waterlogging. Thus there is a sequence of processes that provide positive feedback for increasing moisture levels in the soil that can be seen as a natural succession in areas where the climate and the geological substrate are suitable. Once begun, the cycle may be hard to break, although recovery is possible. Cruikshank and Cruikshank (1981) describe the changes within a woodland in Northern Ireland over the past several millennia and show that the area was deforested on two separate occasions during the last 2000 years. Development of humus–iron podsols occurred, perhaps signalling a change towards incipient peat formation, but the forest recovered once the land-use pressure was reduced and a succession through birch and oak forest took place.

Paludification is heavily modified by small-scale local variability through topography and feedback effects of any existing peatland extent. There are now many examples of the lateral expansion of peatlands which show that paludification continues for thousands of years in most systems and still continues today in some (e.g. Foster and King, 1984; Futyma and Miller, 1986; Nicholson and Vitt, 1990; Charman, 1994). The geomorphological variability of most pre-peat landscapes results in major differences in the time of peat initiation. For raised bogs, there are usually considerable differences between the ages of peats in the centre and at the edges of the site. In a study on an Atlantic plateau bog in New Brunswick, Canada,

Warner *et al* (1991b) show peat which began growth over 10 000 radiocarbon years ago did not spread to some surrounding areas until 1200 radiocarbon years BP, and the peat area probably still continues to expand today. Newnham *et al* (1995) reconstructed the developmental history of Kopouatai Bog in New Zealand over the past 20 000 radiocarbon years and demonstrated the gradual expansion of the peat area, also influenced at various times by changes in drainage patterns and sea-level fluctuations (Figure 4.2). Where the topography is relatively even over a large area, extensive paludification can occur almost simultaneously. For example, on a raised mire in southern Finland, Mkil (1997) found that 20% of the peat extent had begun growth in the first 400 years of development and by 2000 years later almost 50% had been initiated.

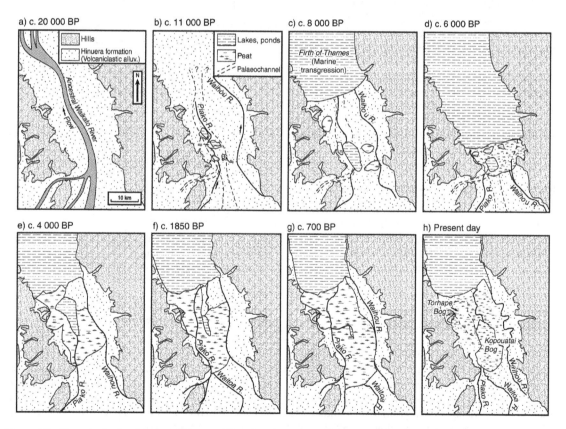

Figure 4.2   Stages in the developmental history of Kopouatai Bog in the Hauraki Lowlands, New Zealand. Early nuclei of peat formation were on the fringes of small lakes. These expanded and new nuclei were created over the early Holocene (c), but then expansion through paludification was limited by sea-level rise (d). The mid to late Holocene saw lateral expansion in all directions, following sea-level fall with the maximum extent of the bog at about 700 radiocarbon years BP. The contraction in the area of the bog (marked by the thick line in h) occurred due to drainage and agricultural activities since European settlement. Redrawn from Newnham *et al* (1995) with permission of Arnold Publishers.

Lateral expansion rates were fastest during the first 400 years at over $1550 \, m^2 \, yr^{-1}$, slowing to $415 \, m^2 \, yr^{-1}$ and then between 140 and 240 $m^2 \, yr^{-1}$ after about 8000 years ago.

While it is conceivable that in stable conditions pedogenetic processes alone could lead to peat formation, there is a range of other influences that can accelerate the process, including climatic change, deforestation and fire. Climatic change is clearly the main driving variable when large spatial patterns of peat initiation are concerned (e.g. Campbell *et al*, 2000) and it can also be related to rates of lateral spread (Korhola, 1994). In other regions and at smaller spatial scales human activities of deforestation and burning may be more important (e.g. Moore, 1993). In the case of climatic change, increased precipitation and/or decreased temperature can increase surface wetness. Deforestation is usually regarded as decreasing evapotranspiration and/ or increasing surface runoff, both allowing more moisture to reach the soil. The role of fire is not always clear. The work by Mallik *et al* (1984) on heathland soils in Britain is usually cited to support the idea that fire can increase susceptibility to waterlogging through the clogging of pore space by small charcoal particles in the upper layers of soil. On the other hand, in North American forests, fire may be more likely to inhibit paludification through the destruction of organic surface material (Viereck, 1983; Davis, 1988).

4.3.3 Primary, secondary and tertiary peat growth

Moore and Bellamy (1973) developed the concept of different pathways of peat growth to identify three broad types of peat: primary, secondary and tertiary. While these concepts are not often referred to explicitly in more recent literature, the concepts they represent are useful in understanding the early stages of the genesis of peatlands. Figure 4.3 shows the basic idea of this concept as applied to three different peatland types. Primary peats are those formed in basins or depressions where water collection is at a maximum and are often preceded by a limnic phase of development. Many peatlands have elements that are based on primary peat formation, including many presently ombrotrophic systems. Secondary peats are those which grow as a result of the tendency of the ground surrounding the primary peatland to become waterlogged and allow the peat to expand beyond the deepest parts of a depression. Again most peatlands have elements that can be regarded as secondary. Tertiary peats are those that develop above the influence of groundwater and conditions allow the peat to cover most areas of the land-

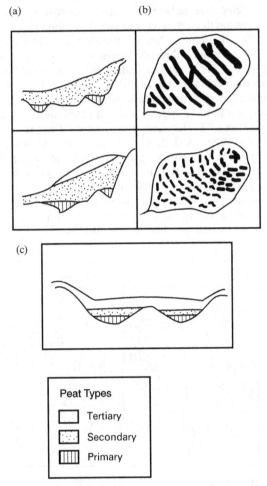

Figure 4.3 Cross-sections of peatlands demonstrating Moore and Bellamy's concept of primary, secondary and tertiary peat formation as applied to (a) a sloping fen, aapa mire or patterned fen, (b) an eccentric raised mire, (c) a blanket mire profile. The layers shown are not necessarily recognisable as stratigraphic units, although they may coincide with these. See text for further details. After Moore and Bellamy (1973) by permission of Peter Moore.

scape. Moore and Bellamy (1973) suggested that this was due to the capillary action of the peat, but we now know that it is low hydraulic conductivity that enables this to occur (Ingram, 1982). These tertiary peats occur primarily in ombrotrophic mires.

Moore and Bellamy also linked their mire types explicitly to climatic conditions, with primary mires occurring throughout many areas, including those with relatively warm and sometimes dry conditions, and secondary and tertiary mires more limited to regions with cooler and more humid climates. In practice the distinction between primary, secondary and tertiary peats in the field or even with microscopic examination is not always straightforward, with the three categories merging into one another in sites that have extensive peat development. Figure 4.3 shows that all three peat types may overlie each other. While the sequence primary–secondary–tertiary may be identifiable in peat composition, the primary–secondary transition is particularly difficult to be clear on. Normally the secondary–tertiary transition is associated with a change from minerotrophic to ombrotrophic peat types but in, for example, poor fen–bog transitions even this may be vaguely defined. However, the concept of primary–secondary and tertiary peats remains a useful idea when visualising peat initiation and lateral spread (e.g. Edwards and Hirons, 1982).

4.4 EVIDENCE FOR THE ORIGINS OF PEATLANDS

The evidence for the origin of peatlands comes from palaeoenvironmental records preserved within the basal layers of peat and the sub-peat sediments and soils. Chapter 6 examines the fundamentals of these methods in greater detail, but we will briefly discuss aspects of the evidence here that are relevant to understanding some of the problems involved in its interpretation.

4.4.1 Defining peat initiation

In the transition from dry land to peatland, the actual point of transition can be hard to identify. In a hydroseral succession, we could identify the change as primarily one of vegetation, from an aquatic community to a terrestrial fen community. Peat composition and perhaps pollen data reflecting wider vegetation change could be used for this. In paludified soils the change is not so much one of vegetation as one of soil status. Rarely does the transition to a peat from a mineral soil appear as a sharp dividing line in the stratigraphy of the profile, and on closer analysis of the content of the sediments the picture often becomes more, rather than less, complex. Often there may be a transition phase, where the soil is becoming increasingly organic but is not yet a true peat. As we have already mentioned, such a change may be reversible if it does not proceed too far (Cruikshank and Cruikshank, 1981), so at what point do we say peat initiation took place?

Definitions could be based on appearance, organic matter content or pollen and/or macrofossil content. There is no easy answer to this question but definitions used need to be borne in mind when attempting to compare sites and ages using different approaches. Since basal peat accumulation rates may be very slow, a few centimetres up or down a profile may make a difference of several hundred or even a thousand years in some cases.

4.4.2 Types of evidence

Besides stratigraphic description, the evidence used in studies of peat initiation is usually from three main sources: pollen, charcoal and radiocarbon data. These may be supplemented by plant macrofossil information (e.g. Charman, 1992a) or geochemistry (e.g. Cruikshank and Cruikshank, 1981) but this information is generally supplementary. The derivation and interpretation of these data have various problems particular to the base of peat profiles. First, pollen analysis is a well-established technique for reconstructing vegetation change around a site. However, in soils, the preservation of pollen may not be complete and differential loss of particular pollen types may occur (Dimbleby, 1985). Second, charcoal data are derived in a variety of ways. Counting on microscope slides is probably the most advanced method (Clark, 1982), but some studies only use visual inspection to identify layers of charcoal (e.g. Smith and Cloutman, 1988). However, it can be difficult to identify clear changes in charcoal frequency using the latter approach because highly decayed basal peats are often very black and greasy in appearance even where charcoal is absent. Third, radiocarbon dates from basal peats can be affected by contamination as a result of root penetration from above and movement of mobile humic acids. These problems often result in unexpected ages, mostly ages that are too young (e.g. Smith and Cloutman, 1988).

4.4.3 Correlation and causation

The final issue to consider in interpretation of peat initiation data is the problem of attribution of cause and effect. There are many examples of correlation between environmental change and peat initiation, some of which are explored below. In most cases, the best we can do is to find evidence for coincidence of change but we cannot easily attribute cause and effect. For example, a decrease in forest cover is often associated with the beginnings of peat growth on paludified ground. There are two possible explanations for this. First, deforestation (for example as a result of human activities) could have caused peat growth by increasing moisture levels in the soil (see above). Second, deforestation and peat initiation could be caused by the same external factor, such as climate change. The identification of the actual cause of peat formation (human activity or climate) is not necessarily easy to identify and we must rely on other independent data to support one or other hypothesis. A further good example of this problem is shown by a consideration of peat initiation and archaeological radiocarbon dates in southern Africa (Meadows, 1988; see below).

4.5 EXAMPLES OF PEAT INITIATION

We have thus far considered the nature of peatland initiation in general terms, briefly examining the underlying causes of peat formation, looking at pathways of development and the evidence of these processes that is left behind. The remainder of this chapter will examine a variety of specific examples of peatland initiation by exploring the range of situations in which peat begins growth. A particular focus will be on the sometimes conflicting evidence for the role of climate and human impact in mire formation. While some mires must be 'naturally' developed under past climate conditions, there is equally strong evidence that others are a product of human activity and would not have developed without this interference. The discussions over the relative importance of human vs. climate in the origins of peatlands are perhaps not helped by the fact that much of the early speculation on this subject was over the origins of blanket mires in the British Isles. In retrospect, it is clear that this is one of the most complex situations that can be found because of the long history of human activity in the landscape and the stratigraphic complexity of the blanket mire landscape. The discussion will begin with a consideration of the genesis of

blanket mires in the British Isles and will then turn to other areas of the world and other peatland types.

4.6 BLANKET MIRE INITIATION IN THE BRITISH ISLES

Blanket mires are by far the dominant type of peatland in Britain and Ireland, covering large areas in Scotland, Wales, northern England and Ireland (Table 4.1). Over 85% of the 1 645 970 ha of total peatland area in England, Scotland and Wales is blanket mire and almost 75% of this is in Scotland. The distribution of blanket mires is strongly related to climatic factors such as rainfall and rain days per annum (Lindsay *et al*, 1988; O'Connell, 1990). There has been much debate concerning the origin of blanket mires in Britain and reviews and discussion of the evidence are given by Moore (1993) and Tallis (1991).

Early work found that peat growth in the Pennines, northern England, occurred during the 'Atlantic' period (approximately 7500 to 5000 radiocarbon years ago) when climate was thought to have been warmer and wetter than the present, and the growth of peat was therefore thought to be entirely due to a climatic cause (Conway, 1954). However, as further data on basal peat ages and pollen contents became available, anthropogenic impacts were strongly implicated as a trigger for peat growth (e.g. Moore, 1973, 1975; Figure 4.4). This work suggested that many peat profiles began growth at or around the time of the 'elm decline' and were associated with the commencement of more intense human activity in the early Neolithic. Figure 4.4 shows a typical basal profile with pollen data showing main changes in vegetation. The detailed contiguous sampling strategy reveals a series of phases of disturbances to the woodland growing on the site. Although there were periods when the woodland recovered to an extent, successive impacts of forest canopy clearance, stock grazing or firing of the vegetation even-

Table 4.1 Areas (ha) of different peatland types in mainland Britain according to a recent estimate by Lindsay (1995). Data are incomplete for Welsh and Scottish fens but the excess of bog over fen peat is real and the dominance of blanket bog in the totals is notable.

Country	Fen	Raised bog	Blanket bog	Intermediate bog	Total
England	131 672	37 413	214 138	981	384 204
Scotland	1 215	27 892	1 056 798	10 653	1 095 958
Wales	2 867	4 086	158 770	85	165 808
Total	135 754	69 391	1 429 106	11 719	1 645 970

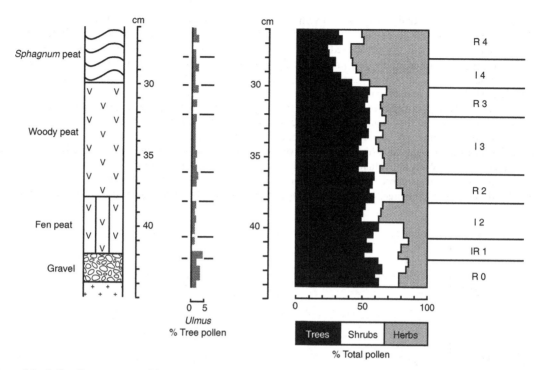

Figure 4.4   Pollen diagram summarising the changes in vegetation during peat growth in a blanket bog in upland Wales, based on Moore (1973). Only selected taxa from the original diagrams are shown for simplicity. The diagram is divided into a series of zones representing disturbance by humans or their animals ('I' phases) and regeneration following disturbances ('R' phases). Redrawn and compiled from Moore (1973) with permission of *Nature*.

tually led to the demise of the woodland cover and the development of *Sphagnum* peat.

Human impact through deforestation was thus suggested as the main cause of blanket peat initiation in Britain through a series of processes (Moore, 1975). Two main lines of evidence were cited in support of this hypothesis. First, archaeological data supported the idea that early human populations were capable of clearing the forest and had the motivation to do it for grazing of wild ungulates and domesticated stock. Second, data from experimental forest clearance in several areas suggested that the decrease in evapotranspiration and increase in runoff as a result of the clearance would produce increased soil moisture and raised water tables. Thus the circumstantial association between human impacts on the vegetation and the growth of peat strongly suggested that the clearance of the forest *caused* the development of peat.

However, there are also sites where forest growth persisted well after peat formation (Tallis, 1975) and other areas where there was much more variability in the timing of peat initiation (e.g. Chambers, 1981, 1982,

1983a). Further detailed research, particularly that of Smith and Cloutman (1988), showed that peat development could be drawn out over several millennia, even within a small area. A total of 16 profiles within an area of about 8 ha showed peat development at various dates from between about 8000 and 4000 radiocarbon years ago. Despite the asynchronous nature of peat initiation, all of these profiles had charcoal in the base of the peat, reflecting burning episodes sometime before peat initiation. Even in areas that might have been affected by lateral paludification such as that seen in peatlands elsewhere (see above), basal peats contained significant amounts of charcoal. Only the very latest peat at 4000 radiocarbon years ago, which was developed directly over rock, may have been caused by lateral spread. Smith and Cloutman (1988) conclude that the first stages of soil changes towards peat growth began with early human clearance through the use of fire from around 8000 radiocarbon years ago. The heathland vegetation that developed was maintained and spread by further clearing, and podsolisation of the soils advanced. This uniquely detailed study is important as it shows that spatial diversity in the timing of blanket peat growth can arise as a result of human activity over an equally protracted period of time. Human impact is often an ongoing process rather than a 'one-off' event.

The overwhelming weight of evidence discussed so far suggests that blanket peat growth in the British Isles resulted primarily from the actions of prehistoric people, but did climate play any role at all, apart from as a 'backcloth' against which human-driven events were enacted? With individual site studies it is hard to make any but the vaguest associations between peat growth and climate change, but climatic factors ought to be apparent if data from a number of sites are looked at together. Surprisingly, there have been few attempts to look for broad spatial or temporal patterns in British peat initiation data. Figure 4.5 shows a plot of basal peat age estimates for northern England and Wales (Tallis, 1991). Data such as these suggest that climatic change did play a role in the development of blanket peat. Tallis (1991) concluded that a wetter climate from around 7500–7000 radiocarbon years BP was important in early peat initiation on more susceptible locations in the Pennines, and that the growth of peat in the Berwyn Mountains after 2500 radiocarbon years BP may also have been triggered

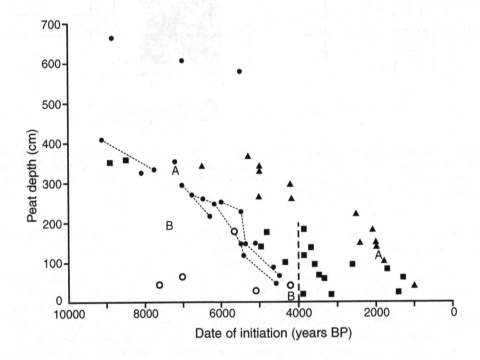

Figure 4.5 Summary of basal peat ages for 57 sites in northern England and Wales plotted as peat depth against age. The southern Pennine dates are generally older than other regions. Dotted lines connect sites within the same catchment showing the diversity of ages within a small area. No obvious clustering of ages suggests that climatic change was less important than local differences between anthropogenic and topographic effects. Redrawn from Tallis (1991) by permission of Blackwell Science Ltd.

by climate rather than deforestation. One of the problems is that much of the evidence comes from the most marginal areas for blanket peat growth in northern England and Wales, rather than from the north and west of Scotland where blanket mire is most extensive and best developed today (see Table 4.1). The limited amount of data show that fire was associated with peat initiation in the Flow Country even from early Holocene peats (up to 9170 ± 90 years BP), although there are none of the pollen indicators of intensive land use (Charman, 1992a). In Shetland and the Faroe Islands it has been suggested, on the basis of pollen data, that some growth of peat occurred throughout the Holocene as a result of natural processes of acidification and leaching (Bennett *et al*, 1992). Climate may also have been important in heathland development on Orkney (Bunting, 1996). On the Western Isles of Lewis and the Uists, blanket peat growth is recorded from the early Holocene and woodland and blanket peat communities coexisted for several millennia until around 5200–4000 radiocarbon years BP, when woodland declined and blanket peat expanded. Initial peat development is attributed to

natural processes in response to a cool wet climate and the acidic bedrock and soils. Although no single factor explains the later expansion, human activities and climate change both probably played a role (Fossitt, 1996).

There is still no clear consensus on the relative importance of human impact and climate for blanket peat initiation in the British Isles. It may never be possible to totally disentangle these two factors. Certainly the data from England and Wales show a strong association between human impact and peat initiation throughout the Holocene. The data from further north are less clear, and climatic change as the primary cause of early peat growth certainly seems likely in many cases. Conceptual models of blanket peat growth can be applied to any of these situations, but the relative weighting of climate and human impacts is likely to vary in space and time (Figure 4.6). As Moore *et al* (1984, p. 232) put it, '... as the climatic limits of blanket mires are approached, their initiation and development become more dependent on human activity'. Perhaps we will not come much further than this in identifying the causes of blanket peat initiation

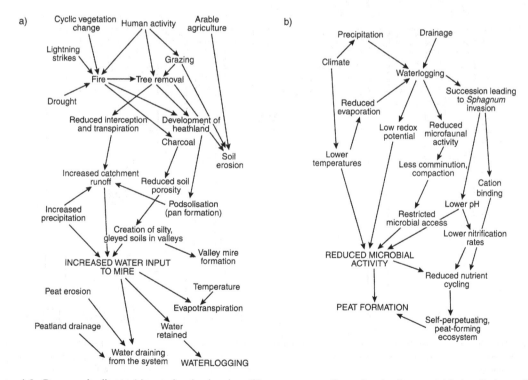

Figure 4.6 Processes leading to (a) waterlogging in mires, (b) consequences of waterlogging for peat initiation. Redrawn from Moore (1993) with kind permission from Kluwer Academic Publishers.

in the British Isles, but we should be wary of assuming similar causes in other paludified mires elsewhere in the world.

4.7 CAUSES OF PALUDIFICATION IN OTHER MIRES

The evidence for the causes of peat initiation in the British Isles is somewhat complicated given the variability of human population history and climatic change in space and time. Other areas of the world have blanket-type peats (see Chapter 1) but are much less likely to have been affected by human activity. Data from coastal blanket mires in Norway show some sites were paludified as a result of climatic conditions (Solem, 1986), while other areas can be attributed to human activity (Solem, 1989). Human effects were clearly still important here and further investigations would undoubtedly reveal sites similar to those in the

British Isles, with a mix of climatic and anthropogenic origins. Further north in oceanic Europe, human impact is less likely. For instance, the steeply sloping peatlands on Svalbard (Lag, 1986) are effectively part of the blanket mire continuum along the fringe of Europe, although little is known of their genesis. In the Northern Hemisphere, one might expect the severity of early human impact to be less in North America, because of the limited development of settled agriculture as a way of life. One area with ombrotrophic mires developed over paludified ground is Newfoundland, Canada. Here dates from various blanket, plateau and raised mires are clustered in the period 4000–2500 radiocarbon years BP with other dates falling within 6000–5000 and a few earlier on. The more recent cluster is also associated with changes from fen to bog conditions in some sites and major pollen changes, especially between 3000 and 2500 radiocarbon years BP (Davis, 1984; Figure 4.7). These peats overlie the remains of forest trees including *Picea* (spruce),

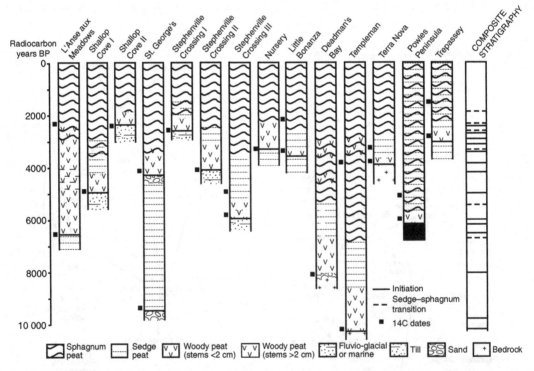

Figure 4.7 Stratigraphies of 14 sequences from ombrotrophic mires in Newfoundland, Canada, shown as changes over estimated time. The composite stratigraphy on the right shows the estimated ages of peat initiation and transitions from fen to bog peat. Accumulation rates are assumed to be linear between radiocarbon dates and the surface, which will inevitably lead to some significant inaccuracies, especially where profiles only have a basal peat age. Redrawn from Davis (1984) by permission of Elsevier Science. See text for further comment.

Abies (fir) and *Betula* (birch), showing that paludification of a forest ecosystem occurred, similar to that for the European blanket mires. Davis (1984) assumes no human activities affected peat growth in Newfoundland and attributes the dates of peat initiation to climate change, modified by pedogenic processes. Since most of the profiles overlie sands and gravels, the impedance of drainage necessary for peat formation required the development of less permeable clay, humus or iron pans and this took some time after deglaciation.

In the Southern Hemisphere, paludification has occurred in oceanic areas of Australia, New Zealand and the subantarctic islands. These peatlands must largely represent climatically determined peats. In Australia, peatlands are limited in extent, but a compilation of data for the state of Victoria provides a useful summary of peat initiation data (Kershaw *et al*, 1997; Figure 4.8). Some of these data refer to successional

peats or those that are later interrupted by renewed minerogenic sedimentation but they do show phases of development which are apparently related to climatic change known from other proxy sources. In Tasmania, Colhoun *et al* (1991) describe a sequence from blanket peats which suggests paludification was taking place from the earliest Holocene or even during the Late-glacial period. It is not clear to what extent these dates are representative of the greater extent of peatland, but they do suggest that initiation was very early here. Human impacts are not discussed, although people were present in this area from much earlier in prehistory. However, in some parts of the New World, human activities can be discounted altogether. A variety of peatlands, including many ombrotrophic peatlands, occur in New Zealand where dates for the first human occupation are probably within the last 1000 years (Newnham *et al*, 1999). Most of the peatlands developed prior to any possible human impact, although later Polynesian and European settlement certainly did alter the vegetation and hydrological balance of peatlands. Some peatlands were initiated as a result of hydroseral succession spread by paludification throughout the Holocene (e.g. Figure 4.2), but largely ombrotrophic peatlands also developed over sloping ground in the South Island from around 7500 radiocarbon years ago in response to increasing rainfall (McGlone *et al*, 1995, 1997). Transitions from wetland to true peatland also occurred around this time (McGlone and Wilmshurst, 1999).

In Africa, peat formation is much more limited as a proportion of the land surface, but there appear to be two main pulses of peat initiation in southern Africa (Meadows, 1988; Figure 4.9). The earlier phase occurred during the Late-glacial and was associated with a well-known increase in moist conditions across much of Africa at this time. The second phase peaked between about 5000 and 3500 radiocarbon years BP and this has also been interpreted as change to a moister climate, following a relatively dry early Holocene. Comparisons with archaeological data suggest that human activity also increased at both these times (Figure 4.9). This leads to an interesting problem of interpretation. Was increased peat growth caused by increasing human activity or did a wetter climate encourage increased human settlement and increased peat growth? Moore (1993) suggests the former, but independent indications of changing climate in Africa perhaps support the latter interpretation (Meadows *et al*, 1988).

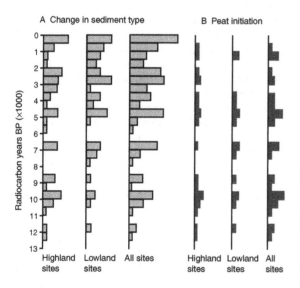

Figure 4.8 Times of stratigraphic change and initiation in peatlands in Victoria, Australia, at 500-year intervals. Phases of peat initiation and change are linked to climate except those in the last 500 years, which are more likely to be linked to European settlement. Redrawn from Kershaw *et al* (1997) by permission of Jack Reiley and Sue Page.

a)

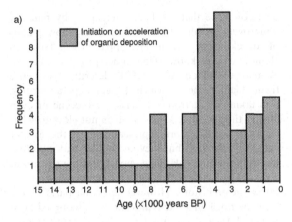

b)

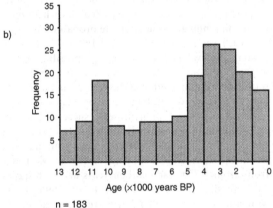

n = 183

Figure 4.9 Frequency of radiocarbon ages on (a) organic deposits and (b) archaeological materials, in southern Africa. Note the correspondence of periods of increasing peat growth and evidence of human activity. Redrawn from Meadows (1988) by permission of Elsevier Science.

4.8 HUMAN IMPACT AS A CAUSE OF PEAT GROWTH IN OTHER PEATLANDS

The origins of many peatlands can be traced to climate but, equally, there are a number of others that are clearly the result of human impacts on the hydrological cycle. One of the clearest examples to emerge is that of the floating bogs of southern Ontario, Canada. These are *Sphagnum*-dominated peatlands that have developed in kettle hole basins, by basin infill where the vegetation mat is wholly or partly floating over a lens of water. They would be referred to as 'schwingmoors' in Europe (see Chapter 1). They are highly varied within this basic definition and are important biogeographically, as they constitute southern outliers

of many 'northern' habitats and the plants that grow there. In the past they had often been considered to be relicts from a time when climate was much colder during the end of the last glacial period (Warner, 1993). However, it had been speculated from ecological principles and observations of the changes in the extent of the *Sphagnum* mat that the vegetation was undergoing rapid changes that must have a more recent origin. Warner *et al* (1989) used peat stratigraphy and pollen data to show that European settlement and deforestation were coincidental with the development of *Sphagnum* bog in four different sites (Figure 4.10). Deforestation would have particularly marked hydrological consequences during spring snowmelt when large volumes of water would run off into the basins. This would increase both the amount of water and the proportion of telluric water (as compared with groundwater) at the surface. This would favour bog shrubs and *Sphagnum* over some of the other taxa on the mire. In particular, the development of a floating mat allows the vegetation to rise and fall with the water level, thus regulating the local water table experienced by the plants. The change to *Sphagnum* bog is the start of peat growth in some cases (e.g. Chesney Bog) whereas in others it marks a transition from previous fen peat accumulation, so it can be seen either as human-induced peat initiation or as successional change forced by human impact.

The initiation of a number of valley mires is also thought to be entirely attributable to human disturbance in several different areas. A number of studies from Britain suggest that the formation of valley peats was aided by anthropogenic forest clearance (Smith and Green, 1995; Mighall and Chambers, 1995), similar to that recorded in upland blanket peats. In areas more marginal for peat development, the human influence may have been even more critical. The stratigraphy and pollen analysis of a valley mire at El Acebron in south-west Spain have been described by Stevenson and Moore (1988). Charcoal layers at the base of the deepest parts of the mire with fluctuations in pollen types such as *Cistus* show that human disturbance was associated with peat development from around 4500 radiocarbon years BP, leading to increased runoff from surrounding slopes. Restricted drainage by eroded material and/or dune movement may also have been involved here, although both would have been aided by the disturbance of soils.

Basin mires in southern Europe were also probably strongly affected by human activity. Cruise (1990) reports new dates from basal peats in basin mires in the Ligurian Apennines in north-west Italy, as well as

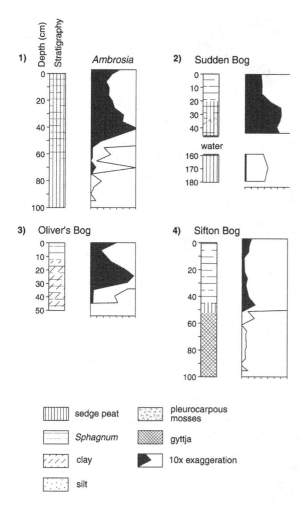

Figure 4.10 Stratigraphy and changing abundance of *Ambrosia* pollen from four floating bogs in southern Ontario, Canada. The rise in *Ambrosia* pollen is a marker for European settlement and forest clearance in the mid-nineteenth century. Note the coincidence between the *Ambrosia* rise and the change to *Sphagnum* peat in all profiles. Extracted and redrawn from Warner *et al* (1989) by permission of *Nature*.

reviewing other data from the region (Figure 4.11). There is a distinct lack of early Holocene basal peat ages in basins that would have been suitable for peat growth had there been an adequate supply of water. While this general pattern is attributable to a change from a drier climate to a cooler, wetter regime in the mid-Holocene for this region, the later peat formation appears only to be triggered when there is disturbance by human populations of the surrounding forest.

4.9 TROPICAL PEAT INITIATION

As with many aspects of tropical peatlands, rather little is known of the nature and timing of their origins. However, several studies suggest that peat development occurred over freshwater marsh or mangrove sediments in Indonesia (see Morley, 1981). In temperate areas, the impact of climate change is often discussed in relation to annual moisture budgets, but seasonality changes may be important in tropical peatlands. The initiation of peatland at Sebangau, Kalimantan Tengah, Indonesia, is attributed to the development of a less seasonal climate during the mid-Holocene, leading to the replacement of topogenous freshwater swamp with ombrotrophic peat swamp forest (Morley, 1981; Figure 4.12).

However, dates for initiation within Indonesian peatlands vary considerably between 9600 and 800 radiocarbon years BP, although most of the ages fall after 6000 BP (Neuzil, 1997). Within individual sites, however, the age range is much more limited, suggesting that these peatlands did not spread slowly but that climatic and topographic conditions became suitable over large areas simultaneously. Lower-altitude sites overlie marine deposits whereas the highest elevation site investigated by Neuzil (1997) has a weathered podsol below the peat. Rieley *et al* (1992) have also noted this difference between peatlands at different altitudes, referring to them as 'high peat' and 'coastal peat', the latter grouped with basin or valley peats as a third category. Figure 4.13 shows a map and cross-sections through an area of peatland in central Kalimantan, Indonesia, with the relationships between these different types of peatland. The high peat areas grade into thin peats and podsolic soils with the basin peats in the intervening valleys, sometimes dominating the entire inter-fluvial area. Rieley *et al* (1992) quote basal ages of between 800 and 4575 radiocarbon years BP for the coastal and basin peats with high peats between 4790 and 6000 years. Neuzil (1997) finds dates as old as 9600 radiocarbon years BP for a higher site. Sites at higher elevations therefore tend to be older than those lower down and it seems likely that peat accumulation was inhibited in the earlier Holocene at lower sites due to higher sea levels than at present.

The initiation and development of other tropical peatlands at low altitudes have also been strongly controlled by changes in sea level. For example, the Black River and Negril swamps in Jamaica developed through a succession from mangrove swamp to freshwater *Cladium jamaicense* linked closely to changes in sea level. Climate change and consequent soil erosion

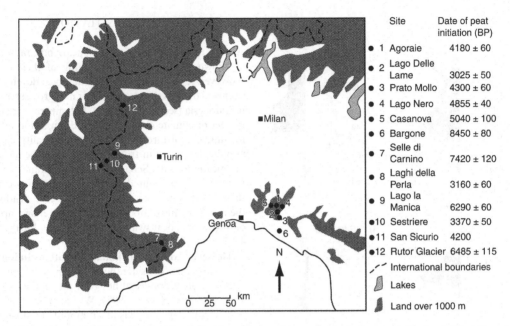

Figure 4.11 Map showing sites with dated basal peats in north-west Italy and surrounding areas (Cruise, 1990). Although most of the basin peats in this region did not begin growth until after 5000 radiocarbon years BP, the exact timing may be related to local deforestation events caused by human populations. Redrawn from Cruise (1990) by permission of Elsevier Science.

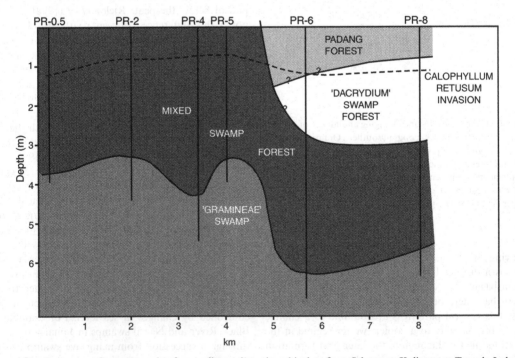

Figure 4.12 Peat swamp forest succession from pollen and stratigraphic data from Sebangau, Kalimantan Tengah, Indonesia. Redrawn from Morley (1981) by permission of Blackwell Science Ltd.

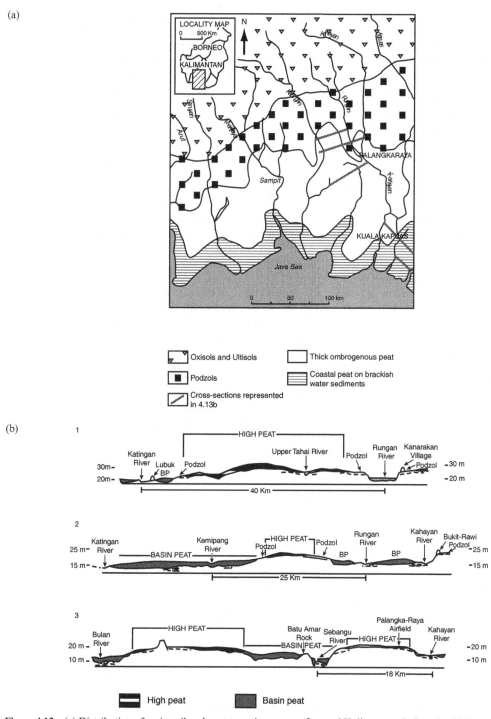

(a)

(b)

Figure 4.13 (a) Distribution of main soil and peat types in an area of central Kalimantan, Indonesia. (b) Cross-section through transects of the basin and high peat areas showing the relationship between the peat types, river valleys and podsolic soils on steeper slopes. The topographic relationships observed here are very similar to those in temperate areas shown in Figure 4.3. Redrawn from Reiley and Page (1997) by permission of the Finnish Peatland Society.

and sedimentation have also affected their development, but human impacts do not appear to have had any effect until clearance of forest in the eighteenth and nineteenth centuries during British colonial occupation (Bjork and Digerfeldt, 1991). Further data on these peatlands will be needed to really discover the links between climate, sea level, topography and underlying substrate in peat initiation, but certainly there is little suggestion that human activity in any form had any bearing on it.

4.10 BEAVERS AND PEAT INITIATION

The main foci of discussion over the causes of peat initiation are climate change in various forms and human activity, together with changes in other local factors such as sea level. Occasionally, animals other than humans have the capacity to modify the landscape considerably, either by sheer weight of numbers or by particular kinds of behaviour. Beavers are one such animal, among a larger group which have been referred to as 'ecosystem engineers' (Jones *et al*, 1994). Ecosystem engineers also exist in the plant kingdom and in peatlands; *Sphagnum* moss is a prime example by the way it builds up over the years, modifying exist-

ing conditions and creating a new habitat. Beavers are common throughout many areas of North American peatlands and also in some areas of Europe, although their original range is much contracted here. The building of dams by beavers, reducing water flow and creating shallow lakes, provides a possible mechanism for the initiation of new areas of peatland. Certainly beaver activity creates much additional wetland in North America, with damming killing extensive areas of forest and providing habitat for other wetland animals and plants (Keddy, 2000). However, the changes appear to be cyclic (Figure 4.14), with a succession from the open water pond created by damming to a pond with aquatic plants. A reversion to dry land occurs when the pond is abandoned by the beavers, normally through loss of the trees on which beavers feed. With abandonment comes dam collapse and the creation of a beaver meadow with wet grassland and marsh species. Eventually a reversion to forest occurs and the area becomes a possible location for renewed beaver activity (Keddy, 2000). It is by no means certain that this cycle will at any stage result in peatland formation as opposed to temporary wetlands. In order for this to happen the dam would have to inhibit water flow significantly over a much longer period of time, probably by surviving in a residual form and by creating adequate siltation in

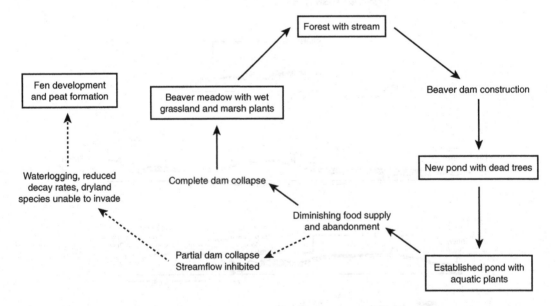

Figure 4.14 The beaver pond cycle with a possible route to peat formation. The normal cyclic progression is renewed unless the right hydrological conditions for peat formation are created and there is no renewed beaver activity, which could flood the incipient peatland and bring the area back into the original cycle. Main stages are outlined with solid lines. Events in open boxes. The original complete cycle is modified from Keddy (2000) and is shown with solid lines, and a suggested pathway to peat growth is shown with dashed lines.

low-gradient valleys. If there is an adequate supply of water but flow is inadequate to re-establish a stream to drain the basin, conditions may become suitable for peat growth.

While in theory this seems like a likely mechanism for peat formation in areas where beavers are, or have been, present, in practice there are relatively few cases where it has been shown to occur. This may be due to the difficulty of locating conclusive evidence of beavers coincidentally with basal peat formation or it may be because it does not occur as frequently as we might imagine. In Europe, beavers were formerly present over much larger areas than they are found now and this has led to speculation that they might have been a significant agent of landscape change and peat growth in the past (Chambers and Elliott, 1989; Gurnell, 1998). Sjörs (1983) goes further in suggesting that this could have been an underlying cause of many instances of paludification in Swedish peatlands. Some direct evidence of beaver activity does now exist in Britain (Coles, 1992), in particular the finds of beaver-worked timber at the base of a peat sequence in north-west England (Wells *et al*, 2000). Macrofossil and pollen remains showed that the area had originally supported a mixed hazel (*Corylus avellana*) and alder (*Alnus glutinosa*) woodland dated to between 3380 and 3000 cal. years BP. This was succeeded by the development of wetland and subsequent peat growth with a piece of beaver-worked hazel wood at the base of the transition to wetland sediments. Radiocarbon dates on the piece of wood showed that the beaver activity dated to between 2750 and 2450 cal. years BP. This find is the first direct evidence of beavers associated with a basal peat in Britain and suggests that other sites may also have been affected by beaver damming at the time of their inception.

4.11 SUMMARY

This chapter has explored how peatlands begin their growth, examining the principles and directions of development that lead to the early accumulations of organic material which eventually become peatland systems. A large part of the discussion has centred on the causes of peat initiation in different systems and in different locations around the world in an attempt to review the many different influences on peat formation. Of necessity, most of the evidence for the nature of peat initiation is palaeoenvironmental data, which represent an imperfect record of change through time. Coincidence and cause are hard to distinguish in many cases and multiple hypotheses to explain the conversion of dry land to peatland are possible. Despite these limitations, enough is now known to begin to see what the main reasons for peat growth are. It is obvious now that paludification does occur as a natural event under prevailing climatic conditions or as a result of climate change in some areas of the world. Pedogenetic processes are often implicated in this and become more important in areas that are suboptimal for peat growth. Equally, in particular regions and peatland types, human activity is critical to the alteration of the hydrological balance that allows peat to develop. We should be wary of attempting to conclude that one or other factor is the only one of importance, but rather we need to consider the balance between climate and human activity, and their interactions with local variations in topography and geology.

5

Peat Accumulation

5.1 INTRODUCTION

Having discovered the ways in which peatlands begin their growth, we next need to look at what keeps peat growing – the accumulation of peat. The build-up of peat on peatlands is impressive; 10 m depth or more is not unusual in many raised mires in temperate and tropical areas, representing gross rates of accumulation of more than 1 cm every 10 years ($1 \, \text{mm yr}^{-1}$). Much higher rates are given in exceptional circumstances. For example, Botch and Masing (1983) give values of up to $3 \, \text{mm yr}^{-1}$ for boreal bogs during certain periods of rapid accumulation and even up to $20 \, \text{mm yr}^{-1}$ for low-land swamps. The end results of this accumulation of organic matter are the peatlands we see today and their ecology, hydrology and functioning are strongly dependent on the amount and nature of the peat that has accumulated. Particularly important in view of the concern over greenhouse gases and global climate change is the role of peat accumulation in global carbon cycling. We will explore some aspects of this in terms of the carbon balance here, but the detail of gaseous emissions and export of carbon from peatlands will be discussed in greater detail in Chapter 9. Peatlands are an important terrestrial store of carbon, and peat accumulation ensures they remain a net sink for atmospheric carbon dioxide.

However, understanding the processes involved in the accumulation of peat is not straightforward. Although theoretically it is a simple equation of 'production minus decay equals accumulation', in practice the measurement of this balance is more complex and needs to consider the peatland system as a whole rather than just the more dynamic surface. In this chapter, we will look at the fundamental processes of production and decay, the long-term accumulation of mass in the system and the models that attempt to describe and predict these relationships. In many mires, sediment accumulation may be a mix of organic material produced at the point of accumulation (auto-

genic sediments) and inwashed inorganic and organic materials from elsewhere on the mire or from the catchment (allogenic sediments). These systems are all fens, especially floodplain mires, and accumulation rates may largely be a function of the input of allogenic sediments rather than processes of *in situ* production and decay. For example, Kosters *et al* (1987) show how the total accretion rate in the Mississippi delta peats is a function of the clastic influx and marine inundation as well as the productivity and decay of the peat-forming plants. In this chapter, we will discuss only accumulation of peat *in situ* although some aspects will apply to allogenic sediments also. Various aspects of peat accumulation are reviewed by Clymo (1983, 1991).

5.2 PEATLAND CYCLES AND PEAT ACCUMULATION

Nutrient and carbon cycles in peatlands are often in a state of imbalance. Inputs to the system are normally greater than the outputs – a situation that results in the accumulation of dead plant material. Animal material is also accumulated but is not generally an important component of the peat in terms of total mass. The three main modes of operation of nutrient cycles identified for wetlands in general also apply to peatlands (Figure 5.1). Nutrients and carbon enter the peatland system either through surface or atmospheric supply and are used within the peatland, or in unusual circumstances may simply flow through the system. Some of the basic building blocks of the ecosystem will be returned to the environment outside the peatland but others will be retained or reused. In many peatlands, the export of material is much less than the inputs and the peatland acts as a 'sink' so that accumulation takes place. In other situations, transformation of inputs takes place (Figure 5.1b). In exceptional circumstances, the export of material may exceed the input of material (Figure

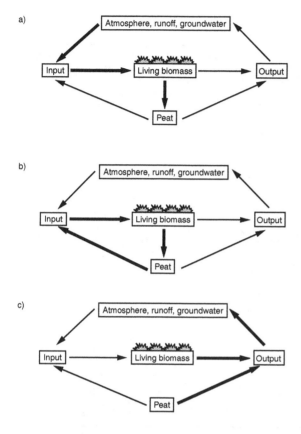

Figure 5.1 Three modes of operation in nutrient and carbon cycling in peatlands. The arrows indicate the relative importance of different transfers between the pools of available nutrients and carbon but are not intended to indicate accurate quantitative amounts. (a) As a sink, with inputs much greater than outputs and a high proportion of the living biomass converted to peat storage. (b) As a transformer, with export of elements in a different form from that in which they entered. Here the emphasis is on the cycling within the peatland system between the pools of biomass, peat and available nutrients and carbon. (c) As a source, with outputs from biomass and peat much greater than inputs. Modified after Mitsch and Gosselink (2000).

5.2.1 Carbon accumulation

Taking carbon as an example, we can see that several different processes operate in the cycling and balance for any particular site (Figure 5.2). Atmospheric carbon dioxide is the main input of carbon, through photosynthesis of the vegetation. Some carbon dioxide is also released through respiration of the vegetation and the peat, but this is often less than the input – thus the peatland is a sink for carbon dioxide. Some of the carbon is altered within the peatland and is produced as methane and exported to the atmosphere or groundwater – a transformation. If the peatland dries out through drainage or perhaps some natural phenomenon such as climate change, respiration of the peat may increase and the peatland may become a net source of carbon dioxide. There are further transformations within the peat which are important considerations in attempting to understand carbon cycling processes. Further details of these will be discussed in Chapter 9, with a particular emphasis on the balance of carbon dioxide and methane with the atmosphere. In terms of global carbon pools, peatlands are an important component, although there is still no agreement on the exact extent of the peatland carbon store. Estimates vary widely depending on assumptions of the extent and depth of peatlands, as well as on the amount of peat-forming or non-peat-forming wetlands. Figures vary between 42 and 1500 billion tonnes (10^{15} g), with most falling in the range 200–500 billion tonnes (Immirzi et al, 1992). Similarly, there is little agreement on the rate of carbon storage currently taking place, although most authors estimate around 100–200 million tonnes (Mt, 10^{12} g) per annum. Since peat is roughly 50% carbon, these figures can be doubled for the total amount of mass stored and accumulated annually. Figures for northern peatlands such as those given by Gorham (1991) are more likely to be correct than those for tropical and Southern Hemisphere peats. However, even here new data on some areas alter previous estimates considerably. In the former Soviet Union, the peat carbon pool was recently estimated as 215 billion tonnes, an increase of 40% on previous estimates (Botch et al, 1995).

Despite these remaining uncertainties, and depending on which estimate one takes, peatland carbon represents around a third to a half of the global soil carbon pool, which is estimated at around 1400 billion tonnes (Table 5.2). Looked at another way, there is almost as much carbon in peat as is present in the entire global atmospheric carbon pool. Separating peatlands from the life zones of Post et al (1982) is difficult, but

5.1c). This process becomes particularly important in damaged peatlands either due to disturbance of the natural processes (such as drainage) or by direct removal of peat for use elsewhere. These processes affect all the elements entering the peatland but in terms of the overall mass of peat, carbon is the most important element to consider and therefore contributes most to peat accumulation (Table 5.1).

Table 5.1 Range and average percentage of important elements in organic soils (after Lucas, 1982; Immirzi *et al*, 1992). Carbon is the dominant element in peat.

Elements	Range for oven dry peat (%)	Typical average for eutrophic peat (%)	Typical average for oligotrophic peat (%)
Elements in organic combination			
Carbon (C)	12.0–60.0	48.0	52.0
Hydrogen (H)	2.0–6.0	5.0	5.2
Oxygen (O)	30.0–40.0	32.0	35.0
Other elements			
Aluminium (Al)	0.01–5.0	0.5	0.1
Barium (Ba)	0.0006–0.3	0.005	–
Boron (B)	0.00001–0.1	0.01	0.0001
Calcium (Ca)	0.01–6.0	2.0	0.3
Chlorine (Cl)	0.001–5.0	0.10	0.01
Cobalt (Co)	0.00–0.0003	0.0001	0.00003
Copper (Cu)[1]	0.0003–0.01	0.001	0.0005
Iron (Fe)[2]	0.02–3.0	0.5	0.1
Lead (Pb)	0.00–0.04	0.005	0.0001
Magnesium (Mg)	0.01–1.5	0.3	0.06
Manganese (Mn)	0.0001–0.08	0.02	0.003
Molybdenum (Mo)	0.0001–0.005	0.001	0.0001
Nickel (Ni)	0.0001–0.03	0.001	0.0005
Nitrogen (N)	0.3–4.0	2.5	1.0
Phosphorus (P)	0.01–0.5	0.07	0.04
Potassium (K)	0.001–0.8	0.1	0.04
Silicon (Si)	0.01–30.0	5.0	0.5
Sodium (Na)	0.02–5.0	0.05	0.01
Sulphur (S)	0.004–4.0	0.5	0.1
Zinc (Zn)[3]	0.001–0.4	0.05	0.005

[1] Cupriferous bogs in Canada contained nearly 0.3% total copper.
[2] Samples with bog iron present could contain more than reported in this estimate.
[3] 6.7% of zinc has been reported present in New York soils containing toxic amounts of zinc.

Billings (1987) suggests 27% of the global soil carbon pool resides in the northern boreal and tundra soils, including some non-peatland mineral soils. Add to this the peatland element in the 'cool temperate forest', the 'tropical moist forest' and the 'wetlands' life zones and the estimate of 35–50% does not seem unreasonable. At a smaller scale, there is increasing interest in assessing the contribution of peatlands to national carbon budgets. Cannell *et al* (1999) estimate that peat accumulation in the UK is occurring at 0.7 Mt C y^{-1}, with losses due to drainage and peat extraction at about 0.5 Mt C y^{-1}, or up to 1.0 Mt C y^{-1} if drainage of all fenland sites is taken into account. Other elements can also be considered on a similar basis to the main exchanges shown in Figure 5.1, although the nature of transformations is of course different from that of carbon.

5.3 PRODUCTIVITY

Productivity is the positive side of the mass balance equation that determines peat accumulation. One might expect that areas with high productivity would also have high rates of peat accumulation, but in reality, decay rates are often also high in very productive systems. It is the sum of a complex of biological and physical factors that results in a surprisingly delicate balance between positive or negative mass balance in the system as a whole. An important consideration in the production–decay balance is the division of peat into two main functional layers referred to in Chapter 3: the acrotelm and the catotelm. The upper acrotelm is aerated at least seasonally and is a much more active environment for both growth and decay. It will obviously include all the above-ground production

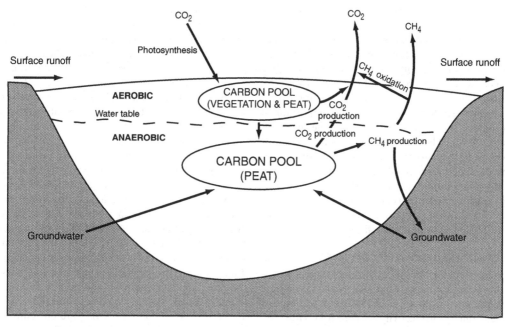

Figure 5.2 Simplified representation of the carbon cycle in a peatland. The peatland is divided into two functional layers (aerobic and anaerobic) but in reality this division fluctuates seasonally and on longer timescales. Supplies from and to groundwater and surface runoff may be dissolved or particulate carbon.

Table 5.2 Estimates of the global soil carbon pool organised by major life-zone groups. Peatlands are not identified as a distinct life-zone group here but are included in several groups including the boreal forests, tundra and wetlands (italicised). Some of the tropical peatlands also fall within tropical forest categories. Data from Post *et al* (1982) and Billings (1987).

Life-zone groups	Area ($\times 10^{12}$ m^2)	Carbon density (kg m^{-2})	Soil carbon ($\times 10^{15}$ g)
Tropical forest – wet	4.1	19.1	78.3
Tropical forest – moist	5.3	11.4	60.4
Tropical forest – dry	2.4	9.9	23.8
Tropical forest – very dry	3.6	6.1	22.0
Temperate forest – warm	8.6	7.1	61.1
Temperate forest – cool	3.4	12.7	43.2
Boreal forest – wet	*6.9*	*19.3*	*133.2*
Boreal forest – moist	*4.2*	*11.6*	*48.7*
Tropical woodland and savanna	24.0	5.4	129.6
Temperate thorn steppe	3.9	7.6	29.6
Cool temperate steppe	9.0	13.3	119.7
Tropical desert bush	1.2	2.0	2.4
Warm desert	14.0	1.4	19.6
Cool desert	4.2	9.9	41.6
Boreal desert	2.0	10.2	20.4
Tundra	*8.8*	*21.8*	*191.8*
Cultivated land	21.2	7.9	167.5
Wetlands	*2.8*	*72.3*	*202.4*
Global soil carbon pool			1395.3

and much of the below-ground production also. Likewise, most of the decay takes place in this zone due to the higher oxygenation. Although there is not nearly so much potential for either production or decay in the catotelm, because it is permanently waterlogged, it should not be considered inert by any means. Decay in the catotelm is slow but because it is acting over a much greater depth of peat, it may be important in terms of the total mass balance of the peatland (Clymo, 1984b). There is much less opportunity for production in the catotelm, although deep root penetration of some higher plants is certainly possible. In the following sections we will consider various aspects of the positive (production) and negative (decay) sides to peat accumulation, and review some of the attempts to provide models of peat growth over time.

5.3.1 Productivity estimates for peatlands

In general, wetlands are highly productive ecosystems and comparisons with other habitats show that they are among the most productive in the world in terms of their net primary productivity (Figure 5.3). However, such general comparisons are based on marsh and swamp wetlands, which are not among the highest peat-producing systems in the world. Productivity in many peatlands is more modest yet peat accumulation is high. Productivity occurs both above and below ground and although the former is easier to estimate than the latter, below-ground biomass of vascular plants is likely to be more important in terms of peat accumulation since it is already partially buried before the death and decay of the plants occur. In some mire plants the proportion of total biomass that occurs below ground is very high. In *Carex rostrata*, up to 90% of the total biomass may be below ground, mostly as fine roots (Figure 5.4; Saarinen *et al*, 1992).

A large number of studies measuring primary productivity of plants in wetlands were reviewed by Bradbury and Grace (1983) and these are summarised in Table 5.3. Even from such a crude grouping of data in a single table, a number of characteristics are apparent. First, the above-ground values for marshes and swamps are the highest, and although values of $>2500\,\mathrm{g\,m^{-2}\,yr^{-1}}$ are probably exceptional, values in the range 1000–2000 g m^{-2} yr^{-1} are not unusual and agree well with other summaries of published data

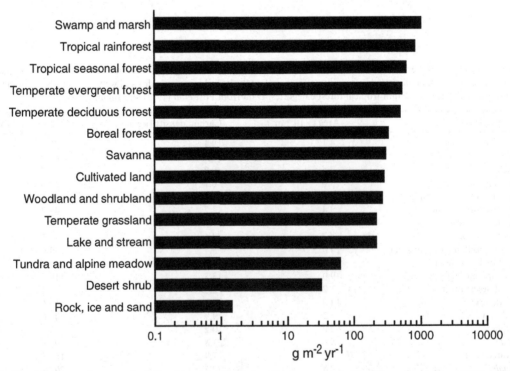

Figure 5.3 Mean net primary productivity of swamp and marsh wetlands compared with other main world ecosystems. Note the logarithmic scale on the x-axis. Redrawn from Keddy (2000) by permission of Cambridge University Press.

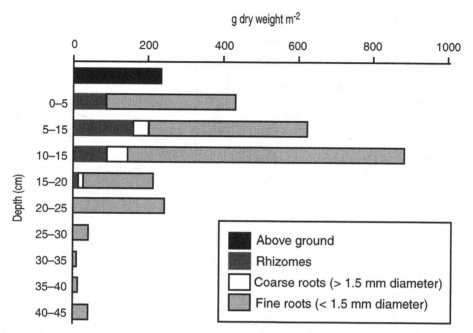

Figure 5.4   Distribution of the biomass of *Carex rostrata* in a sedge fen in southern Finland. Note the large proportion of the total biomass stored under ground, especially as fine roots. Redrawn from Saarnio *et al* (1992) by permission of the Finnish Peatland Society.

given by Mitsch and Gosselink (2000). There are fewer data on below-ground productivity but again values are greater than 1000 g m^{-2} yr^{-1} in a number of cases. Second, although the productivity values of peat and tundra communities are generally lower, typically in the range 300–1000 g m^{-2} yr^{-1} (Bradbury and Grace, 1983), these are still quite high values for plants often growing in relatively cold, poorly aerated conditions. Bryophyte productivity is highly variable, varying with the species and with the habitat, especially in relation to the length of growing season. The highest growth rates included in Table 5.3 are for *Sphagnum recurvum* in Germany. *S. recurvum* is often found in ditches and areas with slow water movement where there is likely to

be some enrichment as well as much less of a problem with summer desiccation. More typical values for temperate Northern Hemisphere *Sphagna* seem to be 300–400 g m^{-2} yr^{-1}. Bryophyte productivity may be a significant contributor (30–50%) to total primary productivity in many poor fens and bogs (Mitsch and Gosselink, 2000), although this is likely to be highly variable spatially within a single site, as some areas will be almost completely dominated by bryophyte cover. In general the contribution of mosses to primary production is greater in wetter areas of temperate patterned peatlands where shrub growth is less vigorous (e.g. Madden and Doyle, 1990).

5.3.2 Factors affecting productivity

The factors that affect primary productivity of peatland plants are likely to be related to the environmental variables discussed in Chapter 3 as principal environmental gradients influencing species and community composition. However, rather few definitive data are available on the main determinants of productivity. Although a certain amount can be gained from comparisons of the diverse range of studies quoted by Bradbury and Grace (1983) or Mitsch and Gosselink (2000), there are

Table 5.3   Ranges of primary productivity estimates from a range of wetland types cited by Bradbury and Grace (1983).

Wetland type	Productivity (g m^{-2} yr^{-1})	
	Above ground	Below ground
Marsh and swamp communities	125–2592	147–1800
Peat bog and wet tundra	42–1118	70–1461
Bryophytes	5–1660	N/A

difficulties in interpreting differences that could be as much related to methodological and site differences as to real environmental relationships. From the limited amount of data available, it is clear that vascular plants and mosses (especially *Sphagnum*) respond in different ways to each other and even within these broad taxonomic groups, considerable variation is revealed by detailed studies. For the trophic gradient, there may be little difference in above-ground productivity between fen and bog sites in the same region (Thormann and Bayley, 1997b), although the picture may be different when below-ground productivity is considered. Within particular peatland types, nutrient concentrations may be more important. For vascular plants on fens in Alberta, Canada, net primary production was strongly related to phosphorus and nitrate levels, whereas in the riverine and lacustrine marshes studied, alkalinity and phosphorus showed the strongest link with productivity (Thormann and Bayley, 1997a).

For the water table gradient, comparisons of similar vegetation types with different mean water tables suggest that productivity increases approximately linearly with depth to the water table (Forrest and Smith, 1975). However, this relationship was derived from a particular vegetation type (*Calluna vulgaris–Eriophorum vaginatum*) on blanket mire in northern England, and considerable variation should be expected in other mire and vegetation types. Responses of different taxa may be more variable than total productivity data imply, and a more detailed level of response needs to be examined to explain whole community relationships. For example, herb productivity is positively correlated with water table on peatlands in western continental Canada, but shrub productivity is inhibited in wetter sites (Thormann and Bayley, 1997a; Figure 5.5).

Rather different relationships with environmental variables are found for bryophytes, particularly *Sphagnum* in northern oligotrophic mires. The increase in productivity in wetter locations has been known for some time, as shown by experimental transplantation of *Sphagnum* species into habitats of differing hydrological conditions (Figure 5.6; Clymo and Reddaway, 1971). From this simple but elegant work, it was also clear that different species do not occupy the niche in which they grow best but the one in which they grow better than all other competing species. Other work by Clymo examined the effects of changes in pH and calcium concentration on *Sphagnum* growth. Raising pH or calcium levels above natural levels had little effect, but raising both together proved lethal for

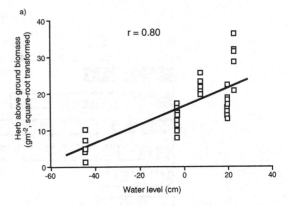

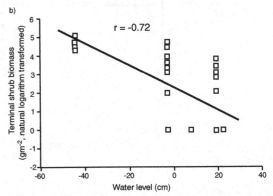

Figure 5.5 Productivity in relation to water table depth on five peatlands in Alberta, Canada, during the growing season of 1994. (a) Peak above-ground biomass productivity (g m^{-2} square root transformed) of herbaceous species. (b) Peak shrub terminal above-ground biomass (new leaves and stems) (g m^{-2} natural log transformed). Redrawn from Thormann and Bayley (1997a) with kind permission of Kluwer Academic Publishers.

most species or at the very least reduced growth considerably (Clymo, 1973). Increased growth of *Sphagnum* has been recorded with increasing nitrate levels from simulated acid rain (Rochefort *et al*, 1990), although the response slowed after the first two to three years of treatment. It might also be expected that increasing temperature would encourage greater productivity in *Sphagnum*, but attempts to discover a relationship between broad-scale climate variation and growth rates in some species suggest that local controls on growth are more important than prevailing climate (Rochefort *et al*, 1990).

Studies of seasonal variability can identify direct relationships with various environmental factors for particular species in a single location, where inter-

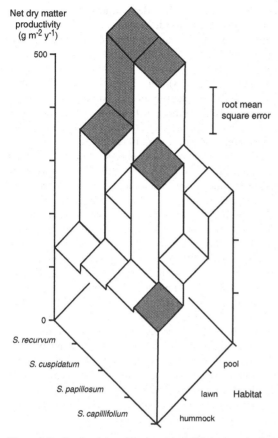

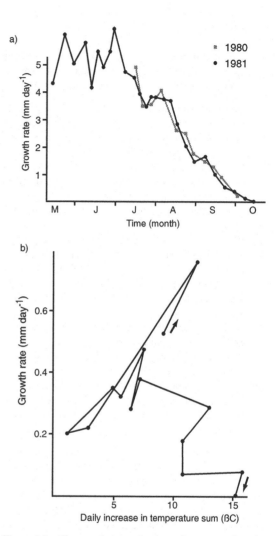

Figure 5.6 Productivity of four species of *Sphagnum* in three different microhabitats on a blanket bog in northern England. Plants were transplanted into the three different habitats and growth rates measured. The shaded blocks show the approximate natural habitats of each species. Redrawn from Clymo and Reddaway (1971) and Clymo (1983) by permission of Elsevier Science.

actions with other variables are minimal. Figure 5.7 shows how the growth rate in two species varies with light availability and temperature over the growing season on a Swedish mire complex (Backéus, 1985). In the case of *Eriophorum vaginatum* (cotton grass), growth rate declined slowly from the end of June, suggesting that light intensities and duration are the major control on production, supported also by fluctuations associated with cloud conditions and snowfall during the longer days in June. For *Betula nana* (dwarf birch), on the other hand, growth rate appears to be determined more by temperature during the first part of the growing season, but to decline after the beginning of July, in spite of increases in the daily temperature sum. Detailed work such as that of Backéus (1985) on a

Figure 5.7 Changes in growth rates of two vascular plant species on a mire complex in central Sweden. (a) Rate of length growth in the youngest leaf of *Eriophorum vaginatum* as a function of time in two years 1980 and 1981. (b) Length growth of *Betula nana* per day as a function of the daily increase in temperature sum. Mean values for 13 measurements on 26 May, 1, 8, 13, 17, 22, 28, 30 June, 6, 15, 21, 25, 30 July, 4 and 10 August in 1982. Temperature sum is the daily increase between two measurements. Points plotted in chronological order. Redrawn from Backéus (1985) by permission of Ingvar Backéus.

number of different species in the same field conditions shows that total productivity on peatlands is likely to be a function of a complex set of factors that may well be different for different species and at different spatial scales.

5.4 DECAY

Productivity is difficult to estimate accurately, yet decay is possibly even harder to study by direct measurement. The main techniques are listed by Clymo (1983) as:

(i) Direct measurement of mass loss. This normally involves using a 'litter bag' where plant remains are kept in a mesh bag and reweighed after a period of decomposition.

(ii) Direct measurement of gas evolution. Estimates of the total amount of decay can be made from measurements of the carbon dioxide and methane produced at the surface if it can be trapped effectively.

(iii) Indirect estimates from standard materials. These are measurements made on materials such as cotton strips, which allow estimates of relative decay rates between different sites, but because the strips are a foreign material, absolute values are hard to assess.

(iv) Indirect estimates from models. Given data on peat ages and cumulative mass of the peat, estimates of the decay rate and the addition of dry matter can be made.

(v) Indirect estimates from simulations. Some simulations to estimate decay rates have been made from temperature, plant substrate and moisture content.

Different approaches are suited to estimates of decay rates over several timescales. Field measurements are most commonly made by litter bag or cotton strip techniques and these produce data on short-term decay rates for different species in different settings. Long-term decay rates can only be reasonably made by indirect estimates from models, which will be discussed in connection with net accumulation rates below. Rather fewer data on decay rates are available than for productivity, especially for fens and for tropical peatlands generally.

5.4.1 Decomposers in peat

Decay is largely a result of microbial decomposition although larger soil animals such protozoa, collembola and nematodes are also important. Estimates of the relative importance of microbial decay compared to that of animals can be made directly using litter bags with different mesh sizes. In litter bags with a coarse mesh, measured decay is due to microbial decay and larger animals, whereas in smaller mesh bags, it can

only be due to microbial decay and some smaller animals including nematodes and protozoa. In a study on a blanket bog in England, up to 43% of the decay was attributed to larger animals, although there was considerable variation between plant substrates (Coulson and Butterfield, 1978). The direct estimation of microbial biomass and examination of community structure is more difficult, although there are now a number of studies that show a range of different organisms are present (e.g. Martin et al, 1982; Nilsson et al, 1992). Dickinson (1983) reviewed microorganisms and their role in peatland ecology and peat decay and more recent techniques have now been applied to peatlands to reveal further details of this 'hidden world' of microscopic activity which is so important to peat accumulation. Sundh et al (1997) used analysis of fatty acid profiles to show that both the total microbial biomass and the nature of the bacterial populations change with depth and microhabitat type in a Swedish mire complex (Figure 5.8). This revealed that different groups of bacteria peak in different levels within the peat, often related to the position of the water table. For example, methane-oxidising bacteria occur at about the same depth as the water table. Sulphate-reducing bacteria peak slightly below the water table. The relative proportions of anaerobic and aerobic bacteria change with depth relative to the water table as one would expect, but anaerobic organisms occur above the water table and aerobic organisms are present below it. This is partly explained by the presence of some oxygen in the upper saturated levels and by anaerobic microsites above the water table. However, where water tables fluctuate most markedly (in the drier sites), particular communities are present where the organisms are either facultative or are at least able to survive fluctuating aerobic/anaerobic conditions. For example, methane-oxidising bacteria survive long periods of anaerobic conditions although they are not active during these times (Sundh et al, 1997).

5.4.2 Pathways of decomposition

The degradative pathways in peat are a slow change from large to smaller molecules (Figure 5.9). The raw materials contributing to the peat mass are the various materials from the dead plants and animals at the surface, including celluloses and lignin (plants) and proteins, carbohydrates and lipids, among others. Mineralisation of some compounds (the labile macromolecules) occurs relatively quickly, but the major constituents of plant and animal material become the

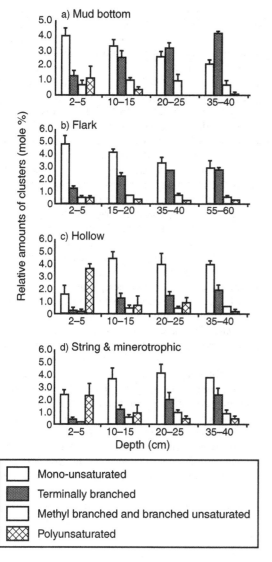

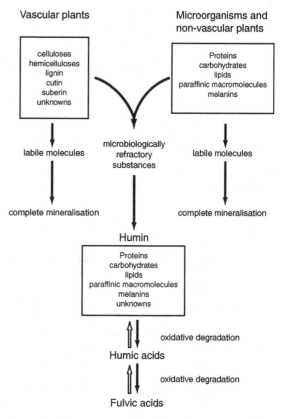

Figure 5.9 Degradative pathway for microbial decay to humic and fulvic acids. Redrawn from Hatcher and Spiker (1988) with permission. © John Wiley & Sons Limited. See text for details.

Figure 5.8 The average relative amounts of phospholipid fatty acid clusters at different depths in four mire microhabitats in a Swedish mire complex. (a) Mud-bottom community with *Sphagnum majus* and liverworts, water table 2 cm depth. (b) Flark with *Carex* spp., *Eriophorum angustifolium*, *Equisetum fluviatile*, *Utricularia intermedia*, water table 7 cm above surface. (c) Hollow with *S. majus*, *E. vaginatum*, water table depth 4 cm. (d) Grouped data from two drier habitats, namely a string (*S. fuscum*, *Oxycoccus quadripetalus*, *Calluna vulgaris*, *Andromeda polifolia*, water table 16 cm depth) and a minerotrophic area (*S. majus*, *S. balticum*, *Carex rostrata*, *O. quadripetalus*, water table 12 cm depth). The different bars represent different groups of fatty acids, which in turn reflect the bacterial community structure. Redrawn from Sundh *et al* (1997) by permission of the American Society for Microbiology.

bulk of the peat, generally termed 'humin'. In turn this is broken down to form humic acids, which are more soluble and therefore more mobile. Further breakdown results in fulvic acids, which have lower molecular weight and are even more soluble than humic compounds. This sequential degradation is also reflected in the distribution of organic fractions down cores. Hatcher *et al* (1986) show how major decreases in holocelluloses and increases in humin occur in the top 20 cm of *Cladium jamaicense* peat in the Everglades, Florida. Humic acids show a big increase in this surface layer but are at low levels at depth, suggesting they are mobilised and removed from the deeper peat layers.

5.4.3 Factors affecting decay rates

The rate of decay varies considerably between different sites and conditions and, as we have already commented, it is a much more important determinant of peat accumulation rate than productivity. As a result, an understanding of the controls on decay rates is crucial to predicting and modelling peat accumulation. The principal factors that affect decay rates in peat are temperature, water content, oxygen supply, microbial and animal populations and the plant material. There are also likely to be many interactions between these variables so that understanding rates of decay is at least as complex as understanding changes in productivity on peatlands. Clymo (1983) provides a detailed review of decay in peatlands which examines these effects. Studies of decay rates are of interest not just for understanding peat accumulation but also for nutrient and carbon cycling research.

Much of the more detailed work on decay rates in peat has been carried out on northern peatlands, especially on *Sphagnum*-dominated communities. Early work suggested that decay was primarily related to water table depth and with decay decreasing with depth, especially below the water table (Clymo, 1965, 1983). It was additionally suggested that lower decay rates at depth could be affected by declining litter quality with age (Clymo, 1965). Clearly one would expect decay rates to be increased with greater depths of the aerobic zone since aerobic decay is faster than anaerobic. However, the effects of species-specific decay rates appear to be extremely important in the surface layers of peat and may override local microenvironmental variables (e.g. Johnson *et al*, 1990; Johnson and Damman, 1991). This idea stems mainly from experiments using litter bags to transplant plant material between different microhabitats on patterned *Sphagnum*-dominated peatlands.

Johnson and Damman (1991) measured mass losses in *Sphagnum fuscum* and *S. cuspidatum* in hummock and hollow microhabitats over a two-year period. Both species were inserted into both microhabitats at different depths, representing oxic, intermittently anoxic and fully anoxic conditions. Mass losses were always greater in *S. cuspidatum* at all depths and in both microhabitats. Although decay rates were higher in oxic conditions, the effect was not marked for *S. fuscum*, which decayed at approximately the same rate under all conditions (Table 5.4). On this basis it appeared that species control of decay was far more important than the effects of water table or microhabitat. However, comparisons of decay rates determined from laboratory incubation showed no difference between different *Sphagnum* species (Hogg, 1993). Instead, it was suggested that the amount of previous exposure to decay was the most important factor. Belyea (1996) undertook experiments to separate the effects of litter quality and microenvironmental conditions on decay rates. She found evidence for both species effects and degree of humification of the peat, as well as interactions with water table position and microhabitat type – a complex picture which nonetheless can be generalised to show the main relationships between species, environment and decay rates on *Sphagnum*-dominated peatlands (Figure 5.10). Interestingly, in relation to water table position, maximum decay rates occur around the position where water tables fluctuate most, rather than in the fully aerobic zone. This may be linked to the specialist microbial communities that inhabit this zone, which are adapted to constant change between aerobic and anaerobic conditions (Sundh *et al*, 1997).

Species differences in decay rates are most evident in comparisons between *Sphagnum* and vascular plants. Decay rates in *Sphagnum* are much lower than for other plants, and because of the dominance of *Sphagnum* in bogs, bogs are found to have generally lower decay rates than fens (Verhoeven, 1990; Aerts *et al*, 1999; Figure 5.11). It seems likely that low decay rates in

Table 5.4 Average mass loss as a percentage of the original mass after 10 and 22 months for *Sphagnum cuspidatum* and *S. fuscum* decomposition in their native microhabitats (hollows and hummocks respectively) and transplanted into foreign habitats. The litter bags used in this experiment were inserted at three different depths, representative of different levels of oxygenation. Figures in parentheses are ± SE and in all cases $n = 6$. Data reproduced from Johnson and Damman (1991).

Zone	Time (months)	Hollow		Hummock	
		S. cuspidatum	*S. fuscum*	*S. cuspidatum*	*S. fuscum*
Oxic	10	–	–	21.8 (1.4)	11.3 (0.3)
	22	–	–	27.6 (1.1)	12.8 (0.0)
Intermittently anoxic	10	16.9 (0.5)	11.5 (1.1)	17.6 (1.4)	11.7 (0.4)
	22	16.5 (1.3)	9.4 (0.8)	20.3 (1.3)	10.6 (0.2)
Anoxic	10	15.1 (0.3)	10.6 (0.4)	15.7 (0.6)	10.8 (0.3)
	22	16.1 (0.6)	9.1 (0.9)	18.1 (0.1)	9.5 (0.7)

		Rate of decomposition				
		Low ⟶			⟶ High	

<table>
<tr><td rowspan="2">Litter quality</td><td>Species composition</td><td colspan="5">Racomitrium hummock < Sphagnum hummock < lawn < hollow</td></tr>
<tr><td>Degree of humification</td><td colspan="5">highly humified < slightly humified </or/> unhumified</td></tr>
<tr><td rowspan="2">Microenvironment</td><td>Microhabitat type</td><td colspan="5">hollow < Racomitrium hummock = Sphagnum hummock = lawn</td></tr>
<tr><td>Position relative to water table</td><td colspan="5">below lowest water table < above highest water table < zone of water table fluctuation</td></tr>
</table>

Figure 5.10 Summary of the effects of four factors on the relative rates of decomposition in *Sphagnum*-dominated peatlands. Two separate factors related to litter quality (species composition and degree of humification) and two related to microenvironment (microhabitat type, position relative to water table) are identified. Redrawn from Belyea (1996) by permission of *Oikos*.

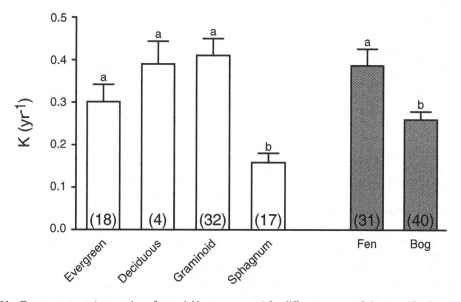

Figure 5.11 Decay constants (proportion of material lost per annum) for different groups of plants and for fens and bogs from a literature review by Aerts *et al* (1999). The numbers of observations are given in the bars. Mean and SE are shown by the main bar and error bars respectively. Note the much reduced decay in *Sphagnum* vegetation and bog habitats (significant at $p < 0.05$). Redrawn from Aerts *et al.* (1999) by permission of the Ecological Society of America.

Sphagnum are mainly related to the presence of phenolic compounds in the plant tissues and excreted into
the surrounding water (Verhoeven and Liefveld,
1997). Similar compounds have also been reported
from tropical peats, although rather little is known
about the preservative properties of these (Katase,
1993). A variety of compounds in *Sphagnum* are
found to have functions related to inhibition of
decay, broadly divisible into two groups: phenolic compounds and carbohydrates, especially uronic acids,
which we have already mentioned in connection with
acidification in Chapter 3. Acidification reduces decay
and the phenolic compounds that are present in cell
fluids and walls also slow decomposition by providing
a defence against herbivores and diseases, as well by
slowing microbial activity. Tannins are released in
deeper *Sphagnum* peats, which further inhibits decay
in the catotelm (Verhoeven and Liefveld, 1997). These
particular qualities of *Sphagnum* peat are important in
the preservation of material incorporated into the peat
such as the 'bog bodies' of north-western Europe, as
well as preserving the plant remains themselves
(Painter, 1991). The presence of decay-resistant materials including phenolic compounds, lipids and waxes
may also reduce decay in *Sphagnum* peats (Johnson
and Damman, 1993). Although in general fens are
found to have higher decay rates than bogs (Aerts
et al, 1999), bogs may sometimes have higher decay
rates if the surface layers are well aerated.
Szumigalski and Bayley (1996) found that decomposition of *Carex lasiocarpa* litter was faster in bogs than in
fens in Alberta, but this may have been due to the lower
water tables in the bog sites studied.

In vascular plants, higher nutrient concentrations
in some plant remains may lead to enhanced decay
rates, especially where the nitrogen concentration or
the carbon:nitrogen ratio is high. However, some
experimental studies suggest the opposite. Aerts
and deCaluwe (1997) measured decay rates in leaf
litter from mesotrophic and eutrophic fens in the
Netherlands and found lower decay rates in the
more productive fen species (*Carex acutiformis*)
than other less productive species. In this case phosphorus content of the plant litter may be more
important than nitrogen, especially in short-term
decay (< 3 months). Nutrient content, pH and temperature differences of the substrate are probably of
secondary importance to moisture status in sites that
have been drained, where decomposition is increased
significantly (e.g. Bridgham *et al*, 1991). Seasonal
changes in decay that have been recorded may also
be a function of water table changes rather than

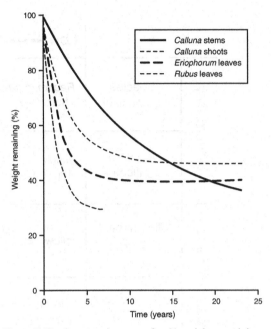

Figure 5.12   Asymptotic curves for % weight remaining of
different litter types over time derived from litter bag experiments at Moor House in the northern Pennines, England.
Rubus leaves show the most rapid initial decay, with *Calluna*
shoots and stems being initially much slower to decompose.
After 23 years, the weight remaining for all the litter types is
similar. Redrawn from Latter *et al* (1998) by permission of
Springer-Verlag.

temperature (Doyle and Dowding, 1990). Very
long-term data on decomposition of vascular plants
in peats suggest that interspecific differences in decay
decline with time so that over periods of >20 years
different litter types make an approximately equal
contribution to peat accumulation (Figure 5.12).
Spatial variability in decomposition appears to
increase over these longer periods so that microenvironmental controls may be more important
than species composition (Latter *et al*, 1998).

5.5 MODELS OF PEAT ACCUMULATION

Although we can learn a lot from examining the
detailed processes of production and decay in peatlands, for many purposes we are interested ultimately
in finding out how fast peat is accumulating. The balance between production and decay need not, however,
be measured directly from first principles by data on

productivity and decay but is often best estimated by other means, such as those mentioned in section 5.4, including gas exchange measurements, models and data on mass–age relationships.

Developing models of peat accumulation that accurately fit empirical data can help to develop understanding and prediction of the processes of peat accumulation enormously. In many ways, the array of data on production and decay over relatively short periods of time (two to three years is typical) has not necessarily furthered our appreciation of the longer-term processes which influence so much of the ecology, hydrology and functioning of peatlands. Professor Dicky Clymo of Queen Mary College, University of London, is pre-eminent in this field and has presented, refined and summarised models

of peat accumulation a number of times in the recent literature, most notably in Clymo (1984b, 1991, 1992b) and Clymo *et al* (1998). Clymo (1992a) provides an excellent summary of the principles and variety of models which can be used in peat accumulation studies.

5.5.1 Structural and functional layers

Fundamental to most models of peat accumulation are the functional layers of the acrotelm and catotelm, which we have already seen differ considerably in terms of their physical, hydrological and biological characteristics (Table 3.1). A more sophisticated representation of functional layers which helps further understanding of these processes is shown in Figure

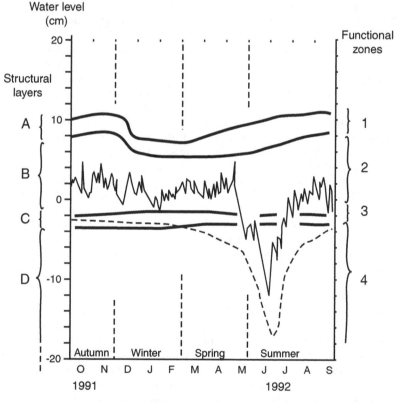

Figure 5.13 The structure and function model for peatlands (Clymo, 1991, 1992a), as applied to Ellergower Moss, Scotland. The structural layers are defined on the left as: A–green, B–litter-peat, C–collapse and D–peat proper. The functional zones are defined on the right as: 1–euphotic, 2–aerobic decay, 3–transition, 4–anaerobic decay. The functional zones move with water table fluctuations through the year and over shorter timescales. The thick solid lines mark the positions of the boundaries between the structural layers. The finer irregular solid line shows water table fluctuations. The functional layer boundaries are defined as follows: boundary 1/2 – as structural boundary A/B, boundary 2/3 – flows water table changes, boundary 3/4 – dashed line, approximately 10 cm below generalised water table changes. Redrawn from Clymo (1992a) by permission of the Finnish Peatland Society

5.13. Here we see a distinction made between structure and function. The structural layers are more or less fixed and only move gradually upward with the growth of the peat surface so that material in structural layer A (the green surface) becomes litter (layer B), then collapses (layer C) and eventually becomes part of the peat proper (layer D). Meanwhile new material is added to the green surface and material gradually shifts down this conveyor belt into the peat. On the other hand, the functional zones do move, particularly with changes in water table through the seasons and over even shorter time periods, especially during warmer and/or drier periods of the year. The euphotic zone (zone 1) is approximately coincident with the green structural layer and is where most of the production occurs. The zone of aerobic decay (zone 2), however, shifts position with water table fluctuations. In dry periods it extends through the litter and collapse layers and even into the upper part of the peat in layer D. Conversely in wet periods, it may be restricted to the green or litter layers. In some circumstances, the water table may be at the surface, in which case functional zone 2 would disappear altogether. The transition zone (zone 3) is the zone immediately below the water table, where oxygen declines although some aerobic decay can still take place. This zone shifts in line with changes in the lower boundary of the aerobic zone. The anoxic zone (zone 4) covers all of structural layer D most of the time, except for dry periods. It may also cover parts of the collapse layer and even some of the litter layer in exceptionally wet conditions.

The critical point here is to see the structures as relatively stable in time but the functions (and hence decay processes) at particular depths as changing through time. The acrotelm–catotelm boundary is normally taken to be where the greatest depth of the water table under normal conditions (the zone 2/3 boundary) occurs. This structure–function model should be generally applicable, although of course specific values will change from those given in Figure 5.13. Clymo (1992b) points to particular variations such as in Antarctic and Arctic peats where low decay in lower peats is a result of low temperature and permafrost rather than saturation. In forested peatlands, larger roots and often a greater depth of the aerobic zone will also require adjustments in the model. Likewise, spatial heterogeneity in many sites means that the size and importance of the functional layers will vary considerably with different water table conditions in pools and hummocks. The applicability of this model and the general idea of a diplotelmic structure in relation to fen systems may also be less clear. The collapse layer is less easy to iden-

tify in fens and in non-*Sphagnum* peatlands generally, and fens also have a less effective means of regulation of the water table due to greater variability of supply and the lack of a water-shedding mire form. The model is helpful in visualising what happens to material produced at the surface in its journey to becoming peat through the upper layers of the peatland in particular – growth, death, structural collapse and descent into permanently saturated denser peat through gradual decay.

5.5.2 Clymo's peat accumulation model

The implications of structure and function for accumulation have also been clearly expressed by Clymo (Figure 5.14). The fundamental issues are: (A) Decay in the acrotelm is fastest by a long way, and loss of the material produced at the surface is high by the time it enters the catotelm. (B) Peat decay does not cease within the catotelm, but continues to decay at a much slower rate, which only becomes significant over long periods of time (millennia). If productivity at the surface remains constant, and the decay rates in acrotelm and catotelm remain constant over long periods of time, the total amount of decay occurring over the whole depth of the peat rises gradually. Ultimately it may reach a level equivalent to the production at the surface and there will be no net accumulation of peat mass. This is the principle upon which the idea of a limit to peat growth is founded (Clymo, 1984b).

Models for peat accumulation based on this principle normally consider the acrotelm and catotelm separately since the rates of change are so different. On the other hand, the basic processes are the same. The acrotelm is receiving material from primary production and the catotelm is receiving material from the acrotelm – although they have a different appearance, they are both inputs of mass to the layer (acrotelm or catotelm) under consideration. Likewise, decay is simplified to a constant value for each of these layers. The obvious potential problem with this as a basis for modelling peat growth is that the assumptions of constant inputs and constant decay are incorrect. We have already seen that production and decay rates can vary quite a lot with different microhabitats, environmental conditions and plant communities. However, when tested against empirical data for both the acrotelm and the catotelm, the fit of the constant rate of decay models is surprisingly good. For the acrotelm, Clymo (1984b) used data from moss banks in the Antarctic to test this idea, as the remains of the plant which forms them (the moss

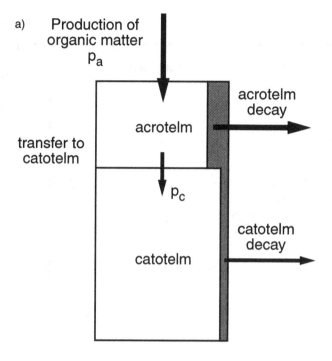

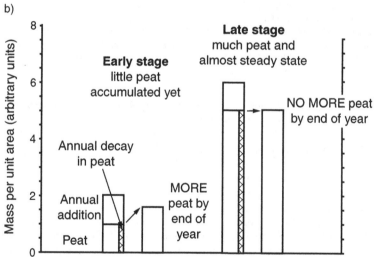

Figure 5.14 Schematic representations of peat accumulation assuming constant production and decay rates. (a) Material is added to the acrotelm at a constant rate through primary production (p_a). A proportion of this is lost before it enters the catotelm through decay. The total addition to the catotelm is the remaining proportion of the acrotelm peat that is passed on (p_c). Decay in the catotelm means that a small proportion of the entire depth of catotelm peat is lost. The net addition to the catotelm is therefore the amount added from the acrotelm minus the amount lost through decay in the catotelm. (b) At an early stage of peat growth, the total amount of decay in the catotelm is small because the total amount of catotelm peat is also small. At a later stage of peat growth, there is a large amount of material in the catotelm so that the total annual loss of material is also large even though the decay rate in the catotelm has not changed. Eventually the decay throughout the catotelm may equal the total additions from the catotelm (p_c). Redrawn from (a) Belyea and Warner (1996) by permission of NRC Research Press and (b) Clymo (1992a) by permission of the Finnish Peatland Society.

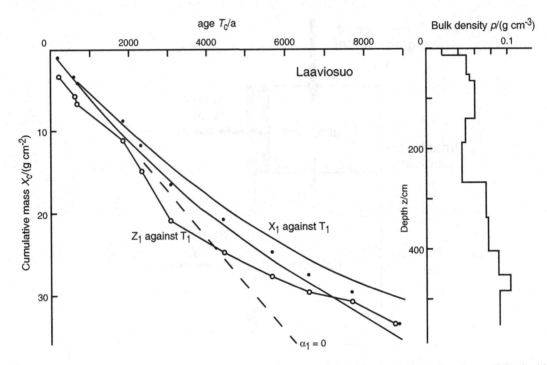

Figure 5.15   Profile of cumulative mass (closed circles) against time (calibrated radiocarbon scale) for Laaviosuo, Finland, with a fitted curve from the constant decay model of Clymo (1984b). The two solid lines show the 95% confidence limits using estimates of addition to the catotelm (p_c) of $5.7 \times 10^{-3} \pm 1.3 \times 10^{-4} \, g \, cm^{-2} \, yr^{-1}$ and constant decay (α) of $1.1 \times 10^{-4} \pm 7.4 \times 10^{-6}$. The open circles show the relationship between age and depth measured as distance for comparison. The dashed line shows the relationship that would result from the same rate of addition but no decay. Redrawn from Clymo (1984b) by permission of The Royal Society.

Chorisodontium aciphyllum) allows a reliable timescale to be derived and the mass remaining can be measured directly (Fenton, 1980). He found that it gave a reasonable representation of the constant rate of decay model, certainly no worse than the more complex exponential model, which would assume a rate of decay proportional to the amount remaining. For the catotelm, the model had to be tested against long-term data of radiocarbon ages and cumulative mass. In most of the cases tested the model proved to be a reasonable fit to the data from a number of *Sphagnum* bogs (e.g. Figure 5.15). Further tests of the model also suggest that it is a reasonable approximation of peat accumulation in most cases (Clymo, 1991). Clymo (1992a) has since tested other possible models which allow for different decay relationships. However, on the basis of the data available it is not possible to distinguish effectively between these models, and the constant decay model is accepted because of its simplicity, not because it is a more plausible portrayal of the complex processes that contribute to peat accumulation variability.

5.5.3 Application and development of the model

A number of studies have used Clymo's model to investigate rates of peat accumulation in different deposits. At the last count, 25 out of 37 profiles had the expected concave age versus mass curves, with 9 approximately linear and 3 distinctly convex curves (Clymo *et al*, 1998). Values for productivity and decay that have been calculated from these studies are quite variable. One extreme example comes from a peatland on the Falkland Islands in the South Atlantic where the long-term decay rate is estimated at $1.1–2.2 \times 10^{-4}$ and the rate of addition of dry matter is $430–720 \, g \, m^{-2} \, yr^{-1}$ (Lewis-Smith and Clymo, 1984). This rate of decay is not unusual but the rate of addition of dry matter is about 10 times that from the profiles in Scandinavia (Clymo, 1984b). Other high rates for dry matter addition are reported from oceanic Canada where it was estimated at $190 \, g \, m^{-2} \, yr^{-1}$, again with a similar decay rate to previously reported values (Warner *et al*, 1993). Data from continental peatlands have so far failed to find results that suggest long-term decay rates significantly different from zero (Charman *et al*, 1994; Belyea and Warner, 1996). One explanation for these results may be that these particular sites are more susceptible to allogenic influences such as fire and drought, which could make production and decay rates more variable (Belyea and Warner, 1996). There are still rather few data for acrotelm peat, probably because the derivation of reliable age–mass data is more difficult. Belyea and Warner (1996) used a variety of dating methods to estimate acrotelm productivity and decay from the Clymo model in a continental peatland which was then used to estimate the overall mass balance of the acrotelm (Figure 5.16). They found considerable variation between different

microsites, especially in the amount of material transferred to the catotelm from the acrotelm. In particular, hummocks had higher positive mass balances than hollows. The implication of these results is that the steady state of production–decay for acrotelm peat assumed by the Clymo model is not realistic, as otherwise hummocks and hollows would develop unsustainable height differences. A more variable rate of decay would be a more plausible portrayal of processes in this zone, especially over the acrotelm/catotelm boundary.

More recently, models based on continuous decay have been used to examine broad-scale patterns of peat accumulation derived from a large data set of basal peat ages and accumulated carbon (Clymo *et al*, 1998). These models also include linear and quadratic

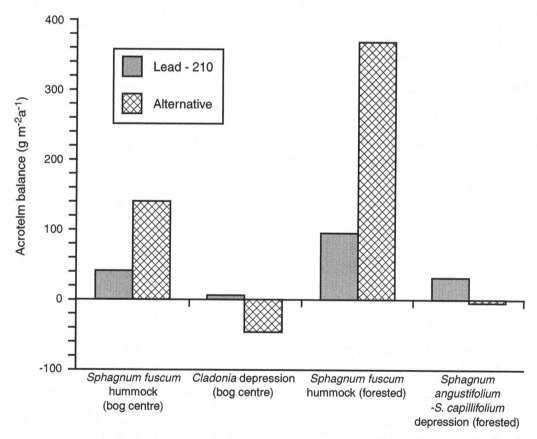

Figure 5.16 The balance between additions and total losses from the acrotelm of four short cores from Rainy River Bog, Ontario, Canada. Cores were taken from one hummock and one depression in both the centre and at the forested edge of the site. Two different estimates are given for each core, one based on dates from ^{210}Pb dating and the other using alternative techniques of moss increment dating (hummock cores) or pollen density dating (hollow cores). Both chronologies made use of some additional age estimates from other pollen and charcoal marker horizons. Redrawn from Belyea and Warner (1996) by permission of NRC Research Press.

decay models, incorporating the idea that decay in the catotelm is not constant but that it declines over time as the remaining peat becomes less decomposable. These more complex models are a good way to understand long-term peat accumulation dynamics and explain the broadest-scale patterns of peat growth quite well. At the scale of the individual peatland, if we want to understand the variability in the accumulation of peat, models that incorporate interaction with environmental variables may be needed. Hilbert *et al* (2000) present a model that incorporates the effects of depth to water table. This model suggests that net peat accumulation may be more variable with shifts in water table conditions and that peatlands may be capable of moving from net sinks to sources of carbon quite rapidly. Other data from 'wiggle matched' radiocarbon-dated peat sequences (see Chapter 6) have emphasised the difficulties of deriving accurate age–depth profiles in peat, and have suggested that species-dependent change in accumulation rates is in fact important in the long-term rate of peat accumulation (Kilian *et al*, 2000). Their reinterpretation of data from the same sites used by Clymo (1984b) suggests that there have been significant hiatuses in peat accumulation in several sites and other changes in peat accumulation that are best explained by stratigraphic changes and botanical composition. They argue against the assumptions of continuity in accumulation and of constant production and decay, suggesting instead that homogenous stratigraphic layers show no sign of the concave relationship between depth and age.

5.6 VARIABILITY IN LONG-TERM ACCUMULATION RATES

An acceptance that constant decay does occur in catotelm peat has important implications for attempts to calculate changes in long-term peat accumulation. Often in palaeoenvironmental studies, it is of interest to look for changes in accumulation rates that may be related to the ecological development of the peatland or perhaps to external factors such as climate. Clymo (1991) has often argued that attempts to calculate accumulation rates purely on the basis of age and depth are overly simplistic as they ignore the constant decay of peat over time. Thus, accumulation rates (calculated as incremental depth per annum) would be expected to be lower for older peats, even if production/decay rates at the surface remained constant. Despite this, many authors have continued the practice of giving accumu-

lation rates in centimetres per year. For very general comparisons, especially for different peatlands over the same time periods, this is still justifiable, especially where decreases in accumulation rates with time are noted (i.e. the opposite trend to that expected with Clymo's model). There are very major differences in net peat accumulation between sites, regions and time periods that reflect genuine differences in accumulation unrelated to long-term constant decay. Changes in peat composition are particularly important and it should be remembered that Clymo's model assumes no change in peat composition over time. Thus the concept of constant decay and the comparison of crude peat accumulation rates are not incompatible.

Comparisons between sites over approximately the same time intervals show a large range of variation. A comparison of lake and mire sediments in eastern North America shows that peat accumulation rates varied between 4 and 500 cm per 1000 years (0.004–0.5 cm yr^{-1}), although the majority of the rates fall around 50–100 cm per 1000 years (Webb and Webb, 1988). Comparisons across large geographical areas also often reveal significant differences in peat accumulation rates. In Canada, major differences in peat depth and accumulation rate can be identified at a continental scale. Regions in the southern boreal and oceanic areas have higher mean accumulation rates than high boreal and subarctic regions (Figure 5.17). These differences primarily seem to reflect the influence of climate (Ovenden, 1990). These broad patterns of change consider the entire Holocene and may mask even greater temporal differences in particular regions. For example, in Arctic peatlands, a number of studies show that besides long-term average rates of accumulation being relatively low, the last few thousand years have seen a considerable slowdown or even cessation of peat growth. This has been reported from the Canadian Arctic (Vardy *et al*, 2000) and several sites in Siberia (Jasinski *et al*, 1998; Peteet *et al*, 1998).

At the other extreme of the climatic gradient, tropical swamp peats show some very rapid peat accumulation rates. Anderson (1983) reports rates of between 2.2 and 4.8 mm yr^{-1}, faster than most temperate peat accumulation rates by two to five times. Interestingly, these accumulation rates show a decreasing trend through time, contrary to the expectation from the constant decay model. This is explained by changes in the forest communities forming the peat. Broader ranges of tropical peat accumulation have been reported elsewhere, with ranges from 0.2 to 13 mm yr^{-1} (Neuzil, 1997). However, the most commonly encountered rates in South-east Asian tropical peats are around

a)

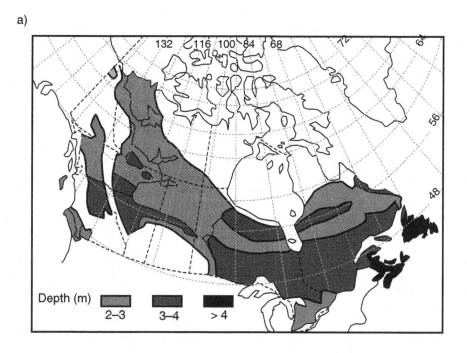

b)

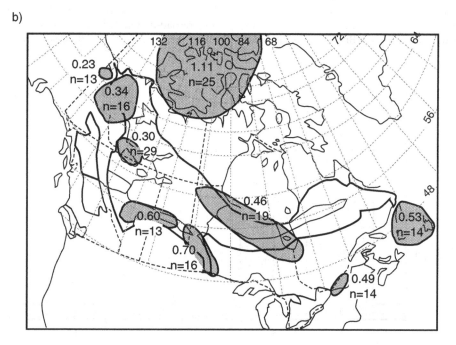

Figure 5.17   Broad-scale patterns of peat accumulation in Canada, as shown by (a) average peat depth and (b) mean rates of long-term peat accumulation (mm yr^{-1}). Peat depth and peat accumulation rates tend to increase in the southern boreal and especially in the oceanic regions. The much higher rates of peat accumulation in the Arctic islands are probably due to high ice and/or mineral sediment content and they are not therefore directly comparable with the other rates, which are primarily of organic material. Redrawn from Ovenden (1990) by permission of Academic Press. © The University of Washington.

2–5 mm yr^{-1} (Sorensen, 1993). The trend of decreasing accumulation rates over time is almost universal in older peats although linear or increasing rates are recorded in a few sites (Neuzil, 1997). In other tropical regions, comparable rates have been recorded, such as those from a spring-fed peatland in Queensland, Australia, which has grown at a rate of between 2.5 and 4.0 mm yr^{-1} for the past 1000 years (Bell *et al*, 1989). Other warm climate peatlands have more variable accumulation rates, which may be partly related to much longer-term climate shifts from glacial to interglacial conditions. The Kashiru peatbog in Burundi, Africa, a highland equatorial peatland, formed peat faster during the end of the last glacial period than in the early Holocene, despite the presumably much greater effects of long-term decay over this time period (Aucour *et al*, 1994).

In temperate peats, the rates of accumulation have been calculated for many sites. An early comparison of raised mires showed that most sites in Europe have accumulation rates around 0.2–1 mm yr^{-1} (Aaby and Tauber, 1975), with a few higher rates in the later Holocene. The general increase in accumulation rates is likely to be a function of the constant decay in the catotelm in these sites, where peat composition is fairly similar (see also Figure 5.15 and related discussion above). Oceanic raised mires in Britain often have accumulation rates of approximately 1 mm yr^{-1} (e.g. Barber *et al*, 1994), giving a convenient rough estimate of 10 years for each centimetre of peat growth. Blanket mire accumulation rates in the British Isles are more variable, with estimated rates of 0.1–1.2 mm yr^{-1} (Figure 5.18). Due to the more complex nature of underlying topography, stratigraphy and development of blanket peats, this is perhaps not surprising. Clearly one would not expect a sloping blanket peat on a relatively well-drained site to grow as fast as peat during a phase of basin infilling in an underlying depression. Fen peats can have much higher rates of accumulation, even if they are not receiving significant amounts of allogenic material. Almost 14 m of peat accumulated over about 7500 calendar years in a kettle hole peatland in Estonia (Punning *et al*, 1995) and approximately 18 m of peat formed in 8000 radiocarbon years in a small peatland in Dorset, England (Waton and Barber, 1987). The latter rate is related to exceptional geological circumstances, where the sediment was subject to continuous slow subsidence, but shows how fast peat can form given the right conditions. In Southern Hemisphere temperate peatlands, accumulation rates do not appear to be much different from those in the north. Newnham *et al* (1995) find overall rates of peat accumulation in a large restiad-dominated raised mire to be approximately the same as those of oceanic raised mires in

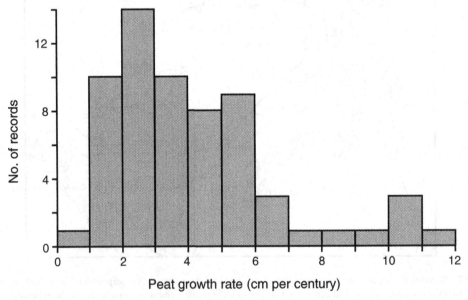

Figure 5.18   Rates of peat accumulation in blanket peats from the British Isles showing the broad spread of crude accumulation estimates in a relatively small region. Differences in growth rates are primarily due to local conditions. Redrawn from Tallis (1995a) with permission. © John Wiley & Sons Limited.

Britain at 0.3–0.9 mm yr^{-1}, with a more recent phase at 1.8 mm yr^{-1}, including the upper acrotelm peat. Other peatlands in New Zealand, perhaps more akin to blanket peats, have more variable rates, but again they are not significantly different from those recorded in the Northern Hemisphere. For example, the Kaipo Lagoon peats in North Island accumulated at an average rate of approximately 0.2 mm yr^{-1} over the Holocene (Lowe and Hogg, 1986).

5.7 SUMMARY

Peat accumulation is probably the most important process in the formation of peatlands. Clearly, peatlands would not be peatlands at all unless it took place! This chapter has explored some of the fundamental processes that occur in all peatlands, as well as touching upon some of the more detailed aspects that control peat accumulation at individual sites. The development of Clymo's models (Clymo, 1984b; Clymo *et al*, 1998) has been a major influence on our understanding of long-term peat accumulation processes. While some may argue about the details of the applicability of the models to specific sites, the fundamental basis of the model, that decomposition of peat is slow but continuous (unless frozen), is inescapable. This has important implications for carbon cycling and gas exchange with the atmosphere, which will be discussed further in Chapter 9. Peat accumulation also strongly influences the development of peatland landforms, for example the development of hummock–hollow systems. Chapter 7 will follow up this influence in discussing autogenic processes on peatlands. Likewise, peat accumulation is strongly affected by human activities, particularly drainage, forestry and peat extraction, a subject for Chapter 8. This is not the last we will hear of peat accumulation by a long way!

PART 3

CHANGES IN PEATLANDS

So far we have considered two main aspects of peatlands, structure and functioning. For the former, this has included a consideration of the many shapes and sizes of peatlands but only a brief, mainly hydrological, consideration of how they became that way. In the case of the latter, the focus has been primarily on the ecological and hydrological processes within the system, together with fundamental aspects of the origins and accumulation of peat growth. While we touched upon some aspects of ecological, hydrological and structural development through time, we skated over the detail of the evidence for the processes involved in these changes. The next section of the book moves on to consider various aspects of changes in peatlands and the ways in which they are influenced by internal and external factors, as well as looking at how peatlands affect the local and global environment. The importance of time in peatland science was pointed out in Chapter 4 when we first entered the peat 'time machine' and started to explore the wealth of evidence for process and change held deep in the ground. Chapter 6 expands on this theme and reviews the various techniques and types of data that are used in peatland palaeoecological and palaeoenvironmental studies. Different forms of this evidence are used extensively in Chapters 7 and 8 to unravel long-term changes in peatlands and their underlying causes. Chapter 9 takes this on a little further and looks at feedbacks from peatlands to the surrounding environment. This encompasses local, regional and global influences of peatlands.

6

The Peatland Archive:
Palaeoenvironmental Evidence

6.1 INTRODUCTION

There is a massive amount of literature on peatland palaeoenvironments and it is not our intention to review it in its entirety here – that would require a separate book in itself. Neither is this intended as a 'how to' guide to techniques of peatland palaeoecology. Again this would be a major undertaking and, in any case, various guides and papers already exist on this. There is a very wide range of texts covering palaeoenvironmental methods and applications. Some are mainly concerned with methods (e.g. Birks and Birks, 1980; Berglund, 1986; Warner, 1990b), others are general Quaternary environmental change texts (e.g. Bell and Walker, 1992; Lowe and Walker, 1997; Roberts, 1998; Williams *et al*, 1998). There is no text specifically dedicated to peatland palaeoenvironments but these books all cover key aspects of palaeoenvironments in general. The main aims of this chapter are twofold. First, to provide a brief overview of the range of evidence that can be used for understanding palaeoecological change in peatlands. Second, to show how some of this evidence has been used to study particular issues in peatland science. Much further work will be referred to throughout the book, especially in Chapters 7, 8 and earlier in Chapter 4, but we hope that this chapter will provide a reasonably balanced overview of the range and scope of peatland palaeoecological studies. Some of the more established and widely used techniques are given less attention simply because they are covered elsewhere. Other techniques that are more exclusively used on peats or which are recent developments with further potential have been given greater space than their application to date would strictly warrant.

6.2 THE RANGE OF EVIDENCE AND SOME GENERAL PRINCIPLES

The major groups of evidence that are routinely used are divisible into four main areas (Table 6.1). First, to describe and understand broad-scale, gross changes in peatlands, various survey and stratigraphic techniques are used. Second, a wide range of biological techniques is used to gain greater detail of changes through time of the peatland and its surrounding environment. These involve the identification and estimation of the changing abundance of different fossil groups, especially of plants and animals. Third, various chemical and physical parameters of the peat are used for the same purpose. These are becoming increasingly important and many new techniques and ideas have been explored over recent years. Finally, while we can use the above techniques to discover the sequence of change, we need a suite of methods to provide a timescale for these sequences. This part of the process is often the most difficult, not only in terms of the level of technology and instrumentation for analysis but also in the interpretation of results. Radiocarbon dating in particular has been looked at more critically in recent years. The following pages will go on a whistle-stop tour of key aspects and applications of these techniques as they apply to peatlands. The selection of material and especially of examples is personal – there are many more to be found in the literature.

6.2.1 Uniformitarianism and time

One of the overarching principles in the reconstruction of past environments is that of 'uniformitarianism'.

Table 6.1 Summary of palaeoecological techniques and their main purposes in peatland palaeoecology. Sections on biological remains and chronology modified from Warner and Bunting (1996). * Indicates that biological remains are not commonly preserved in peat deposits due to their calcium carbonate content. Many of the more unusual biological groups are not well known and are not often used.

Type of evidence	Main purpose
Survey and stratigraphy	
Peat depth and topography:	
– depth probes and standard survey equipment	Morphology of peat deposit
– radar	Morphology and general stratigraphy
	Stratigraphic description
Manual coring techniques:	Sample extraction for description and analyses
– Russian D section	
– Piston	
– Various short corers	
– Excavation	
Stratigraphy:	
– visual description	Detection of main changes in peat composition
– X-ray	Identification of particular features (e.g. tephra layers)
Biological	
Pollen and spores	Changes in plant communities in peatland and surrounding areas
Plant macrofossils:	Normally derived from growth at sample location
– tree stumps and logs	Tree-ring studies
– other higher plants	Changes in peat-forming plant community
– moss leaves and stems	Changes in peat-forming plant community
Charcoal	Fire history on site and in surrounding area
Algal:	
– diatoms	Hydrological changes
– other (desmids, etc.)	
Testate amoebae shells	Hydrological conditions
Fungal remains:	Various, including identification of specific hosts, hydrological
– hyphae, spores, macrofossils	characteristics
Arthropod	Cladocera, crustacea* and ostracoda* exoskeletons
Insecta:	
– Chironimidae head capsules	
– Coleoptera exoskeleton	
Other remains:	Various details of past ecology and environment
Mollusca*	
Porifera (sponges),	
platyhelminths egg capsules,	
Rotifera egg capsules,	
Tardigrada egg capsules and exoskeletons	
Vertebrate bones* and soft tissues	
Physical and chemical characteristics	
Mineral content:	Identify allochthonous input
– loss on ignition	Total in organic and/or carbonate content
– magnetic susceptibility	Ferromagnetic mineral content

(Continued)

Table 6.1   (*Continued*)

Type of evidence	Main purpose
Bulk density	Accumulation rates, compaction estimates
Humification	Degree of decay
Inorganic elemental chemistry: – various	Mainly pollution studies, land use and human impact
Organic biomolecules: – various	
Isotopes of oxygen, hydrogen and carbon	Past water sources and climatic conditions, balance between C_3 and C_4 plants in tropical peats
Others, including: – near the infrared spectra – luminescence	Used as rapid measurements for other parameters or may indicate other factors – more work is needed on these
Chronology Radiometric: – radiocarbon (^{14}C) – lead (^{210}Pb) – other radionucleides (^{241}Am, ^{137}Cs, ^{3}H)	 For organic sediments up to 40 000 years old Recent sediments (up to 200 years old) Mostly related to development and testing of nuclear weapons in the last 50 years
Time markers	Provide markers of various ages that allow correlation of different sequences. Strongly dependent on region of study
– volcanic ash (tephra) – pollen markers – fires – spheroidal carbonaceous particles (SCPs)	 e.g. elm decline in NW Europe at *c.* 5000 BP Post-industrial record in Europe
Incremental counting: – dendrochronology – moss growth increments	 Provides accurate age on larger timbers in peat Recent peat accumulation in suitable peats (50–100 years)
Accumulation: – pollen density – peat accumulation models	

Often referred to by Charles Lyell's phrase 'the present is the key to the past', this is the idea that by looking at processes and patterns observed today we can interpret the evidence of the past by assuming similar basic relationships existed then. This is particularly apposite in the consideration of biological evidence. We will also refer to some terms that are widely used in palaeoenvironmental science but which may not be familiar to some readers. Critical among these are those of major time periods and the scales used in the measurement of time. The Quaternary is the geological time period covering approximately the last 2 million years ('the ice ages'). The Holocene is the most important time period for peatland growth and it covers approximately the last 10 000 radiocarbon years. We also refer to the Late-glacial, which is the last few thousand years of the last glacial period (approximately 15 000–10 000 radiocarbon years ago). Time is expressed in either radiocarbon years or in calendar years before present. There is a difference between these two timescales that gets greater over longer time periods for reasons that will become apparent below. For example, the start of the Holocene is about 10 000 radiocarbon years ago, which is the same as approximately 11 500 calendar years ago. The text will normally make it clear which timescale is being used and where possible calendar years will be used. Any historical dates are given as years AD or BC.

6.2.2 History of peatland palaeoenvironmental study

Peatland palaeoecology has a long history – the accumulation of plant matter to impressive depths did not escape the attention of early writers on landscape and nature and they often speculated on the age of peat and the remains found there. Many observers did not recognise that peat is almost entirely made up of plant material, but William King described peat formation by mosses in Ireland as early as AD 1685 (King, 1685; Du Rietz, 1957). Shortly after this, Linnaeus attempted to characterise peat in terms of its botanical constituents (Du Rietz, 1957). Extensive discussion of peatland palaeoecology was under way by the mid-nineteenth century. For example, Geikie (1866) debated the phenomena of buried trees in Scottish peat and the causes of their destruction, including possible climatic changes and human impacts in the past. He also considered feedback effects of the demise of forests, noting (p. 375), 'Did the fall of tree by choking the drainage, only then bring about the requisite conditions for the increase of *Sphagnum* and its allies?' Although we now know much of this speculation is entirely wrong – such as the supposition that most of the Scottish peats are approximately Roman in age – it is fascinating that such detailed consideration was given to peatlands at this early stage. The Scottish peats received a lot of attention at the turn of the century, especially extensively discussed by Lewis (1905, 1906, 1907, 1911). Probably the most famous of all the early work on the fossil record in peats is that of Blytt (1876), which, together with the paper by Sernander (1908), formed the basis of a widely accepted climatic division of postglacial time in north-west Europe for almost 100 years. Known as the Blytt–Sernander scheme, this was subsequently shown to be invalid in many respects (see Chapter 8). Much of this early work was concerned with peatlands as geological archives of broad-scale regional changes, in particular as a source of information on changing climatic conditions over time. Although peat sequences are still used as sources of palaeoclimatic information, the applications of peatland palaeoecology have expanded considerably.

6.3 REASONS FOR PALAEOENVIRONMENTAL STUDIES ON PEATLANDS

In the simplest terms, there are two main categories of peatland palaeoenvironmental research. First, we can view the peat as a receptacle of information about the surrounding environment. Here we are interested merely in peat as a receiver and preserver of evidence. Climatic inferences from changes in the peat are a good example of this, as are studies that use pollen analysis to record regional vegetation change over time. Second, we may be more interested in the peat as a recorder of the ecological and hydrological development of the peatland itself. Thus, changes in stratigraphy that arise from local phenomena may be of greater interest than broad-scale shifts representative of a wide area of non-peat environment. However, while the prime interest might be in the external or internal peatland environment, in reality it is not possible to ignore the influence of one on the other. A good understanding of any changes on peatlands depends on an appreciation of the whole peatland system, including the key external forcing factors. Some of the particular applications of palaeoenvironmental research on peatlands have already been mentioned in the course of Chapters 4 and 5. More will come in Chapters 7 and 8 especially. However, I provide a brief overview of the most common general categories of applications here, to place the specific studies in a broad framework.

6.3.1 Peatland development

In most palaeoecological work, the development of the peatland itself is implicit, even if it is not the explicit purpose of the study. Thus, even if we are aiming to obtain information on climate or land-use history from the peat archive we *must* consider peatland development too. It is perhaps not given enough attention in some studies. However, determining the development of the peatland may be our major aim, perhaps for understanding processes of hydroseral succession, the development of particular vegetation stands or the evolution of surface patterns.

6.3.2 Vegetation history and landscape development

Much of the early work on peatland palaeoecology in the 1940s to the 1970s principally focused on pollen records in peatlands as archives of past vegetation history. Such studies are still undertaken, although it is less common simply to aim to acquire another record of vegetation history, except in under-recorded areas of the world (e.g. Blyakharchuk and Sulerzhitsky, 1999).

6.3.3 Human activities and land-use change

Many changes in pollen diagrams, charcoal records and geochemical data relate to human activities. Prehistoric impacts were principally a result of the use of fire and forest clearance for agricultural activities. These manifest themselves in records of vegetation, charcoal and in geochemistry from inputs of soil dust. Some of these early records and those for later periods contain evidence for different land management practices: pastoral and arable land use, different crops and woodland management. In historic times, the impacts of pollution, drainage and erosion become more apparent, again through a variety of the evidence to be discussed below.

6.3.4 Climatic change

There has been an interest in the link between peat and climate change since the earliest observations of peat stratigraphy. The 1990s have seen a renewed interest in palaeoclimatology generally, and reconstructions from peat evidence have been a part of this. Since climate and hydrology are such strong controls on the ecological status of peatlands, they are often implicated in palaeoenvironmental change in peat profiles.

6.3.5 Other reasons

There are a number of other very specific reasons for undertaking palaeoenvironmental studies that do not fit easily into the above categories. Studies of modern peatlands are undertaken to help understand the development of coal deposits (Cohen *et al*, 1989; ChagueGoff and Fyfe, 1996). Sometimes peat deposits become archaeological sites with excavations of particular features where the palaeoenvironmental evidence in the peat is a crucial part of the data. Examples of this are discussed below.

6.4 MEASURING TIME – PEATLAND CHRONOLOGIES

Although it is often one of the last things to be determined of all the analyses in any specific study, the timescale for a palaeoenvironmental record is clearly critical. As increasing amounts of data become available and interest in discovering spatial and temporal patterns of change grows, the need for reliable chronologies in peatlands and other deposits becomes ever more important. Most texts on Quaternary science such as those listed above include sections on these tech-

niques and more specialised works are those of Smart and Frances (1991) and Mahaney (1984). However, while the Quaternary science literature covers many aspects of the dating of peat deposits quite well, the coverage of some techniques specific to peatlands is poor, although Averdieck *et al* (1993) provide a review of some of the methods used. There are four main types of dating techniques that can be used in peat (Table 6.1): radiometric, time markers, incremental counts and methods based on accumulation. Radiometric techniques provide an absolute timescale that can be compared with other sequences and types of sedimentary record. However, there are normally errors resulting from depositional processes and statistical factors that limit the precision of these techniques. Time markers provide a relative timescale so that we can compare very precisely between sequences that contain the same marker. For example, a peat core can be readily correlated with a record from a lake core, if tephra (volcanic ash) from the same volcanic eruption is present in both. However, absolute dates for these time markers are derived from other techniques. Incremental techniques rely on the regular addition of material from a known reference point (including the surface), similar to counting the annual rings of a tree. These approaches are not widely used in peatlands but there are several interesting possible approaches, especially for more recent peats. Peat accumulation models can also be used to infer ages, but they are usually based on a set of ages from the other techniques. They were discussed in Chapter 5 and we will not say any more about them here.

6.4.1 Radiometric dating

Radiometric dating depends on the radioactive decay process. The basics of this are as follows. Radioisotopes are deposited at the surface, contained in dead plant tissues, from dry deposition or in water sources. Once in the peat column, they begin to 'decay', changing to other isotopes. Since the rate of this decay for particular isotopes is known, we can estimate the age of a deposit by the amount of the original radioisotope remaining. We must also know the amount or proportion of the radioisotope that was originally deposited to do this. The rate of 'decay' is expressed as the half-life of the isotope, or the time it takes for half the original isotope to change to a new form.

Radiocarbon dating is the most widely used dating technique for peat, and various aspects are reviewed in Lowe (1991). Since peat is approximately 50% carbon, it is an ideal material for radiocarbon dating. The carbon is of course derived from fixation of atmospheric carbon dioxide, which is incorporated into the peat when plant tissue dies. The amount of the isotope ^{14}C (carbon-14) is measured as a proportion of the total carbon. The half-life of ^{14}C is approximately 5700 years, so only very small amounts of ^{14}C remain after long periods of time (Figure 6.1). Although the limit of radiocarbon dating is about 40 000 radiocarbon years, beyond which there is so little ^{14}C left that it is almost impossible to detect, most peatlands are much younger than this and radiocarbon dating can be used. For peat or any other organic material older than this, it may be possible to use uranium series dating, although further work is needed on this before it can provide reliable age estimates (van der Wijk *et al*, 1988; Smart, 1991).

Measurement of radiocarbon can be done on large samples by so-called 'conventional' methods in which natural decay is measured over a period of time, or by accelerator mass spectrometry (AMS) in which individual atoms are counted. AMS techniques have the advantage of requiring about 1000 times less material than conventional techniques. In both methods radiocarbon dates are quoted as radiocarbon years plus or minus one standard error to indicate the level of uncer-

tainty. Ages are quoted as BP (before present), where 'present' is AD 1950. In essence radiocarbon dating is a simple process, but there are a number of problems in using it associated with (i) assumptions of the atmospheric concentration of ^{14}C in the past, and (ii) depositional and post-depositional processes.

First, atmospheric ^{14}C has not remained constant over time. If it had, then we would expect to see a linear relationship between 'real' calendar time and radiocarbon content of organic materials of known ages. We know it has not from comparisons of the ^{14}C content of tree rings and their known calendar ages (Figure 6.2). Thus radiocarbon years and calendar years are not the same, but we can use the calibration curves from tree rings to correct for this effect and estimate calendar ages. These are usually quoted as cal. years BP (again where present is 1950) or converted to the AD/BC timescale. Although calibration appears to solve the problem, particular sections of the calibration curve have unusual shapes due to very rapid changes in atmospheric ^{14}C levels in the past. Particularly problematic are periods where there is a plateau on the calibration curve. For example, samples from between 2400 and 2800 calendar years BP have the same radiocarbon age (Figure 6.2). This effect limits the usefulness of radio-

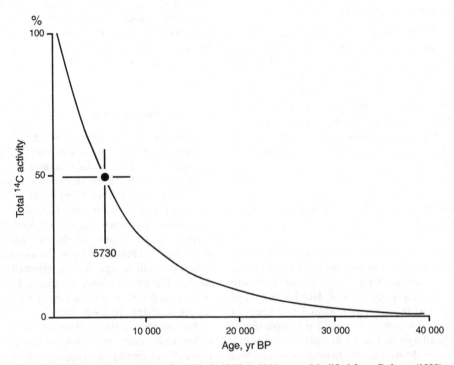

Figure 6.1 Plot of the decay of radiocarbon over time. The half-life is 5730 years. Modified from Roberts (1998).

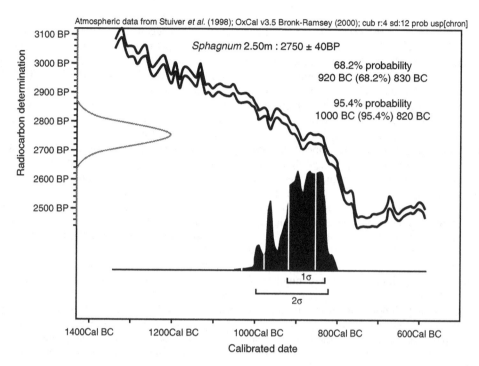

Figure 6.2   Results of calibrating a radiocarbon date to calendar years BP using one of the calibration programs (OxCal v3.5, Bronk-Ramsay, 2000). Calibration uses a data set based on the radiocarbon content of tree rings of known calendar age such as that of Stuiver *et al* (1998) to estimate calendar ages from a measured radiocarbon age. Here a date of 2750 ± 40 BP from *Sphagnum* peat from 2.5 m depth has been calibrated. The calibration curve for the period between 1400 cal. BC and 600 cal. BC is shown. The probability distribution of the radiocarbon age is shown on the left by the smooth curve and the probability distribution of the calendar age produced from calibration is shown in the lower solid curve. The calibration curve shows that the relationship between radiocarbon age and calendar age is not linear and that for some periods the same radiocarbon age may actually relate to widely spaced calendar ages (e.g. after *c*. 800 cal. BC).

carbon dates in particular time periods, especially where only a few dates are available.

The second problem is the relationship between the event that we want to date and the material we are actually dating. For example, if we want to date a change in a pollen profile, we should ideally date the pollen itself. While this is possible (e.g. Brown *et al*, 1992), it is more common to date the peat from which the pollen came. However, peat contains material from various sources, some of which will be of different ages. After peat deposition, roots may penetrate the 'fossil' stratum, bringing younger material to the horizon being dated. Equally, older material could be inwashed to minerotrophic peatlands. The 'hard-water' effect may also introduce older carbon from carbonates in the water supply and dissolved organic carbon may shift younger or older material around the peat system. More recently, Kilian *et al* (1995) have suggested that there may be a reservoir effect created by 'old' carbon

dioxide emitted from depth being fixed by plants growing at the surface. Clearly, it is important to consider these sources of error in sampling and pretreating materials where some problems can be reduced. For example, sieving can remove larger roots and acid washing is used in standard pretreatments to remove carbonates. Intensive investigation of the difference in ^{14}C ages between different pretreatments on blanket peat samples suggests that there is not necessarily any consistent relationship between them (Shore *et al*, 1995). However, any such problems are likely to be increased on shallow peats with low accumulation rates, such as those used by Shore *et al* (1995). AMS dating offers a potential solution in some cases, as specific components of the peat can be dated separately, as mentioned for pollen above. *Sphagnum* or other moss leaves and stems are potentially some of the best components to date separately, where they are available. Their lack of roots means they cannot translocate old carbon from depth to their growing parts and they are large enough to be 'fixed' in position in the profile. However, they may still be affected by fixation of 'old' carbon dioxide emitted at the surface. However, in AMS dating of any sample, it is important to be sure that it is pure and not contaminated by other material to achieve a reliable date. Even small amounts of contamination may represent large proportions of the carbon dated and thus adversely affect the derived age.

'Wiggle-matching' is a recently developed technique for higher-precision estimates of peat age that could help overcome some of the uncertainties associated with conventional approaches to radiocarbon dating in peat (e.g. Kilian et al, 1995). It uses the non-linear relationship between radiocarbon age and calendar age to match the shape of the calibration curve with a series of closely spaced peat dates. However, it has not been widely used in routine palaeoecology, probably because of the high costs of obtaining an adequate number of closely spaced dates.

Other radiometric techniques are largely those used for providing dates for more recent peats. Radiocarbon dating is not particularly useful for the last 300–400 years partly due to ^{14}C activity being very similar to modern ^{14}C values and partly because of the large fluctuations in the calibration curve for this period. Additionally, nuclear weapons testing since AD 1950 has increased atmospheric levels of ^{14}C dramatically. This means that the ^{14}C activity of peat deposited in the last 50 years is much higher than the 1950 'modern' reference value. While it is possible to use these unusual fluctuations to wiggle-match very recent radiocarbon ages in peat (Clymo et al, 1990), a number of other radiometric, time marker and incremental techniques are more commonly used to estimate ages. The isotope ^{210}Pb is potentially well suited to dating recent peat and other sediments since it has a half-life of 22.26 years and is detectable in sediments up to around 200 years old. Normally a constant rate of supply of ^{210}Pb over time is assumed in calculation of age from measured activity (Appleby and Oldfield, 1978). In general, chronologies derived from hollow sites seem to be more reliable than those from hummock sites (e.g. Oldfield et al, 1979), although others have found that hummock profiles of several different species appear to provide reasonable chronologies (Pakarinen and Tolonen, 1977a, b). The problems of obtaining reasonable chronologies from ^{210}Pb are related to possible differences between the total amounts accumulated at different microsites in peatlands (Urban et al, 1990) and various processes displacing Pb within the profile after deposition (Oldfield et al, 1995). Oldfield et al (1995) compared

^{210}Pb dated profiles with a number of other age estimates and suggest that ^{210}Pb chronologies should always be used in conjunction with other techniques as a cross-check. This is sound advice for any chronological technique, but especially for ^{210}Pb, given the variable results obtained in the past. It should certainly be followed wherever possible, particularly for peatland types and geographical regions where ^{210}Pb dating has not been applied before.

Several further radiometric techniques are worthy of mention, particularly as they are often carried out at the same time as ^{210}Pb analyses. ^{241}Am (americium-241) and ^{137}Cs (caesium-137) are both radionucleides from post-1954 weapons testing detectable in peat profiles. The year of peak fallout for both of these isotopes was 1963, but measured ^{137}Cs activity shows a much greater range of variation than ^{241}Am due to its greater movement in solution and active biological uptake by plants (Oldfield et al, 1995). As a result, only ^{241}Am is recommended for dating peat, although ^{137}Cs has also been used in the past (Urban et al, 1990). Figure 6.3 shows a good example of the use of ^{241}Am to constrain ^{210}Pb chronologies in three peat profiles in Britain. In all cases, pollen markers are also superimposed on the age–depth curves as a further cross-check. Finally, a recently developed tool is the ratio of ^{206}Pb:^{207}Pb, reflecting different sources of historical lead pollution used successfully to help date a peat profile in Switzerland (Shotyk et al, 1998).

6.4.2 Time markers

There is a wide range of time markers that in some circumstances can provide a complete chronology on their own. More often they are used in conjunction with other techniques, as there are too few markers to provide a complete chronology for the profile being studied. Radiometric techniques are in principle applicable in any area of the world. In contrast, time markers are highly variable regionally and locally. Thus an understanding of the local environmental history is vital to gaining maximum benefit from the use of time markers. The three markers we discuss here are volcanic ash layers (tephras), pollen markers and spheroidal carbonaceous particles. The principle in all cases is that an event is detectable in the peat and that we can use this to correlate between cores and sites. If the date of the event is known from other sources, we can then assign an age to the layer where the event occurs.

Tephra layers from volcanic activity are common in soils in many regions of the world, although they do not

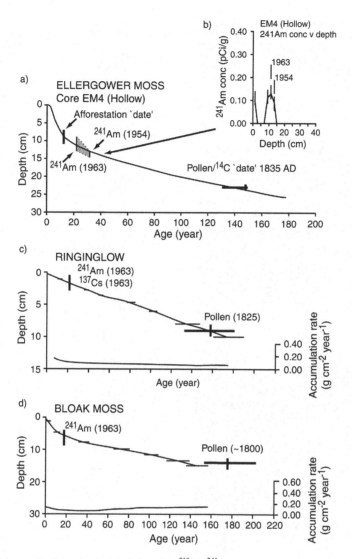

Figure 6.3   Age–depth curves for three sites in Britain based on ^{210}Pb, [241]Am and pollen markers. In the case of Ellergower Moss, the [241]Am–depth curve is illustrated to show how the 1954 start and 1963 peak of weapons testing are represented in the record. The pollen markers are different for each site as the vegetation histories are slightly different for each area but are known from documentary records. Redrawn from Oldfield *et al* (1995) by permission of Arnold Publishers.

always coincide with the areas of peatland distribution! Some areas, such as New Zealand, have a number of late Quaternary tephra layers, including many visible to the naked eye, which have been used for dating and correlating peat and other deposits (Lowe, 1988; Newnham *et al*, 1999). There are at least 36 tephra layers over the last approximately 50 000 years in New Zealand (Froggatt and Lowe, 1990). A striking example of multiple tephras in a peat deposit is that

of the Kaipo Bog sequence in North Island, New Zealand (Lowe *et al*, 1999; Figure 6.4). Here, 16 tephras throughout a sequence covering 15 000 radiocarbon years of peat accumulation were identified. The youngest tephra is the Kaharoa from 665 ± 15 radiocarbon years BP and the oldest tephra is the Rerewhakaaitu from $14\,700 \pm 95$ BP. In other areas, the detection and recognition of very thin tephra layers from distant volcanic sources are increasing. These tephras are not

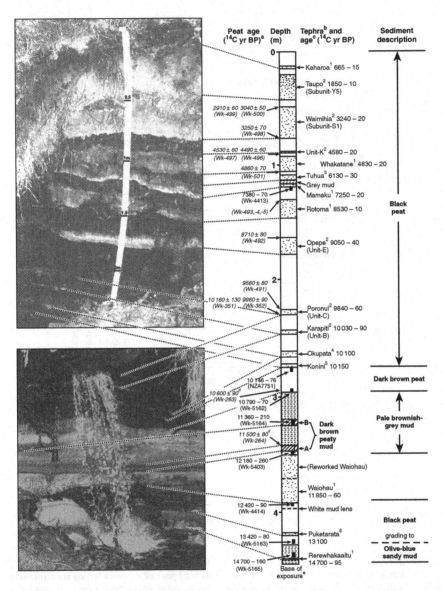

Figure 6.4 Tephrostratigraphy and chronology of the Kaipo Bog sequence, Te Urewera National Park, North Island, New Zealand. The photographs on the left show two separate field sections of tephra layers interbedded with peat and mud layers. The two sections have been linked to form a complete tephrochronological sequence for the site covering the last 15 000 radiocarbon years BP. The radiocarbon ages are those derived from this sequence, and the tephra ages on the right are based on multiple radiocarbon age estimates from other sites where the same tephras occur. Reproduced from Lowe *et al* (1999) by permission of the Royal Society of New Zealand.

visible in the peat except with treatment to remove much of the organic matter. Microscopic tephras are becoming especially important in peatland palaeoecology in north-west Europe, where there are a number of tephras of Icelandic origin throughout the Holocene.

Research in Scotland and Ireland has revealed around a dozen tephra layers (Dugmore *et al*, 1995; Pilcher *et al*, 1995), some of which are not identified with specific known eruptions but which still form useful tephrochronological markers for correlation purposes. Even relatively small eruptions in the historic period are known to have produced ash that reached some distance from source (Dugmore *et al*, 1996). It therefore seems very likely that other regions will also possess tephra layers from more distant volcanic sources, and there is much potential for research in this area. Certainly as work progresses on sites more distant from volcanic centres, the distribution of known tephras expands regularly. For example, Icelandic tephras have now been detected from as far afield as England and Germany (Van Den Bogaard *et al*, 1994; Wells *et al*, 1997).

The use of tephras in dating requires both the detection and characterisation of the glass shards that form the tephra. The latter is usually achieved by geochemical analysis of a number of individual tephra shards using electron microprobe techniques. Often this is to confirm a best guess of identification based on radiocarbon ages or stratigraphic position, but it is essential in most applications, especially where there are a number of possible candidate sources and eruptions. Post-depositional dissolution (Hodder and de Lange, 1991) and the analysis techniques used (Hunt and Hill, 1993) may be important considerations here too. Therefore, the identification and correlation of tephra tend to be based on multiple criteria, including field properties, stratigraphic interrelationships, geochemistry and radiocarbon dating. A consideration of all of these aspects is necessary to unravel complex sequences such as that shown in Figure 6.4.

Pollen markers have already been mentioned in connection with their use in constraining ^{210}Pb chronologies. The most useful pollen markers are those which show major changes in the pollen composition over very short periods and which are either known to be consistent across large areas or for which there are detailed data on the local variability of ages. Older pollen markers include those of the elm decline around 5300 radiocarbon years BP in Britain and Ireland, and the hemlock decline in North America, which occurred around 4800 radiocarbon years BP. For the historic period there are a number of changes in pollen diagrams that can be referred to known historical events. For example, the major expansion of European settlement in North America is marked by a very rapid rise in *Ambrosia* pollen (ragweed). Of course this can vary considerably between areas, but historical archive data can be used to constrain ages for particular sites quite well. The same is true of afforestation events in Britain, where a rise in pine pollen was associated with planting on estates in the late eighteenth and early nineteenth centuries and a further rise is found with the establishment of twentieth-century plantations (Oldfield and Statham, 1963; Oldfield *et al*, 1995).

The third type of marker that is now being used in recent peats is counts of so-called spheroidal carbonaceous particles (SCPs). These are microscopic spherules that are derived from the burning of coal and oil. They begin to occur at the start of the industrial revolution and particularly after 1950, with a peak in Britain and Ireland in the later 1970s or early 1980s (Rose *et al*, 1995). As a result, they are particularly useful in dating very recent peats. They have not yet been used extensively for peat but there are a few published examples (Barber *et al*, 1999; Chambers *et al*, 1999). They have been much more widely applied in lake sediments, and data exist for sites throughout much of Eurasia and, to a lesser extent, North America (Flower *et al*, 1997; Boyle *et al*, 1998, 1999; Lan and Breslin, 1999; Rose *et al*, 1999; Yoshikawa *et al*, 2000). Again, documentary data on the changes in the volume and nature of industrial activity in particular areas are required to estimate the dates of particular features in the record. Figure 6.5 shows a typical pattern of change in a number of lake sites in northern Britain, compared to that found in a peat profile from Northumberland. It seems likely that SCP counts could be more extensively used elsewhere in the world, and certainly could be much more widely employed in peatland research.

6.4.3 Incremental techniques

There is a wide variety of incremental dating techniques for Quaternary sediments, including annual laminations in lakes, tree rings (dendrochronology) and ice cores. One of the few direct incremental techniques that has been used successfully in peat is that of moss increment counting (Pakarinen and Tolonen, 1977b). It is only applicable to recent peats as it relies on well-preserved stems of *Sphagnum* or other mosses (*Polytrichum* species) being present. When stems of these taxa are carefully separated from a horizontal slice of peat (normally 4–5 cm in depth), individual growth years can be identified by changes in colour, branching patterns and stem orientation. The length of the stem formed each year can thus be measured directly. The time of formation for each slice can be estimated by dividing the total stem length by the average increment length. It has been successfully carried out in a number of sites, although it generally only provides useful data for the last 40–50 years (El-Daoushy *et al*, 1982; Belyea and Warner, 1994).

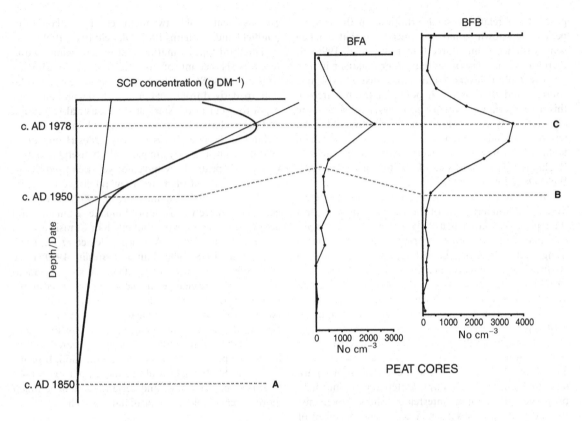

Figure 6.5 Changes in the concentration of SCPs with depth in profiles from different regions in the UK compared to those recovered from peat profiles in Northumberland, England. The vertical scale on the peat cores is omitted for clarity but both cover 50 cm. Horizons A, B and C are estimated ages of AD 1978, AD 1950 and AD 1850. Adapted from Rose *et al* (1995) and combined with unpublished data held by the author.

Several other techniques exist for peats, which are essentially incremental methods, although they do not count increments of equal time in the same way that tree rings and lake sediment laminae are counted as annual layers. Estimates of age using the constant bulk density method (Ilomets, 1984, cited by Belyea and Warner, 1994) and pollen density methods (Middeldorp, 1982) both work up from a known basal age, assuming either constant addition of material at the surface or constant influx of pollen, respectively. Bulk density is very quick to measure even in the contiguous samples required to develop a cumulative mass curve, but there is much more work involved in estimating pollen concentrations for a large number of samples. In both of these techniques, it is important that the basal reference age is reliable but also that the assumptions are reasonable. Comparisons of various techniques, including all three of these incremental techniques, were carried out by Belyea and Warner

(1994). They concluded that moss increment counting worked well but was susceptible to systematic bias as deeper depths were reached, even where moss stems appeared to be intact. Bulk density dating worked well for rough age estimates, as long as no major stratigraphic changes were encountered. Likewise pollen density was useful, although did not appear to work as well as bulk density dates on these particular cores.

The well-known incremental technique of dendrochronology is applicable in peat palaeoecology if timbers of sufficient size are available for study. The applications of this technique will be discussed along with plant macrofossil studies generally, as it is often more informative about the palaeoecology of the trees themselves than it is about the peatland record as a whole and it is rarely used for dating of the peat. However, some recent studies have used estimation of the age of living trees to establish vertical growth rates of *Sphagnum* hummocks over the last 150 years

(Ohlson and Dahlberg, 1991; Okland and Ohlson, 1998). Here the age of living trees was estimated from counts of tree rings. The depth of the surface of the peatland when the tree began growth was found by excavating to the original base of the stem so that the depth of peat that had accumulated during the growth of the tree could be established.

6.5 SURVEY AND STRATIGRAPHY

Many studies require the collection of basic data on the surface and subsurface profiles of a peatland and major stratigraphic changes. Samples of peat are also collected for laboratory analysis using the techniques reviewed in this chapter. The particular methods used vary according to peatland type, the requirements of the study and sometimes depend on the tools favoured by the investigator! Survey of the overall shape of the deposit is usually achieved by standard topographic survey techniques (e.g. Pugh, 1975), together with depth probing using rods such as those from coring devices (see below). This allows a three-dimensional picture of the peat deposit to be built up. This simple approach works well in most cases, but geophysical techniques are a possible alternative and may have the added advantage of providing basic stratigraphic data as well. However, the only examples of its use appear to be radar techniques to detect the peat–mineral contact with additional details on major stratigraphic changes, occurrence of tephra layers and large subsurface features such as tree stumps (Lowe, 1985; Warner et al, 1990).

Data on the overall shape of the peat deposit are of limited use in most cases, and for most applications a stratigraphic description is necessary and samples of peat are essential. A variety of devices for extracting peat have been designed over the years, all with various advantages and disadvantages. The most common devices used in peat are piston corers or the so-called 'Russian' corer. The detail of these is described elsewhere (e.g. Barber, 1976; Aaby and Berglund, 1986b) but the main principles are as follows. For piston corers, the principle is that the sample tube is pushed down into the peat and then pulled out. For Russian corers, the idea is that a half-cylinder-shaped head is pushed down into the peat and then a sample is enclosed by rotating the half-cylinder, enclosing a section of peat which is then pulled up to the surface. Variations of these basic designs have developed for more specific tasks in peat studies, especially for the extraction of larger and less disturbed samples (e.g.

Barber, 1984; Wright et al, 1984). Both approaches suffer from potential difficulties with compaction, distortion and contamination of the samples, but these are usually minimal. The choice of sampling technique ultimately depends on a balance of factors, including peat type and the amount of sample required. Neither of these common devices is particularly effective at retrieving samples from unconsolidated near-surface peats. Alternative sampling devices have been designed for surface peats, including box corer devices (Wardenaar, 1987) and large-diameter tube devices (Clymo, 1988). In some situations, sampling from an exposed face may be possible. This approach is well suited to sites that are being subjected to peat extraction or which have large drainage ditches. It can also be achieved by excavating shallow pits on some sites.

Stratigraphic descriptions are essential for most applications and they should be carried out in the field or as soon as possible thereafter. On exposed sections, stratigraphic changes are sometimes clearer in a weathered section than in a cleaned fresh face. Stratigraphic descriptions can give a rapid general understanding of changes in time and space in a peatland and often provide essential information for the next phase of investigation such as plant macrofossil or pollen analysis. The selection of the location of samples for more detailed analysis is important and an inappropriate location could mean a lot of wasted effort later. A standard system for description of sediments was developed by Troels-Smith (1955), and this is still widely used today, although it is often supplemented by other descriptions and analyses. The principal idea is that any sediment, including peat, can be described in terms of basic physical properties, degree of composition and the relative proportions of its constituents. The physical properties include colour expressed as degree of darkness, the degree of stratification, elasticity and dryness. The characteristics and components are given Latin names to avoid confusion in different languages. For peat, the most common components in the Troels-Smith system are those of *Turfa*, for example any peat composed predominantly of moss remains is termed *Turfa bryophytica* or Tb for short. More complicated peat composition can also be described using Troels-Smith's shorthand formulae. There is also a set of standard symbols for the representation of stratigraphic change that most authors use wherever possible. Aaby and Berglund (1986a) provide a useful guide to the use of the Troels-Smith system since the original publication is not always easy to obtain.

6.6 BIOLOGICAL EVIDENCE OF PAST CHANGES

There is a vast range of biological evidence to be gained from peat deposits, because the preservation of so many organisms is possible under the special conditions found in peat. Detailed descriptions of all of the techniques is not possible here, but we can provide an overview of some of the most widely used evidence, as well as pointing to other areas that are less commonly employed but which may nevertheless be useful in certain settings. The remains of plants are the most widely used data from peatlands, principally because they are ubiquitous, abundant and provide a good picture of the changes on the peatland, as well as those in the surrounding landscape. For all biological evidence there are certain factors that should be considered in the acquisition and interpretation of the data:

(i) *Preservation*. What is the quality of preservation and has it affected taxa differentially?

(ii) *Identification*. What taxonomic precision can be achieved?

(iii) *Fossil assemblages vs. living assemblages*. Are the relative proportions of the fossil remains likely to be the same as those in the living assemblage that they represent?

(iv) *Taphonomy*. What processes led to the deposition of the fossils and where did they come from? Are they predominantly autogenically deposited (i.e. at the place where they lived) or allogenically (away from the place they lived)?

(v) *Uniformitarianism*. Are there likely to have been changes over time in the relationship between the organisms and the environment?

Some of these factors are part of broader considerations of palaeoenvironmental evidence, which are discussed in greater depth elsewhere (Lowe and Walker, 1997). We will explore some of these in relation to specific groups of remains in the following pages.

6.6.1 Plant macrofossils

The division between plant 'macrofossils' and 'microfossils' is not based simply on size, as the names would suggest, but is related to the nature of the remains themselves. Plant macrofossils include the vegetative parts as well as the seeds and fruits of plants. Plant microfossils are the pollen and spores. The terms 'macrofossil' and 'microfossil' are not very easily applied to other groups of organisms, although 'micro-

fossil' tends to be used for any remains that need high-powered light microscopy ($> \times 100$ magnification) to be easily visible. The size of plant macrofossils varies enormously, between tiny seeds less than 1 mm in length to entire tree trunks. There are thus no standard methods for their study, and each group requires particular techniques from careful sieving and mounting on microscope slides (moss leaves) to a chainsaw (tree trunks)!

Leaving aside very large tree and shrub remains, most plant macrofossils need to be sieved and separated for identification and assessment of abundance, using microscopic techniques. Several papers on plant macrofossils provide useful ideas and guidelines on the methods used in peat and other sediments. Good guides to the study of seeds, fruits and vegetative remains include those of Wasylikowa (1986), Grosse-Brauckmann (1986) and Warner (1988b). In addition, bryophytes are especially abundant in many peatlands, and several different approaches have been used for their study (Dickson, 1973, 1986; Janssens, 1983, 1989; Barber *et al*, 1994).

In most cases, plant macrofossils in peat will be deposited autogenically, so they represent the plant community at the sample point. Although there are rather few taphonomic studies on peatland plant macrofossils, it seems likely that even most seeds and fruits will not travel very far (Greatrex, 1983). However, some more dynamic environments such as floodplains, where water flow is stronger, may have a significant component of allogenic macrofossils. In general, though, plant macrofossils provide excellent evidence of the changes in mire plant communities through time. They thus underpin many studies that examine the relationship between mire development and both internal and external environmental changes. An area where there has been particular interest in recent years is that of plant macrofossils on ombrotrophic peatlands and climate change in north-west Europe, following the work of, for example, Barber (1981), and similar work has begun in New Zealand (McGlone and Wilmshurst, 1999). While ombrotrophic peatlands have perhaps been the focus for many plant macrofossil studies, fen peats are also suitable for the preservation of macrofossils. Wells and Wheeler (1999) reconstruct vegetation change using multiple cores from the Ant Valley in Norfolk, England. The peat here contains a diverse mix of reproductive and vegetative macrofossils showing a series of phases of development in response to sea-level change, climate and human impact (Figure 6.6). The phase of marine influence with development of saltmarsh (zone SM1b)

Sedge Marsh 1
Norfolk UK

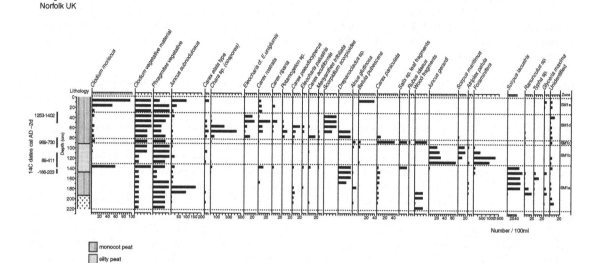

Figure 6.6   A plant macrofossil diagram from a floodplain mire in the Ant Valley, Norfolk, England. This is one of five cores from the area and reflects a sequence of change from a wooded marsh ('carr') environment (SM1a), through to a period of saltmarsh development (SM1b), a short phase of shrub growth (SM1c), a flooded marsh (SM1d) and then a dry phase dominated by *Juncus*, *Phragmites* and *Cladium mariscus* (SM1e). Redrawn from Wells and Wheeler (1999) by permission of Arnold Publishers.

reflected in the presence of foraminifera is unusual in peatlands, but demonstrates a more general point that many other biological remains are usually also found, identified and counted, along with the plant fragments in 'plant' macrofossil analysis.

Large tree remains are a feature of many peatlands throughout the world. They are particularly well known from areas where erosion and disturbance have exposed them, such as in Ireland and northern Britain, but they are also well known from Scandinavia, parts of North America and New Zealand. Extensive tree remains must also exist in other northern boreal peatlands and in tropical peatlands as they are such important components of the modern vegetation, but they have not been extensively studied. Buried trees can yield information on forest history but are also the principal source of dendrochronological data for the radiocarbon calibration curve discussed above, and for much of the palaeoclimatic data from tree-ring studies. Work such as that by Kullman (e.g. Kullman, 1989; Kullman and Engelmark, 1990) on a variety of tree species in Sweden is probably some of the most detailed, showing how tree lines have changed during the Holocene, often in response to climatic change. For example, Kullman (1999) suggests summer temperatures were up to 2.4 °C higher than today at 8500 BP.

6.6.2 Pollen and spores

The term 'pollen analysis' includes both pollen of higher plants and spores of ferns, mosses and clubmosses. They are the most commonly studied fossils in peat, probably because they are present in every deposit and are almost always highly abundant. Pollen analysis is a powerful tool, yielding information about the vegetation of the peatland itself as well in the surrounding landscape. There are many general palynological guides to preparation and identification of pollen, and peat is one of the easiest sediments to work with for pollen extraction (e.g. Faegri and Iversen, 1975; Moore *et al*, 1991). Indeed, the study of palynology really began on peat with the work of Erdtman (1934). It is critical to consider taphonomic issues in any study using pollen, but especially in peatlands. Pollen is distributed by wind, water and insects but the wind-borne component is usually the most important in peatlands. There are thus many potential source areas for pollen in a peat deposit and all possibilities should be considered in interpretation of pollen data from peat. Tauber (1965) developed conceptual models of these processes and these can be adapted for particular situations. The local component (from plants growing within a few metres) of the pollen rain is much more important in peat than many other sediments. Wind directions and dominant air movements may also result in particular patterns of pollen deposition in peatlands (Figure 6.7). It is also known that the proportion of local components of the pollen rain increases rapidly from mire edge to mire centre (Caseldine, 1981), and other modern pollen rain studies suggest strong local variability (e.g. Bunting *et al*, 1997). As a result of processes such as these, one might expect the local variability of the peatland pollen record to be quite high. Edwards (1983) suggested that there was greater variability in the detail of multiple cores than is normally recognised, as general patterns and presence/absence of taxa are often the same. Depending on the objective of the study, this can be viewed as an advantage or as a disadvantage. The peat pollen record may be able to detect subtle spatial differences but requires a more careful approach to reconstruction of broad-scale vegetation change.

Pollen analysis of peats is most often carried out to determine the vegetation history of the surrounding area, although in recent years lake sediments, where available, have often been preferred for this purpose. This reflects the taphonomic problems of peat where the influence of local pollen on the total assemblage is difficult to disentangle. Despite these difficulties, peatland pollen records are still a major contributor to regional syntheses of vegetation history (see e.g. Berglund *et al*, 1996) and mapping of broad-scale vegetation and climate change over time (e.g. Huntley and Prentice, 1988, 1993). One advantage of peat is that it is more widely distributed in many areas than lakes and there is thus more choice of location and greater

potential for examining detailed spatial variability of vegetation history (e.g. Turner *et al*, 1993; Dumayne, 1993). The potential for high-resolution studies has also been explored by several authors, and work on peat and other sediments is reviewed by Green and Dolman (1988).

More importantly for our purposes of examining changes through time on peatlands themselves, palynological data are important in reconstructing spatial and temporal patterns of vegetation change at the scale of microtope and mesotope. These often need to be related to other data such as plant macrofossils and charcoal counts to provide a full picture of events. There are many examples of such applications; here are just a few that show different approaches to the subject. It is common to use pollen data to reconstruct both vegetation history of the surrounding area and bog development simultaneously. Table 6.2 describes the developmental phases of a raised bog in France as well as the vegetation changes in the surrounding area (Barbier and Visset, 1997). Here the data from a single core are subjectively interpreted in spatial terms, from knowledge of where the taxa detected were most likely to be growing. Alternative approaches use multiple cores to differentiate the spatial patterns in past vegetation on the site (e.g. Bunting *et al*, 1998). The use of pollen data together with information from plant macrofossils is another way to help separate local and regional vegetation change, as well as detecting additional taxa or providing higher taxonomic precision than some pollen data allow (e.g. Birks, 1970).

6.6.3 Charcoal

Pollen and plant macrofossils provide the main evidence for vegetation change on mires but charred plant fragments are an additional valuable source of information for the reconstruction of fire history, which is often closely related to vegetation change. Reviews of various aspects of charred particle analysis are provided by Tolonen (1986a) and Patterson *et al* (1987). In peats, charcoal reflects both burning of local vegetation and more widespread fire, and some of the considerations involved in interpretation of charcoal records are the same as for pollen. In general, larger charcoal particles will reflect fire closer to the site than small particles. We have already discussed the wide distribution of industrially produced SCPs for dating; these particles clearly show how far small charred particles can travel. Beyond this simple concept, there has been much discussion of charcoal source–sediment

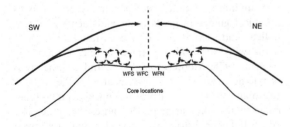

Figure 6.7 Suggested movement of air masses and pollen deposition on a flat-topped, steep-sided mountain area in Wales. Here, there is a sharp division between the two sides of the peat-covered hilltop, with the pollen rain dominated by one valley or the other but with little mixing between the two. This is one example of the need to consider pollen taphonomy in the selection of sites and interpretation of data for pollen analysis. Taphonomic processes may be very different for different sites. Redrawn from Price and Moore (1984) by permission of Peter Moore.

Table 6.2 Vegetation history from pollen analysis used to reconstruct main phases of bog development and changes in the surrounding mineral soils from palynological data from Logne, western France. The conditions on the peatland itself are separated into those thought to be occurring at the mire margins (E) and those in the mire centre (B). Changes in the surrounding mineral soils are indicated as H. The pollen zones are given down the left-hand column. After Barbier and Visset (1997).

Zones	Phases	Main vegetation types	Hydrological events and other factors
n		B: *Calluna* and *Erica* moor invaded by *Myrcia* and birch	
m	VI	E: Colonisation by birch and hazel	Rise in water level and anthropogenic influences
l		H: Clearance of oak	
		Intensive arable farming but declining in the most recent period	
k		H: Oak + abandoned cultivation	Open conditions at edge of bog
j		B: Filling-in of the bog	
		E: Residual alder woodland	
		H: Oak + cultivation	
I	V	B: Increase in *Sphagnum*	Rise in water level
		E: Grasses, helophytes	Closed conditions at edge of bog
		Reduction in *Osmunda*-alder carr, willows and buckthorn	
h		H: Oak + lime	
		B: Increase in *Sphagnum*	Rise in water level
g	IV	E: Domination of helophytes and *Carex*	Closed conditions at edge of bog
		Sudden renewed development of *Osmunda*-alder carr	
		H: Oak + lime	
		B: Increase in *Sphagnum*	Maximal rise in water level and possible anthropogenic intervention
f	III	E: Grasses, helophytes, *Carex*	
		Near-disappearance of *Osmunda*-alder carr	
		H: Oak + beech + cultivation	Open conditions at edge of bog
		B: Development of *Sphagnum*	Rise in water level
d	II	E: Colonisation by Cyperaceae, helophytes and grasses	
		Decline of *Osmunda*-alder carr	
		Development of willow carr with buckthorn and *Myrcia*	Closed conditions at edge of bog
c		H: Oak woodland	
b		B: Free water at the time of colonisation by *Sphagnum* mosses and other pioneer species	Constant water level
		E: *Osmunda*-alder carr	
a		H: Mixed oak + lime and hazel	Closed conditions at edge of bog

relationships (Patterson *et al*, 1987; Clark, 1988), although it remains difficult to use these ideas in any very certain way to interpret changes in fossil charcoal content. Various methods exist for the estimation of charcoal content (Robinson, 1984; Winkler, 1985), but the most commonly used technique is to count charcoal abundance on pollen slides – the so-called point-count technique (Clark, 1982). In peat, there is the potential problem that severe fires could remove peat and disrupt the sedimentary sequence, and even subsurface formation of charcoal has been proposed (Boyd, 1982), although this seems unlikely (Moore, 1982b). In drier conditions, fire can be a severe problem on both tropical and temperate peatlands, and the destruction of the peat record is a possibility.

It is common to combine pollen and microscopic charcoal analysis, especially where fire is hypothesised to have played a significant part in the vegetation history of an area. In western oceanic Europe, the development and abundance of heathland communities are thought to have been heavily influenced by early human populations and especially their use of fire. Edwards *et al* (1995) show that there is considerable spatial diversity in charcoal records and heath development on the island of South Uist in Scotland. Here, heathland had begun to develop before significant charcoal was recorded but later burning may have encouraged its expansion. The attribution of fire to human or natural causes is difficult in most records, but where multiple profiles reveal such spatial diversity it seems unlikely that the fires would have been a result of periods of much drier climatic conditions and increased natural fire frequency.

6.6.4 Protozoa

There is a large group of other microfossils detectable in peat and many are commonly observed in pollen preparations. These are predominantly of fungal or algal origin (see below) but also include remnants of animal origin and protozoa. The protozoa that are preserved in the peat are those that possess an external shell and are known as rhizopods or testate amoebae. While they can be observed in pollen preparations, only a proportion of the tests survive the chemical treatments, and special preparations are needed for reliable counts. They are particularly abundant in oligotrophic peatlands and have been extensively studied in Europe, Scandinavia and North America, with some work in New Zealand peats. Reviews of the methods, identification and applications are given by Tolonen (1986b),

Warner (1988a) and Charman *et al* (2000). They are mainly indicators of the moisture status of the peatland and have been extensively used for palaeohydrological reconstructions. However, they also show major changes along the pH gradient and may have potential as indicators of trophic status.

The relationships found in ecological studies have been used for many years in the reconstruction of past peatland environments. Much of the earlier work was in Europe and Scandinavia (e.g. Tolonen, 1966; Grospietsch, 1967; Beyens, 1985) and interpretations were made subjectively. More recent work has gone a step further in using quantitative taxa–environment relationships with the application of 'transfer functions'. This is a technique now widely used in Quaternary ecology (Birks, 1995), but only applied in a limited number of peatland studies. Here the known relationship between taxa and an ecological parameter (such as depth to water table) is used to reconstruct the same variable from a fossil assemblage of organisms. This has been done for a number of profiles in Canada (Warner and Charman, 1994), Britain (Hendon *et al*, 2001) and Switzerland (Mitchell *et al*, 2001). Figure 6.8 shows the results of such a process on a profile from Northumberland, England, using the transfer function of Woodland *et al* (1998). The principal benefit of this process is in summarising the complex changes in assemblages as a single parameter of interest. It has also been used for bryophytes, notably by Janssens (1988). This will be discussed further in relation to surface moisture reconstructions and climate change in Chapter 8.

6.6.5 Other microfossils: fungal, algal, animal remains

Of the 'other' microfossils often recorded in peat samples, fungal and algal remains are most common. In addition, there are various pieces of evidence from other organisms that are found in pollen preparations or other microscopic examinations such as simple 'squashes' of untreated peat samples. The potential of many of these organisms is still unexplored but the work by van Geel and his co-workers at the Hugo de Vries Laboratory in Amsterdam is the most extensive (van Geel, 1978 is an especially useful paper, together with the review by van Geel, 1986). Other papers summarising aspects of work on these microfossils are those of Cronberg (1986), Pirozynski (1990), Smol (1990) and Warner (1990). One of the factors emphasised by van Geel (1986) is that many remains that can be seen have

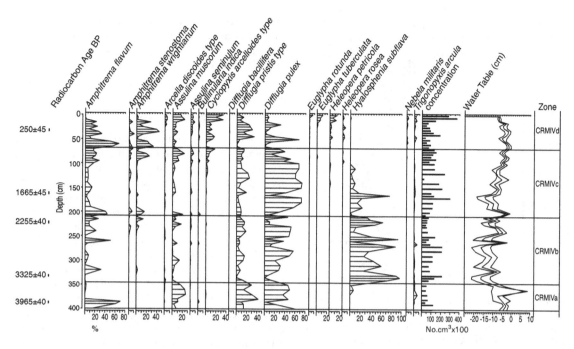

Figure 6.8 A testate amoebae diagram from Coom Rigg Moss, Northumberland, showing main changes in taxa and the reconstructed depth to water table based on a transfer function from modern ecological data. The complex changes in taxa are summarised as a single parameter of interest, making interpretation of the data much simpler, although the limitations of the transfer function need to be considered. Redrawn from Charman *et al* (1999).

not yet been identified as particular species and have instead been given 'type' numbers, many of which have subsequently been identified. Even without identification, types can be useful for characterising different sediments and a certain amount of palaeoecological information can be gained from knowledge of other sediments in which they occur.

6.6.6 Diatoms

Diatoms are widely used in studies of lake palaeoecology but are not reported from peatlands so often. Despite their presence on the surface of the wetter parts of peatlands (e.g. Kingston, 1982), they may suffer dissolution in the peat column (Bennett *et al*, 1991). However, they certainly are preserved in some peats and have been used to reconstruct development of peatlands in a variety of locations including temperate (Beyens, 1985) and Arctic (Ruhland *et al*, 2000) areas. The main indicator value of diatoms on peatlands is changing trophic status, with a secondary role of dryness, since a number of taxa are aerophilic. There may be further scope for using diatoms as indicators of

other variables such as diatom analysis on some peatlands where it is known they occur in abundance at the surface, such as phosphorus concentrations, as suggested for the Florida Everglades (Cooper *et al*, 1999).

Diatom analysis can also be carried out on lake sediments from water bodies within or next to the peatland, to reconstruct mire development. Pogonia Pond is a small lake within a peatland in southern Minnesota, USA, almost at the southern limit of peat distribution in the area (Brugam and Swain, 2000). Analysis of diatoms and the application of a transfer function shows that there was a major change in the acid-neutralising capacity of the lake after 2200 radiocarbon years BP (Figure 6.9). This is linked to the development of the peatland following regional cooling in the climate.

6.6.7 Macroscopic invertebrates

Various invertebrates occur in peatlands. Coleoptera (beetles) are commonly encountered in many peat deposits and are noticeable because of their relatively large size. Others are smaller and/or less abundant, especially in acid peats where calcareous shells and

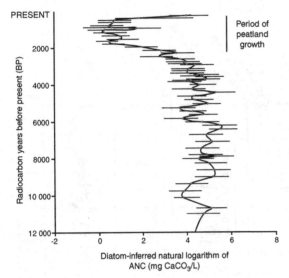

Figure 6.9 Variation in the acid-neutralising capacity (ANC) of Pogonia Bog Pond, Minnesota, USA, reflecting the growth of peatland in the area. The variation in ANC is inferred from a transfer function developed from modern samples in the region. Redrawn from Brugam and Swain (2000) by permission of Arnold Publishers.

exoskeletons do not survive. However, molluscan remains do occur in peats with a higher pH (Miller and Bajc, 1990) although they are not widely reported. Other taxa such as Cladocera and Chironimidae are predominantly aquatic and likewise have not received much attention in peatland studies, although they are occasionally reported from specific contexts.

Beetles are best known as indicators of temperature changes in the Quaternary, especially at the glacial–interglacial transition (Lowe and Walker, 1997). However, they may also be useful indicators of temperature and other environmental changes during the Holocene when most peat formation has occurred. There are still rather few major examples of the study of beetles in Holocene peats as the principal indicators of environmental change over long time periods, but several studies have been carried out in Canada (Lavoie *et al*, 1997a, b). One might expect beetle faunas to reflect climatic changes in the Holocene as they have been shown to do so effectively for earlier periods. However, the evidence here is equivocal. For example, in the tundra of Arctic Canada Lavoie *et al* (1997b) found a high proportion of boreal forest species, suggesting mean July temperatures were 2.8–5.5 °C higher than today during the period approximately 6000–2000 radiocarbon years BP. In southern Quebec, there was

little change in the fauna in response to climate over the last radiocarbon 7000 years, and the beetle fauna were mostly dependent on local ecological conditions (Lavoie *et al*, 1997a).

6.6.8 Archaeological remains

Other biological remains related to human activities are sometimes recovered from peat deposits. The study of wetland environmental archaeology is a large field on its own and much of the evidence is based on the techniques already described. The unique aspect of wetlands as an archaeological resource is that they preserve many organic materials that are not preserved in other contexts. Two particular cases are worthy of mention: bog bodies and trackways.

Finds of human bodies in bogs have been made for some time and were well publicised by the book *The Bog People* (Glob, 1969). Probably the most intensively investigated of recent bog body finds are those discovered during peat extraction on Lindow Moss, a remnant raised mire in Cheshire, England. A series of human finds named Lindow I, II and II was found in the 1980s. The investigations into the evidence of the body of Lindow II (known as Lindow Man) and the environment in and around it are reported by Stead *et al* (1986), and a review of recent work on bog bodies with further information on Lindow III is provided by Turner and Scaife (1995). The detail and diversity of evidence from the finds at Lindow Moss are startling: all aspects of the anatomy of the body were looked at, the gut contents were examined in detail for plant residues and parasites, and the surrounding peat was studied for palaeobotanical and entomological remains.

The second important area of peatland archaeology is the trackways discovered in many areas of Europe. These date from a wide range of times but were built by prehistoric peoples from at least as far back as the Neolithic onwards for access to and transport across wetland areas. There are now quite large numbers of these reported, but the greatest density and most intensively studied are those in the Somerset Levels in England. Much of the work is reported in the *Somerset Levels Papers* and also summarised by Coles and Coles (1986). Some of the most impressive evidence comes from the timbers themselves, which preserve the woodworking marks of tools as well as holding information on woodland management practices and dates of tracks. Equally, the surrounding peat holds related palaeoenvironmental evidence that can be used to help understand the relationship between ancient human populations and the environment of the time.

6.7 PHYSICAL AND CHEMICAL CHARACTERISTICS

Besides the ecological information locked up in the preserved remains of various organisms, the peat archive contains a wealth of data represented by physical and especially by chemical attributes. There is a rapidly increasing range of characteristics that can usefully be measured and this will almost certainly continue to increase with new developments in analytical chemistry. Basic measurements such as bulk density are now regularly carried out, mainly in response to the increased interest in mass balance and peat accumulation, discussed in Chapter 5. Other basic data such as loss on ignition are also easy to collect and of great interest in determining the proportion of the inorganic contribution to the peat. Beyond these basic measurements, characteristics fall into five main groups:

(i) Measurements related to the degree of decay (humification).
(ii) Inorganic elemental chemistry.
(iii) Organic chemistry.
(iv) Isotopic analyses, excluding those made for chronological purposes.
(v) Other novel techniques of characteristics that are proxies for some physical and chemical characteristics of the peat.

6.7.1 Humification

Humification is the degree of decay of the peat. Although there are a number of methods for estimating it, there is no single scale on which it measured and no accepted definition of what it really means. The simplest method is a visual assessment to provide a relative score on a scale of 1 to 10 (von Post, 1924), also described by Aaby (1986). More precise measurement by digestion and colorimetric assessment of the extract is now the standard way of assessing humification in palaeoecological studies (Blackford and Chambers, 1993). However, it seems likely that although this provides a reasonable relative measure of humification, the digestion process is itself responsible for much of the breakdown of the peat and changes the composition of the organic acids present considerably (Caseldine et al, 2000).

Humification data are widely used to indicate changes in hydrological conditions on mires: greater degree of humification suggests dry, warm conditions with increased decay, whereas low humification suggests wet, cool conditions with reduced decay. This is most often used on ombrotrophic mires to reconstruct climatic changes. One of the best-known examples of this application is that of Aaby (1976) on Draved Mose, Denmark. There are numerous more recent examples, especially from the British Isles (e.g. Blackford and Chambers, 1995; Anderson, 1998) but also from New Zealand (McGlone and Wilmshurst, 1999). Humification data have also been used in valley mires to detect periods of increased wetness arising from greater surface runoff as a result of human activities (Chambers, 1983b).

6.7.2 Inorganic elemental chemistry

In any sediment, there is a large number of elements that can be determined, but many of them will be present in only very small quantities. As more sophisticated machines and techniques are developed, the detection limits decrease and the number of elements that can be studied rises, although new approaches are also being explored that require very simple chemical techniques. In this section we will primarily consider those data that have been shown to be of significance in palaeoenvironmental changes. No doubt in the future, these data will grow considerably. A special issue of the journal *Water Air and Soil Pollution* (volume 100, 1997) provides a useful series of recent papers on this topic.

One of the main areas of interest has been in heavy metal detection, often to reconstruct pollution histories of particular areas. Although some work has suggested that there could be problems with mobility of lead and other metals in the upper parts of peat profiles (Pakarinen and Tolonen, 1977a; Stewart and Fergusson, 1994), there are now numerous records of lead and other metals from a number of peatlands in Europe. More detailed work on lead isotopes also demonstrates that mobility of lead in the profile has not blurred the historical record (Shotyk et al, 1997). Lee and Tallis (1973) provided the first indications that the record was a realistic assessment of historical changes in pollution levels and recent work has taken this basic idea much further. Shotyk et al (1998) showed how lead concentrations changed in response to both climatic change and human disturbance in the early to mid-Holocene. Lead pollution from mining and smelting activities began around 3000 radiocarbon years BP and were detected from changes in the ratio of ^{206}Pb : ^{207}Pb and lead : scandium. Records such as this also allow an estimate of natural background lead levels to be made, which can be shown to be up to 1500 times lower than during the peak of lead pollution in recent years (Figure 6.10).

The detection of human impact using geochemistry of peat has also been the subject of much interest. Disturbance of the landscape and an increase in dust levels through forest clearance and soil tillage appear to be responsible for increased lead levels in the prehistoric period (see above), also shown by comparison of lead concentrations with anthropogenic pollen indicators for the period between approximately 1000 cal. BC to cal. AD 1000 (van Geel, 1989). These authors also suggested other elements (cobalt, iron) reflected human impact on vegetation in line with the pollen data. Titanium is another element that appears to be strongly linked to human land disturbance throughout Europe (Kempter *et al*, 1997; Hong *et al*, 2000). Steinnes (1997) shows how data on a number of metals from surface and short cores reflect spatial differences in historical pollution in Norway, as well as the confounding influence of oceanic sources of elements such as selenium.

Because different influences can be responsible for changes in a single element, it is often important to look at the changing ratio between elements to identify which of the possible factors is responsible. For example, as we have already mentioned, lead can increase as a result of inputs of soil dust or from mining and smelting activities. Gorres and Frenzel (1997) found increasing lead concentrations that could be attributed to either soil disturbance or mining pollution. These two effects were separated using the ratio between lead and titanium, as the latter is primarily a function of soil dust input alone.

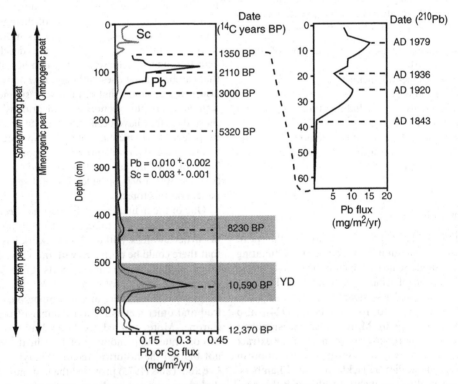

Figure 6.10 Rates of atmospheric deposition of Sc (scandium) and Pb (lead) in a Swiss bog profile. The lowest peak occurs at the same time as very cold stadial conditions existed over north-west Europe, and the peak around 8230 BP could also be associated with colder conditions or volcanic activity. Pb deposition increases after 5320 BP, reflecting forest clearance and farming activities. A further rise and change in the proportions of Sc : Pb occurs after 3000 BP with the start of mining and smelting. Major increases are recorded after the start of the industrial revolution with peak values in AD 1979. Redrawn from Shotyk *et al* (1998) by permission of the American Association for the Advancement of Science.

6.7.3 Isotopes of oxygen, hydrogen and carbon

Isotopic analyses have become increasingly important to the reconstruction of past environmental conditions, especially over the Quaternary period (see e.g. Lowe and Walker, 1997). Oxygen isotopes from deep ocean and ice cores have been particularly influential in our understanding of long-term climatic changes. Isotopes of oxygen, carbon and hydrogen can and have been measured in peat, and although still relatively under-exploited and requiring much further work to understand the signals, they have great potential as palaeoenvironmental indicators. The main interest in isotopes is as indicators of past climatic changes, although we know they are likely to be strongly affected by local variations in vegetation too. The main isotopes of hydrogen and oxygen of interest are ^{2}H (deuterium, sometimes written as D) and ^{18}O (oxygen-18). For carbon, we are primarily concerned with changes in ^{13}C, unrelated to ^{14}C, which we discussed as a tool for dating earlier in the chapter.

The key processes linking hydrogen and oxygen isotopes with organic matter growth and climate are explained by Brenninkmeijer et al (1982). First, the D/H and ^{18}O/^{16}O ratios in precipitation depend on temperature. In ombrotrophic bogs especially, the isotopic composition of water taken up by the plants is therefore also dependent on temperature. Second, evapotranspiration means that some enrichment of the heavier isotopes takes place and this enriched water is used in photosynthesis. The degree of enrichment depends on the humidity of the air. Third, isotopic fractionation occurs within the plants, resulting in a depletion of ^{2}H and an enrichment of ^{18}O. Although these basic processes are known, there is a lot of uncertainty over how ^{2}H and ^{18}O data from peat should be derived and interpreted, since there will be a strong species effect given the mix and temporal changes in many peat profiles. There are particularly large differences that arise between mosses and vascular plants because of their very different mechanisms for moving water to the growing tissues.

Despite these problems and by attempting correction for species effects, Dupont (1986) was able to produce temperature curves for three profiles in the Netherlands based on deuterium analyses (Figure 6.11). There is reasonable agreement in the timing of the major changes, and the calibrated temperature scales suggest the magnitude of changes is similar. Some independent corroboration for the temperature shifts was also found from ice core data, suggesting similar direction of changes at about the same time,

and this was used to convert the species-corrected 'relative deuterium temperature' scale to an approximate real mean annual temperature scale.

Direct use of ^{18}O has also been made in climate reconstructions. Hong et al (2000) avoided major difficulties of species effects in using peat of relatively uniform composition from China to reconstruct climatic change over the last 6000 years (Figure 6.12). They interpreted changes in ^{18}O as reflecting surface air temperature changes, a link supported by other palaeoclimatic data. In addition they were able to show that ^{18}O fluctuations showed a strong link with solar forcing by examining the periodicities in the data.

Carbon isotopes in peat have also been used to reconstruct changes in atmospheric carbon dioxide (White et al, 1994). Again species effects are a consideration, but in this study on South American peat, the two major components (Sphagnum moss and sedges) were analysed separately. Differences between the two sets of measurements were used to separate out the effect of variations in the water availability from the atmospheric carbon dioxide signal. Another approach is to separate organic compounds that can be shown to be associated with a particular taxon, thus avoiding species-specific signals (Xie et al, 2000). Other approaches to isotopic analyses of peat include measurement of D and ^{18}O in the 'fossil' pore water held in frozen Arctic peats (Wolfe et al, 2000).

6.7.4 Organic geochemistry

There have been a number of attempts to use organic molecules from plant tissues as biomarkers in peat. The detailed chemical structure of lipids in particular should be related to the group of plants that formed the peat. Lipids are insoluble in water and are found in vegetable oils, plant waxes and animal fats. Several studies have attempted to use lipid biostratigraphy in peat to provide palaeoenvironmental reconstructions. Some of the first work to be done on this was on tropical peatlands (Dehmer, 1993; Lehtonen and Ketola, 1993). Ficken et al (1998) attempted to link lipid chemistry to plant macrofossil changes in a Scottish mire but found inconsistencies between the plant and chemical data. Farrimond and Flanagan (1996) compared lipid records with pollen data and, perhaps less surprisingly, also found divergence between the two records, with the lipids presumed to be reflecting the peat-forming vegetation rather than broader vegetation patterns. Although the approach has some promise, at the moment it cannot provide more reliable data on any

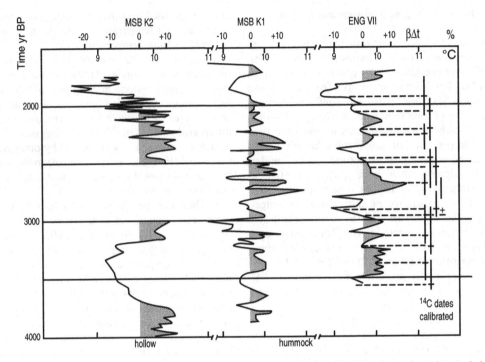

Figure 6.11   Relative deuterium temperature from three sequences in the Netherlands. The relative deuterium scale is derived from deuterium measurements corrected for species composition of the peat. The temperature scale is estimated from comparison of part of the record with ^{18}O data from the Greenland ice core. Redrawn from Dupont (1986) by permission of Elsevier Science.

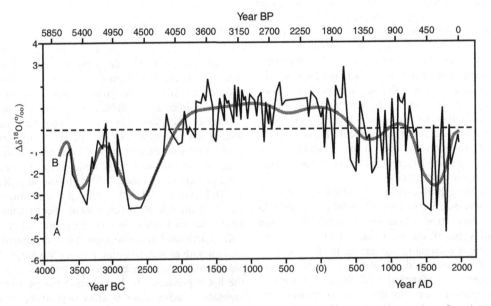

Figure 6.12   Changes in the ^{18}O content of cellulose in a peat bog from China, interpreted as changes in surface air temperature from 6000 cal. years ago to the present. Redrawn from Hong *et al* (2000) by permission of Arnold Publishers.

useful parameter than other more traditional techniques.

6.7.5 Other techniques

There are several other techniques that have been tested on peat, mainly as potential proxies for other characteristics such as humification or macrofossil content. Luminescence measurements on peat or peat extracts may provide more detailed data on humification and decay products (Caseldine *et al*, 2000), although there is some way to go in interpreting the data adequately. Near infrared spectroscopy (NIRS) is another technique with intriguing possibilities for the future (McTiernan *et al*, 1998). In this non-destructive process, the NIRS spectra are measured and a proportion of them are calibrated using conventional measurements of various physical and biological characteristics. The results of these experiments are surprisingly good (Figure 6.13).

6.8 MULTI-PROXY APPROACHES

Much of the palaeoenvironmental evidence from peatlands is complementary and some is overlapping. As a result, different sources of information can be used together for two main purposes. First, if two sources of data show the same changes, it gives us greater confidence in our results. Second, there may be more than one explanation for changes in one set of data, but the addition of a different data source may differentiate between these two possible explanations. A simple case would be the use of pollen, plant macrofossil and charcoal data from the same profile. Pollen spectra reflect vegetation changes over wide areas as well as local changes. Plant macrofossil data come from mainly local sources and can confirm the local presence of plants identified in the pollen record, helping to separate local and regional pollen components. Several hypotheses may be erected concerning changes in the pollen record, including the possibility that fire was the cause of forest reduction, for example. The charcoal record could then be used to test this particular hypothesis directly. The use of a series of palaeoenvironmental indicators is known as multi-proxy palaeoenvironmental reconstruction, and it is now probably more commonly used than single-indicator methods. Of course, for many studies, peat deposits will also be used together with other palaeoenvironmental evidence such as that from lakes and tree rings – the peat record

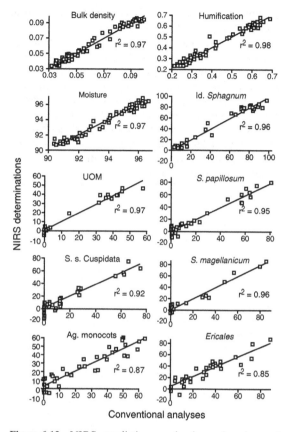

Figure 6.13 NIRS predictions and observed values of some physical and biological characteristics of a peat core. Redrawn from McTiernan *et al* (1998) by permission of Arnold Publishers.

is only a part of the jigsaw of palaeoenvironmental reconstruction.

6.9 SUMMARY

This chapter has provided a broad-brush synopsis of most of the key aspects of using peat to reconstruct changes over timescales where direct observation is not possible. It will be clear from this that the 'peat archive' holds an enormous variety of physical and biological fragments of the past and geochemical evidence of change. New techniques and advances on well-established methods are constantly being developed and the depth and breadth of information we can obtain are unprecedented. Study of peat palaeoenvironments has many different purposes. A large proportion of such research is focused on the peat record as a repository of information about the wider environment (climate, regional vegetation, human settlement), and some of the work relating to past climate change will be reviewed in Chapter 8. Equally, the fossil record provides us with an insight into the long-term development of the peatland system itself. The insight into peatland development has few parallels in other ecosystems. Lakes are the only other habitat for which similar long-term records are normally available through the study of their sediments, but the level of spatial detail is often poor due to sediment mixing and inputs from many sources to any single core location. In Chapter 7 palaeoenvironmental and other evidence is used to discuss how peatlands change through time due to natural ecosystem processes ('autogenic' change). This is followed by a discussion of their reaction to external changes ('allogenic' change) in Chapter 8. Of course for any specific peatland, both types of change need to be considered together to get a full picture of how it has changed over millennia. However, whatever the scale and scope of the research, the only constant result is that change is continuous!

7

Autogenic change

7.1 INTRODUCTION: LONG-TERM CHANGE

Long-term changes in peatlands arise from multiple external forcing factors and boundary conditions (allogenic factors) as well as from a variety of internal processes (autogenic factors). All of these factors act at a variety of scales in time and space, interactions between them are the norm, and it is common for several factors to contribute to the same sequence of change. Given this complexity, there is a variety of possible ways of beginning to explain long-term changes in peatlands. Here I have elected to discuss autogenic and allogenic factors separately, and to restrict most of the discussion to factors arising from natural processes. Although some anthropogenic effects have acted over very long periods of time, such as the landscape disturbance involved in the initiation of peat discussed in Chapter 4, the most significant effects from human activities have occurred over much shorter timescales. These are principally related to the exploitation of peatlands through agriculture, forestry and peat extraction and will be discussed in Chapter 10. In Chapter 7 we will discuss the ways in which peatlands have developed in response to autogenic factors and in Chapter 8 we will explore the relationship with external forcing factors. There are rather few comprehensive reviews of these aspects of peatlands, and the papers by Tallis (1983) and Frenzel (1983) are the most recent available. The latter is especially good in reviewing allogenic causes of changes in peatlands and early European work on palaeoclimate and peat development. The former is more comprehensive and takes a particularly close look at hydroseral succession. Casparie and Streefkerk (1992) provide a detailed account of changes applicable to a restricted area of the world but touch on many more generally applicable aspects as well. The period 1980–2000 has seen a much improved, more detailed understanding of long-term change in peatlands, but there is no reason to alter the opinion of

Heinselman (1975, p. 102), who suggested '... perhaps there is no "direction" to peatland evolution – only ceaseless change'. If anything, we now appreciate there is an even greater diversity of change in space and time than was previously suspected. On the other hand, we understand the peatland system and its responses to change rather better and can use this understanding for much improved reconstruction of past environmental changes in and around peatlands.

7.2 AUTOGENIC AND ALLOGENIC CAUSES OF CHANGE

Changes from autogenic processes always occur against background external conditions ('boundary conditions'). Interactions between autogenic and allogenic factors occur in three main ways:

(i) Stable external conditions as a background against which autogenic processes take place. For example, in the case of paludification of dry ground, the climatic conditions clearly have to be suitable to create sufficient soil moisture for peat growth. However, peat growth is not associated with a *change* in climate. The pedogenic changes are an autogenic process and the development of peat simply requires sufficient time under stable climatic conditions.

(ii) Changing external conditions accelerate an autogenic process. In the case of peat initiation, pedogenic processes towards peat formation may be accelerated by an increase in soil moisture as a result of climatic change or human disturbance of the vegetation cover.

(iii) Changing external conditions inhibit an autogenic process. At the start of peat growth, pedogenic processes may proceed to a certain point (e.g. podsol formation) but then a shift to a drier climatic regime prevents the development of a peat soil.

General conceptual models of disturbance and response in soils can be adapted to illustrate these ideas more succinctly. Figure 7.1 shows a variety of possible interactions between autogenic and allogenic change and further develops the three main forms of interaction already described. In this case the external forcing is represented as a series of disturbances, but in some situations the forcing may be a continual pressure, which would further modify these models. Concepts of equilibrium and thresholds for change are equally applicable here (Butzer, 1982). Figure 7.2 shows how the models can be applied to blanket mire growth in Britain, following the example used above. Here it is hypothesised that pedogenic processes were tending towards podsolisation and the development of a surface organic layer in the soil under a suitable climatic regime. Accelerations of this trend are shown as resulting from deteriorating climate and human impact at various times, increasing the rate of change. Similar concepts can be used to describe many other examples of changes in peatlands, and the models shown in Figure 7.1 are useful to bear in mind when attempting to visualise and understand these processes.

The three types of autogenic–allogenic relationship and the models in Figure 7.1 are obviously simplifications of what occurs in practice, because there may be two or more significant autogenic factors and two or more allogenic factors interacting in different ways to produce an observed effect. Most of the evidence for the relative roles of autogenic and allogenic influences on peatlands comes from the palaeoenvironmental record held within the peat. This record may provide an excel-

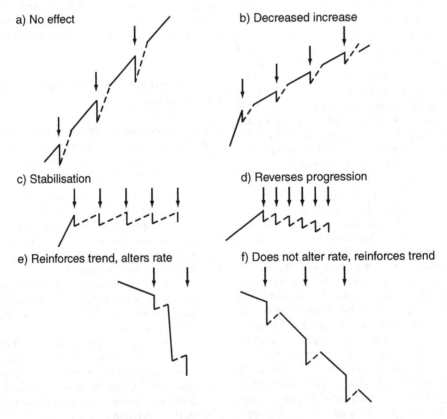

Figure 7.1 Interactions between autogenic and allogenic factors in peatlands. Time runs left to right and the general trend of an autogenic process (e.g. towards terrestrialisation) is the vertical axis. Here the effect of the allogenic factor is shown as a series of events impacting on the rate and direction of autogenic change. In (a) to (d), the external factor is acting against the prior trend, either having no effect (a), retarding the rate of change (b), stabilising the system (c) or reversing the trend (d). In (e) and (f), the external factor is reinforcing the trend, either by accelerating the rate of change (e) or by curtailing the recovery of the system (f). Redrawn from Trudgill (1988).

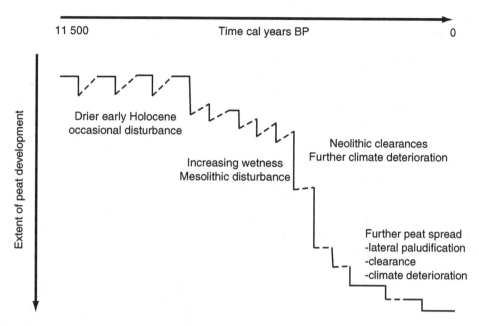

Figure 7.2   Hypothetical schematic of the postglacial development of blanket peats in Britain illustrating interaction between autogenic and allogenic factors. The interaction between a natural pedogenic trend and the possible effects of climate and human disturbance of the hydrological regime are the main factors shown. Adapted from Trudgill (1988).

lent account of what happened, but it requires careful consideration and design of data collection to establish realistically how and why it happened. This is the challenge for peatland palaeoecologists if we are to understand the long-term changes observed in peatlands throughout the world. This is particularly important where we want to use the peat record as an indicator of a particular external factor such as climate. This will be discussed further in Chapter 8, where palaeoclimatic records from peat will be explored in some detail.

There are a number of different types of autogenic processes of importance in peatland change, broadly categorised as follows:

(i)   Biotic processes of growth, decay and accumulation. Spatial variability of accumulation at different scales has a significant effect on the size and shape of the mesotope and can determine the arrangement of small-scale features such as pools and hummocks on the mire surface.

(ii)   Biotic processes associated with plant growth and form, and with animal behaviour. A number of plants have particular growth strategies that impact on other processes in the mire. Animal behaviour can also be important in some situa-

tions – beaver activity is an obvious example of this.

(iii)   Physical processes associated with movement of sediments and soils.

(iv)   Physical processes of freeze–thaw cycles in cold climates.

(v)   Hydrological processes causing an increase or reduction in surface water.

7.3 HYDROSERAL SUCCESSION

Changes in peatlands occur at a variety of scales; those occurring over most of the area of a peatland (mesotope) or at the landscape scale (macrotope) are often the most obvious in peat stratigraphy. More subtle changes at the microtope scale, and especially within individual microforms, may account for as much or even more of the observed variability. We will begin by discussing the large-scale changes and move on to consider the smaller-scale dynamics towards the end of the chapter.

Hydroseral succession is by far the most important and commonly encountered example of autogenic change in peatlands. Almost all peatlands will have been through some changes related to this process. The concept of hydroseral succession was conceived initially to account for the spatial zonation of aquatic and terrestrial communities at lake margins and in peatlands, most famously by Clements (1916) and later made widely known in Britain by Tansley (1939). The essentials of the proposed succession were a gradual change from open water to aquatic macrophytes, followed by colonisation by rooted aquatics, emergent plants, a terrestrial fen with a further transition to raised bog and/or a dry woodland community given adequate time and a suitable climate. The idealised patterns suggested by observations of extant plant communities were later proved to be only applicable in a very general sense and some aspects, such as Tansley's proposed transition to oak forest, were shown to be entirely false.

The best-known and most comprehensive evaluation of hydroseral succession using palaeoecological data is that of Walker (1970). Reviewing the data available at the time, Walker found that actual hydroseral succession sequences in Britain were much more variable than Tansley had proposed. In almost all Walker's examples, raised bog vegetation was the final stage of the hydrosere and there were a number of different possible pathways of change observed in the sequence (Figure 7.3). Tansley's suggestion that woodland would ultimately develop through hydroseral succession was completely unsubstantiated. Since Walker's work, there have been summaries of observed hydroseral pathways for a number of other regions, and a much greater body of data could certainly be added to these now. Tallis (1983) summarised the changes in a number of North American peatlands and found an even greater variety of transitions, confirming that diversity of development in hydroseral succession is the norm rather than the exception. There may be some uniformity of succession within restricted regions (e.g. Rybnicek, 1984), but it seems likely that the early stages of succession are highly variable in terms of the specific plant communities and rates of development, depending on factors such as basin morphology, geological setting and hydrochemistry. More detailed analyses reveal further subtle differences in the precise sequences recorded in similar mire types within a region. Ilomets *et al* (1995) analysed the transitions in Estonian mires; those from topogenous mires are shown in Figure 7.4, and clearly show a very complex set of possible changes when detailed macrofossil data are used.

The rates of peat accumulation for different hydroseral phases can be estimated using stratigraphic methods. Although these often only consider depth–age rather than mass–age relationships and do not

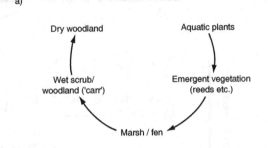

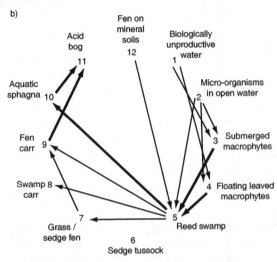

Figure 7.3 Hydroseral succession in the British Isles (a) as proposed by Tansley (1939) and (b) Walker's (1970) results showing the 159 stratigraphic transitions recorded at 20 sites. Only transitions where there is more than one record are shown and thicker lines denote dominant course of succession. (b) redrawn from Walker (1970) by permission of Cambridge University Press.

normally account for constant decay over time (see Chapter 5), it is clear that different stratigraphic units can show very large differences in accumulation rates. The successional stages of development of a forested bog in Saskatchewan, Canada, are shown in Table 7.1. Large shifts in accumulation rate, expressed as both depth and mass per year, are associated with major stratigraphic changes. In aquatic phases, there may be significant allogenic input and accumulation rates are often highest. In this case, total mass is up to 15 times greater than some later phases but total organic accumulation is only 2–3 times as much. Fen peat growth rates are also typically higher than later bog phases, although this is not always the case. While forest growth on a peatland clearly increases

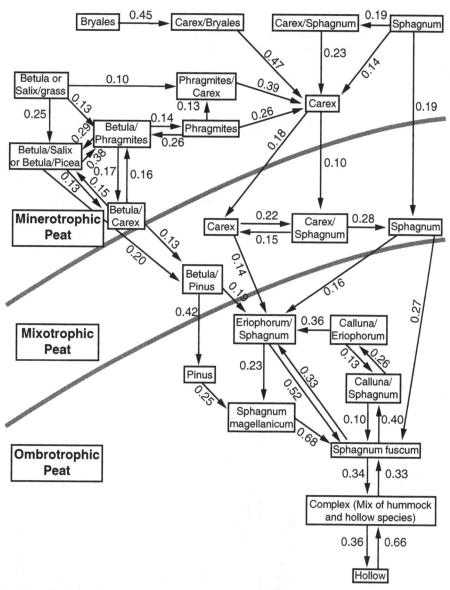

Figure 7.4 Sequence of peat types in topogenous mires in Estonia showing the successional change between different stages. The probability of each transition is shown as a figure next to each arrow. Redrawn and adapted from Ilomets *et al* (1995).

the standing biomass, accumulation rates in treed phases may be lower due to dryness and outbreaks of fire (Kuhry, 1997). An alternative method to stratigraphic estimates for measuring accumulation rates of different hydroseral phases is suggested by Bakker *et al* (1997), who measured the mass of peat accumulated in old peat cuttings with different successional sequences. Here, all the cuttings dated from AD 1937, and the top

of the old cut surface was identified in short cores. The mass accumulated since cutting could therefore be measured and related to the site history, known from air photographs taken at different times since cutting. The accumulation rates for different successional phases could then be calculated (Figure 7.5). Interestingly, the rates recorded were much higher than those estimated from stratigraphic approaches, which could be

Table 7.1 Mean accumulation rates for a core from a forested peatland in Saskatchewan, Canada, expressed in a variety of different ways. The differences between the total dry weight and organic rates of mass accumulation emphasise the role of minerogenic input in the aquatic phases. Differences in later phases can be attributed to variations in productivity related to nutrient availability and moisture levels. Subtotals and totals in italics, individual stratigraphic units in upright text. Extracted from Kuhry (1997).

Radiocarbon age (yrs BP)	Depth (cm)	Vegetation type	Deposit	Height (mm yr^{-1})	Dry weight (g m^{-2} yr^{-1})	Organic (g m^{-2} yr^{-1})
0–1870	0.105	Treed dry bog	Peat	0.56	47.3	44.2
1870–2790	105–166	Treed moist bog	Peat	0.66	87.6	82.9
0–2790	*0–166*	*Total bog*	*Peat*	*0.59*	*60.6*	*57.0*
2790–3260	166–221	Treed fen	Peat	1.15	107.6	98.0
0–3260	*0–221*	*Total peatland*	*Peat*	*0.67*	*67.4*	*62.9*
3260–4330	221–311	Pond with fen	Gyttja and sand	0.85	173.9	94.2
4330–4890	311–412	Pond with marsh	Sand and gyttja	1.80	1015.5	154.1
3260–4890	*221–412*	*Total pond*	*Sand, gyttja*	*1.18*	*463.0*	*114.8*
0–4890	*0–412*	*Total wetland*	*Sand, organic*	*0.84*	*199.3*	*80.2*

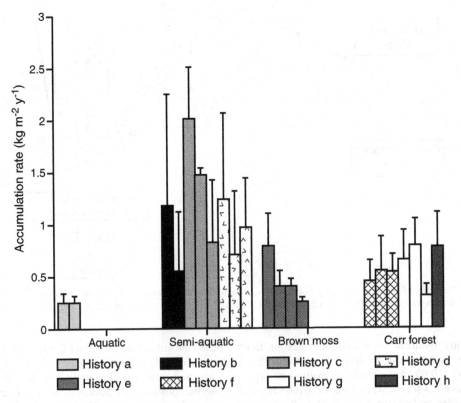

Figure 7.5 Average accumulation rate of organic matter during four successional phases in peat cuttings in the Netherlands. The different shadings represent different types of site history, but the results show that there are major differences between successional stages independent of the exact sequence of change or local situation of the profile. Redrawn from Bakker *et al* (1997) with kind permission from Kluwer Academic Publishers.

explained by the fact that these are artificial habitats in the first phases of renewed peat growth, and nutrient availability and surface wetness are likely to be much higher than in natural long-established communities (Bakker *et al*, 1997). Long-term decay in older peats would also help to explain this difference.

While it is possible for hydroseral succession to take place under stable climatic conditions, in reality, because of the time it takes to progress through multiple phases (usually several thousand years), the likelihood of this happening is rather small. It seems likely that climatic influences (and probably other factors too) affect the rate of change and the timing of transitions. For example, in a study on two *Sphagnum* peatlands in southern Wisconsin, both sites showed initial transitions from lake to shallow ponds with floating aquatics at around 6300 BP due to a regional decrease in precipitation (Winkler, 1988). A later transition to *Sphagnum* bog was also synchronous at the two sites, suggesting a common climatic forcing for the change, perhaps pushing the peatlands through the next phase of development. In the absence of the climatic influence it is possible that the change would have occurred, but it would be much less likely to happen at the same time. Multiple coring strategies are a good way to establish whether climatic forcing was adequate to accelerate (or decelerate) hydroseral succession at all locations in a particular site or in a number of sites within the same climatic region. Sometimes different locations within the same site show very different timing for similar types of change, suggesting climate has been unimportant in controlling succession here (e.g. Hu and Davis, 1995).

Much of the work on hydroseral succession has been carried out in northern temperate mire systems but there are an increasing number of examples from northern cold-climate and southern temperate and tropical peatlands. For arctic peatlands, hydroseral succession was apparently important in the development of some but not all sites (Vardy *et al*, 1998). Although the composition of plant communities is different, the phases of aquatic–fen–bog are at least present in some locations. Vardy *et al* (1998) found a change from open water with aquatic plants through to gradual colonisation by *Drepanocladus*-dominated moss and sedge mats in the early to mid-Holocene. A subsequent change to *Sphagnum* peatland may be linked to the development of permafrost owing to regional cooling, which would have the effect of raising the surface above the influence of groundwater. Similar sequences of general change from aquatic to *Sphagnum*-dominated vegetation have been recorded

from elsewhere in the North American high Arctic but the timing of these is different (Ovenden, 1982), suggesting that local factors can exert a greater influence than regional climate change. Other arctic peatlands have more complex developmental histories. For example, the stratigraphy of a Siberian peatland showed a series of fen communities following an early Holocene aquatic phase but these do not follow a simple basin infill sequence, but reflect fluctuating hydrological conditions from a mix of local and regional factors (Peteet *et al*, 1998).

7.4 REVERSALS AND OTHER SUCCESSIONS

Hydroseral succession frequently follows the basic pathway of aquatic–fen–bog ecosystems, although as we have seen there is considerable variability around this general directional change. Particular circumstances can strongly modify this general trend. We have already alluded to possible climatic influence on the pace of change, but local processes can also delay or reverse the 'normal' hydroseral succession. In order for this to occur, there has to be an unexpected increase of surface water in the peatland. The causes of such a hydrological change are varied and the following examples are by no means exhaustive.

Within blanket mire systems, areas of poor fen vegetation are common in parts of the system where the surface water flow is focused. These are often known as 'flushes' (see Chapter 3) and vegetation is often dominated by sedges and rushes rather than the *Sphagnum* and dwarf shrubs most common elsewhere. In the north of Scotland, these have developed into patterned fens with ridge–pool systems. Stratigraphic studies have demonstrated that these are superficial features that have developed over the top of ombrotrophic blanket mire as a result of local hydrological change (Charman, 1994, 1995). In this case, the ombrotrophic blanket mire developed through a combination of hydroseral succession and lateral spread by paludification, but the succession was reversed back to fen with a switch in water supply to the mire.

Delayed successional change has also been shown to result from slow ground subsidence in the strata underlying a peatland. Hughes *et al* (2000a) show how swamp and fen phases of a mire in Cheshire, England, were extremely protracted by comparison with similar basins elsewhere in Britain. Here, subsidence of the salt-bearing rock underlying the mire would explain the very slow rate of transition from fen to bog, which occurred within 500–1000 years at sites with similar morphometry. The gradual transition over almost 4500 cal. years was punctuated by periods of apparent standstill and reversal, presumably where there were temporary changes in the balance between the level of groundwater and the peat surface. Likewise Bunting *et al* (1996) recorded a reversal of succession on a cedar (*Thuja occidentalis*) swamp in southern Ontario, Canada. The normal process of terrestrialisation had taken place in the early Holocene and the site was a forested tamarack (*Larix*) swamp until a reversion to a much wetter reedswamp took place at around 5600 radiocarbon years BP. The causes of this were not clear,

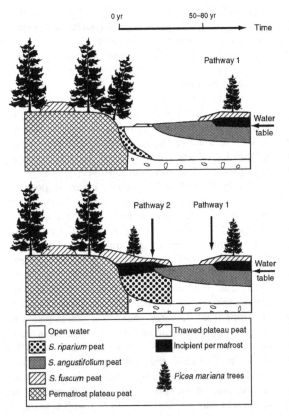

Figure 7.6 Schematic diagram of successions in a boreal peatland following the formation of a collapse scar at the edge of a permafrost plateau. (a) Recolonisation of the collapse scar over time by aquatic *S. riparium*, then lawn *S. angustifolium* and finally, terrestrial *S. fuscum*, suitable for the growth of black spruce (*Picea mariana*) and the beginnings of renewed permafrost development. (b) Re-establishment of *S. fuscum* and *P. mariana* from a collapsed plateau hummock. Redrawn from Camill (1999b) by permission of *Écoscience*.

but were probably a combination of local autogenic factors, augmented by beaver activity and/or climate change.

Very localised natural disturbances can bring about new successional sequences in small areas of a mature peatland that has already formed through terrestrialisation or paludification. In the boreal peatlands of Manitoba, Canada, Camill (1999b) has shown that collapse scars on the edge of permafrost plateaux go through a succession from aquatic, lawn and then hummock communities suitable for renewed permafrost formation (Figure 7.6). Similar examples would be recolonisation of bare peat surfaces in eroded areas of blanket peat or in drained pools.

Reversals or changes in succession can of course also occur if there is a major shift in the external conditions. I do not know of any cases where climate change has been clearly responsible for the reversal of a succession, but other factors can produce such a change. Sea-level rise can interrupt or break a successional sequence. The low-lying land around the Severn estuary in England provides a good example of this, where a full succession from reedswamp through carr woodland to ombrotrophic *Sphagnum* bog during the early and mid-Holocene is truncated by a marine transgression after about 3400 cal. BP.

7.5 PROCESSES OF TERRESTRIALISATION AND THE TRANSITION TO BOG PEAT

The processes of terrestrialisation seem simple at first sight: a depression collects water, gradually fills in and plant communities change according to water depth, nutrient availability and stability of the rooting zone. This is presumably the essential mechanism in most hydroseral sequences, but there are two aspects of terrestrialisation that are worth considering a little further. First, basin infill is explicable as a biologically driven physical process of sedimentation, but it does not explain why a final change to ombrotrophic conditions is found in so many sites. Second, is a gradual silting up of the basin an oversimplification of what actually takes place? One of the features of basins that are apparently midway through succession is an extensive mat of floating vegetation (as in a schwingmoor). This appears to be a mechanism by which the rate of terrestrialisation could be increased, over and above the filling in of the basin with solid sediment throughout the profile.

Many emergent plants spread by vegetative reproduction using rhizomes and for many taxa these can extend over open water. Once such a floating mat is established on the fringes of open water, it provides a location for other species to colonise. Mats may break off the fringes of open water and become islands of floating vegetation. Over time these may expand and coalesce to form very extensive vegetation mats overlying a lens of water. A floating mat of vegetation provides unique conditions because the plants are not subject to the same fluctuations in water level as in 'fixed' locations, because the mat moves up and down with these changes. The mats can build up their own layers of peat, forming up to 5 m in Wybunbury Moss, Cheshire, England, for example (Green and Pearson, 1977). Because the mat moves up and down with the water level, it provides a potential site for colonisation by *Sphagnum* as groundwater only penetrates the lower levels of the mat, depending on its buoyancy. A 'mini-succession' can take place on an individual mat, with the development of patches of floating bog vegetation, raised above the influence of groundwater (Mallik, 1989). As the mat grows in thickness, pH levels at the surface decline and *Sphagnum* increases in abundance (Figure 7.7).

Kratz and Dewitt (1986) suggest a conceptual model for peatland development associated with the extension of a mat from the edge of a lake with a distinct series of zones outward from the edge of the basin (Figure 7.8). Horizontal encroachment of the mat into the lake results from growth of the leatherleaf (*Chamaedaphne calyculata*), which provides a basis for mat growth and thickening. The thickening mat overlies debris peat and open water and eventually forms continuous sediment with the underlying debris peat. Once these layers are

compacted by continued peat accumulation at the surface, no more peat accumulates, as there is inadequate moisture supply to sustain it. In more oceanic conditions, on might expect to see continued growth of peat through continued growth of *Sphagnum* communities. The growth rate of an extending mat can be considerable: average rates of approximately 25 mm yr^{-1} over the last 9000 years are estimated for the *C. calyculata* mat in a Wisconsin peatland (Kratz, 1988). Clearly rates such as these are significant over timescales of decades and certainly for the 1000–10000 years of growth represented at some sites.

The process of mat growth and thickening may be important in driving the succession from open water to bog vegetation in some settings, but it is more common to find a significant depth of fen peat between aquatic and bog sediments in the succession. In such circumstances, what causes the transition from fen to bog peat? In long-term observations on Swedish mires, acidification of the fen mesotopes within a mire complex over the past 50 years may be attributable to natural acidification by the vegetation itself, linked with gradual expansion of the mire system (Gunnarsson *et al*, 2000). In *Sphagnum*-dominated bogs, a combination of factors appears to be responsible for the rapid spread and dominance of this taxon once it has gained a foothold, and groundwater supply is reduced to a minimum. The structural and biochemical attributes give *Sphagnum* its ability to retain water and scavenge nutrients efficiently, as well as its influence on acidity and resistance to decay (van Breemen, 1995; Chapter 3). Therefore, once *Sphagnum* is established, it seems unlikely to disappear as long as climate and hydrological

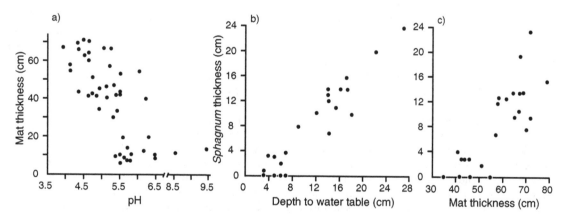

Figure 7.7 Relationships between different variables on floating organic mats in Tintamarre Marsh, New Brunswick, Canada. (a) Mat thickness and pH, (b) thickness of *Sphagnum* peat and depth to water table and (c) thickness of *Sphagnum* peat and overall mat thickness. Redrawn from Mallik (1989) by permission of NRC Research Press.

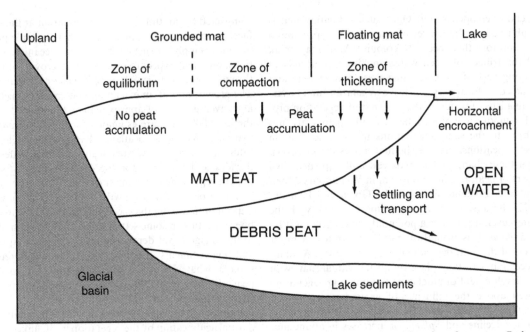

Figure 7.8 Conceptual model of mat extension and peat growth around lakes. See text for explanation of processes. Redrawn from Kratz and Dewitt (1986), who presented it as a model for northern temperate latitudes, although it could apply equally to many other areas of the world, with variations in the particular taxa and growth habits. Redrawn with permission of the Ecological Society of America.

conditions remain stable. On occasion of course the external conditions do change and *Sphagnum* will disappear. Wheeler (1992) suggests that colonisation by *Sphagnum* is the cause of a gradual acidification of a eutrophic fen in East Anglia, England, but a transition to full raised bog is prevented by a rising groundwater table.

However, *Sphagnum* may not be essential to the development of bog from fen. Hughes (2000) suggested that in many north-west European mires, *Sphagnum* is preceded by a phase of *Eriophorum vaginatum* and *Calluna vulgaris*. He points out that drier or fluctuating hydrological conditions often occur before the fen–bog transition and that these plants would be favoured by a fall in the water table, allowing the separation of surface meteoric water and groundwater. Peat formed by these pioneer plants would reduce water flow and provide suitable conditions for the subsequent establishment and expansion of *Sphagnum*, aided by the mechanisms already mentioned. The relatively dry phase of *Eriophorum* dominance appears to last as much as 2000 years in a number of profiles in northern Britain (Hughes *et al*, 2000b). In other raised mires, *Sphagnum* is the first main colonist of areas of fen. For example, at Store Mosse in Sweden, the fen–bog

transition is principally one from *Carex* sedge to *Sphagnum fuscum* dominated communities, although *Eriophorum* phases are also sometimes recorded (Svensson, 1988a). Certainly it seems likely that there may be more than one mode of transition from fen to bog even for peatlands within north-west Europe.

Sphagnum was clearly not important in the development of other major bog types in the world such as the tropical domed peatlands and the restiad-dominated mires of New Zealand. Rather little work on mire development processes has been carried out in these areas, but gradual development of small areas of restiad vegetation from an early stage in mire development occurred at Kopouatai bog in North Island, New Zealand. The main peat-forming species, the wire rush (*Empodisma minus*), may have some similar properties to *Sphagnum* in terms of its ability to reduce evapotranspiration (Campbell and Williamson, 1997; see Chapter 3), and its dense mat-forming ability may give it a competitive advantage over other taxa.

In tropical peats dominated by the remains of a variety of tree species, there may not be any good reason to infer specialised characteristics of the plants, other than an ability to survive soil conditions impoverished by leaching and low pH levels. Swamp forest peat overlies early topogenous mixed sediments in some sites (e.g. Morley, 1981) and a clear fen–bog transition is not apparent in the few published profiles of tropical peats. Geochemical characteristics suggest that ombrotrophic peatland was present for at least 10 000 years in some sites (Page *et al*, 1999) and was initiated over only minimal underlying minerotrophic sediments.

7.6 LATERAL EXPANSION AND THE DEVELOPMENT OF PEATLAND LANDSCAPES

The lateral expansion of peat was discussed in relation to peat initiation in Chapter 4. Although rates of lateral spread are affected by climate change and anthropogenic impacts on the hydrological balance, the mechanism is essentially an autogenic process. At the edge of a peatland, there is a tendency for adjacent mineral soils to become wetter due to surface water flow from the peat area, impedance of drainage into the peatland, and an increase in humidity from evapotranspiration from the relatively wet peat and vegetation. There is therefore a tendency for peat to grow outwards if all other conditions are stable in time and space; on a flat plain with uniform soils and a stable climate, one would expect the peat to continue outward growth at a steady pace. In practice there are fluctuations in climate, hydrology, vegetation and topography that will determine whether the peat deposit is able to grow and the rate at which it can do this.

Detailed analyses of macrofossils and microfossils in a profile from Amtsven, north-west Germany, reveal the changes at the edge of a peatland occurring as a result of pedogenic processes in the soil and the gradually approaching 'front' of an adjacent peat deposit (Kuhry, 1985). At first, a decline in major tree taxa (*Tilia* and *Quercus*) is recorded, followed by an expansion of more acid woodland and shrubs (e.g. *Betula*, *Cornus suecica*, *Rubus*) and eventually there is a transition to carr woodland with a distinct peat layer. This then forms the lagg zone of the encroaching raised bog with a detailed succession of sedge, aquatic and reedswamp communities that ultimately become more acid with thicker peat and *Sphagnum* growth. Kuhry (1985) suggests that these developments are all driven by autogenic processes without any noticeable influence of climate change or human impact.

While it extends outwards, the peat body is also growing upwards and it may be doing this at different rates in different parts of the peatland, related to variations in hydrology, nutrient status and vegetation

types, for instance. This is particularly striking in areas of peat-dominated landscapes where the evolution of different elements of the peatland can be tracked through time with multi-core studies. Figure 7.9 shows the development of peatland types at Mariana Lakes in northeastern Alberta, Canada, reconstructed from 10 cores representing different areas (Nicholson and Vitt, 1990). Here, poor fen dominated by sedge communities developed around the fringes of the lakes from 8000 radiocarbon years BP and lake infill via hydroseral succession began. Poor fen subsequently spread laterally from 6500 BP, probably accelerated by gradual climate change, particularly after 5000 BP. While this process of lateral spread took place, established areas of sedge fen became more acid, and *Sphagnum* growth began. In areas of strongest water flow in the centre of the water track, poor fen vegetation was maintained, but slightly elevated areas at the edge of the peatland were dry enough for the establishment of forest bog communities in the last 2000–3000 years. This example is a good illustration of the development of a peatland landscape over time and shows a typical combination of autogenic and allogenic controls on peat growth and differentiation of peatland landforms.

Topographic influences are often of key importance for the rate and direction of lateral expansion in areas predisposed to peat formation. Where topography varies around the edge of an extending peat body, peat tends to form fastest in the lower, flatter areas than in the more steeply sloping parts of the landscape. Blanket mire spread is particularly strongly affected by topography, as shown by models predicting bog distribution (Graniero and Price, 1999a, b), although this approach could be applied to the spread of other mires as well. Graniero and Price (1999a) found that slope gradient was the most important factor in controlling the occurrence of blanket peats in Newfoundland, Canada, but that the upslope area providing surface runoff also played a role. Upslope extension of a mire is often evident when multiple cores are taken from a transect within the same microtope area of a mire. Basal radiocarbon dates from profiles from a transect on a fen in Sweden illustrate this process very well (Figure 7.10). The age of the peat is successively younger further up the slope, showing that gradual expansion of the peat area has taken place throughout almost the entire Holocene period. The stratigraphic profiles of all cores suggest a rather similar basal peat for most locations but a succession to sedge and *Sphagnum* fen in the older parts of the mire (Foster and Fritz, 1987). Together these results demonstrate paludification of

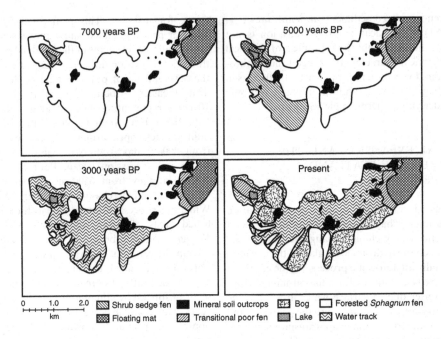

Figure 7.9   Reconstructed peatland development at Mariana Lakes, north-eastern Alberta, Canada at 7000, 5000, 3000 radiocarbon years BP, and the present day. Redrawn from Nicholson and Vitt (1990) by permission of NRC Research Press.

forest primarily by autogenically determined lateral spread of the peat initiated early in the postglacial period. Other data from mires in Europe and North America also reflect this upslope paludification process in the past (e.g. Foster and King, 1984; Charman, 1994). In a model of bog development, Almquist-

Jacobson and Foster (1995) also emphasise the importance of topographic controls on both the establishment of the initial fen peat phase and the subsequent lateral expansion of both fen and bog peats. Furthermore, they find the process of lateral expansion to be largely independent of climate for their study area.

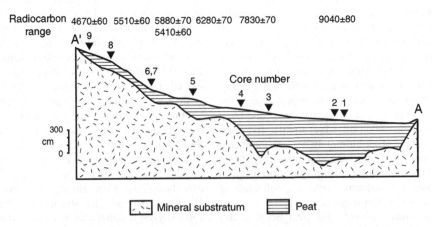

Figure 7.10   Cross-section of Kräckelbäcken fen, central Sweden, with locations of cores and basal radiocarbon ages. A further age towards the edge of the present fen was reported as 1640 ± 50 years BP. The sequence of ages reflects the gradual upslope extension of the peat over a long period of time. Redrawn from Foster and Fritz (1987) by permission of Blackwell Science Ltd.

7.7 'MATURE' PEATLANDS AND EROSION

So far in this chapter we have largely discussed the growth of peatlands, but a reduction in peat extent and depth can occur. Often this is attributable to direct human impact, especially in the last few hundred years (see Chapter 10), but it seems likely that at least some of the degradation of peat has occurred as a result of natural processes. A large body of very wet peat is held together by a relatively fragile web of internal structures and the 'skin' of vegetation on the surface. Physical stresses within the peat can develop, especially in well-established mature peatlands. Hypothetically, instability could be caused by a number of factors, including sheer depth of the peat, slope of the underlying terrain, oversaturation, and disturbance or destruction of the vegetation, and there are a number of cases when some or all of these may have contributed to peatland erosion.

Some of the areas where eroded peatlands are most common are the blanket mire regions of Britain and Ireland. Here, erosion channels dissect many large blanket peatlands and in badly affected areas, upstanding peat blocks (known as peat 'haggs') are the only remnant of the original surface. On the Isle of Lewis in the Outer Hebrides, most of the blanket mire is affected by erosion, with steeper slopes having much better developed gullies cutting into the peat (Goode and Lindsay, 1979). The southern Pennines are another area where erosion is particularly evident, as shown by Tallis (1964, 1965). The distribution and degree of erosion are undoubtedly affected by recent human activities, including grazing with domesticated animals, pollution reducing *Sphagnum* cover and impact of recreational activities, but there is also strong evidence that some degree of erosion of these areas is naturally occurring as a result of autogenic processes. In Ireland, erosion of upland blanket peats has been dated to 3000 and 1500 radiocarbon years BP, showing that recent industrial pollution and increased stocking levels of domestic animals are not the cause of erosion here (Bradshaw and McGee, 1988; McGee and Bradshaw, 1990). Likewise, on the Pennines, erosion first occurred in the prehistoric period, rather than in the last few hundred years when anthropogenic factors came into play (Tallis, 1985; Figure 7.11).

The exact causes of natural erosion are not fully known but a number of factors are thought to be important, especially relating to the physical build-up of deep peats in upland environments. The palaeo-environmental record of erosion has been most thoroughly exploited by Tallis (1985, 1987), who

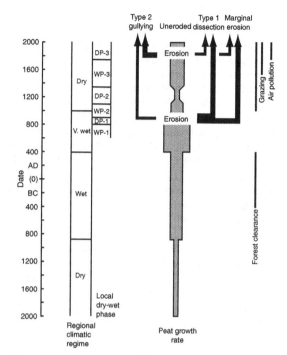

Figure 7.11 Time course of blanket peat erosion on Featherbed Moss, southern Pennines, England, with other potential influences. The central shaded column shows the relative growth rate of the peat and the arrows show the main periods and relative amounts of erosion. The left-hand part of the diagram shows the timescale and main climatic regimes. More detailed surface wetness changes are also given for the later period, represented as wet phases (WP) or dry phases (DP). On the right, the main periods of forest clearance, grazing and burning are indicated. The first phase of erosion (c. cal. AD 800) is unrelated to climate or anthropogenic activities. The second phase (c. cal. AD 1800) occurs during more intensive human activity. Redrawn from Tallis (1985) by permission of Blackwell Science Ltd.

suggested that the peat mass at Featherbed Moss became unstable as it reached the topographic limits of lateral spread and then built up unhumified *Sphagnum* peat on the surface. Tallis saw individual events such as bog-slides and bog-bursts as important in the initiation and acceleration of erosion in the Pennines. At Holme Moss, he even pointed to an individual storm event in AD 1777 as a potential cause of one major erosion episode, possibly linked to a severe burn on the site in the preceding few years. Burning is clearly one possible way in which the surface vegetation could be damaged sufficiently to weaken the resistance of the peat to erosion. Mackay and Tallis (1996) suggest that a major fire in around AD 1921 was the most important contributory cause of peat erosion in the forest of Bowland, Lancashire.

The steepness of the slope is also important in determining which areas of peat are likely to be eroded. Where erosion is present, it is always most advanced on the steeper ground, often at the edges of the site (Pearsall, 1956; Tallis, 1985). Hydrological factors may also play a role, especially the development of subsurface piping within the peat. These pipes develop within peat and other soils, often following desiccation of the surface and the development of cracking in the uppermost layers. Once established they can form networks within the peat and provide potential points of weakness when stress is exerted on the peat (Gilman and Newson, 1980). Certainly extreme hydrological conditions are important in the removal of large amounts of peat. There is some evidence that major summer storms, preceded by very dry conditions, cause severe erosion (Hulme and Blyth, 1985). Drying out of the peat surface, especially where it already lacks significant vegetation cover, can crack the surface and form a crust of dry peat, which is then removed by a severe storm event in a very short period of time.

7.8 CYCLIC REGENERATION

Throughout most of this book I have attempted to provide an up-to-date summary of current facts, ideas and hypotheses concerning peatland ecosystems. As a result, I have not felt justified in exploring in any depth the ways in which understanding of peatlands has changed over time. However, some areas of peatland science achieved such wide attention and the changes in opinion have altered so dramatically that they should be discussed in greater detail. In terms of autogenic changes on peatlands, cyclic regeneration is one such topic. It serves as an excellent example of the way in which an attractive and plausible theoretical idea can become an established truth, without any real attempt to test it experimentally. In the case of the cyclic regeneration theory, when experimental work was undertaken more than 50 years subsequently, it was falsified conclusively, yet lingered on in more general ecological texts for some time after. The idea and its refutation have been reviewed by several authors, including Barber (1981) and Backéus (1991), where the complete nuances of the story are told in full.

Early observations of peatlands noted two things: the hummock/hollow topography and changes between darker and lighter bands of peat in exposed profiles. Two early pioneers of peatland science, von Post and Sernander, proposed the cyclic regeneration theory to account for these observations (von Post and Sernander, 1910). In essence they proposed that hollow and hummock microforms would succeed each other over time in a cyclical way, due to differential accumulation rates in the two microhabitats. It was proposed that hollows would accumulate peat rapidly while hummocks would accumulate more slowly, especially as they became higher. Thus the growth of hollows would eventually exceed the height of the hummocks and then a new hollow would form where a hummock had been previously (Figure 7.12a). Over time, layers of alternating hummock and hollow peat would be found in a peatland where this had occurred (Figure 7.12b). This seemed an attractive idea that would explain the observations of surface features rather well and could account for the relatively few crude stratigraphic observations that had been made. What is surprising is that the idea was accepted as fact very quickly and with little question for so long (Backéus, 1991).

Later work on open peat sections where the spatial variability of stratigraphy could be properly appreciated has shown that the predicted alternating sequence of hummock–hollow peat was rarely present (Walker and Walker, 1961; Barber, 1981). The theory was thus falsified and various other data now also suggest that there is not normally any strict alternation of hummocks and hollows but that hummocks tend to stay in the same locations over millennia (e.g. Moore, 1977; Svensson, 1988b). Occasionally, elements of the succession expected from the theory are encountered but these are more easily explained as related to broader-scale hydrological changes, probably related to climate (Svensson, 1988b). In addition, it now appears that although peat growth rates do differ between microhabitats, there is no simple relationship such as hollows having higher growth rates than hummocks (see below). Instead, it seems more likely that changes in peat stratigraphy are primarily a function of climate changes and that during wet phases the total area of hollow expands, whereas during dry periods the total area of hummock extends. This has been termed the 'phasic theory' of peat growth by Barber (1981) and is important in relation to the interpretation of peat stratigraphy as a palaeoclimatic record (see Chapter 8).

7.9 PATTERN DEVELOPMENT

One of the main features of peatlands in many areas of the world is the intricate patterning formed by the

a)

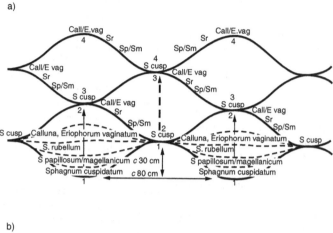

b)

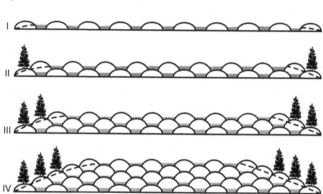

Figure 7.12   Two illustrations of the theory of regeneration at different scales. (a) Idealised cycle of the regeneration cycle, showing the succession of plant communities filling in two hollows and eventually overtopping a hummock where peat growth has effectively ceased. The cycle is then repeated with infill of the new hollows formed by the previous hummocks. The numbers represent the surface at different stages, i.e. at stage 2, the central location is a hollow and the two outer locations are hummocks. Sr = *Sphagnum rubellum*, Sp = *S. papillosum*, Sm = *S. magellanicum*, S cusp = *S. cuspidatum*, Call = *Calluna vulgaris*, E vag = *Eriophorum vaginatum*. Redrawn from Barber (1981). (b) Hypothetical cross-section of a bog showing cyclic regeneration, with 'lenticular regeneration' stratigraphy. Black denotes hollows and white represents hummocks. Hummock–hollow regeneration has resulted in the build-up of a highly structured peat of alternating light and dark peat, which should be observable in the field, but in practice this is the exception rather than the rule. Redrawn from Barber (1981) after Kulczynski (1949). See text for discussion. Figures redrawn by permission of A.A. Balkema.

arrangement of the hummocks and hollows on the surface. The undulating surface of many sites is especially visible where the lowest microforms are pools, but there is a wide variety of microforms and an even greater variety of terms used for these microforms which occur in both Southern and Northern Hemisphere mid- and high-latitude mires as well as in high-altitude tropical mires (e.g. Backéus, 1989; see Chapter 1). Despite this variety of form and terminology, the processes that lead to their formation are likely to be universal. Perhaps because of their prominence, or because of a general intrigue with patterns in nature, the development of surface patterns on peatlands has received a disproportionate amount of attention from peatland scientists for 100 years or more. Fortunately, several reviews provide an excellent summary of this work: Boatman (1983), Seppälä and Koutaniemi (1985) and Glaser (1998) are particularly useful references, while the volume edited by Standen *et al* (1998) is a useful collection of recent work on this phenomenon.

In terms of autogenic processes on peatlands, we are most concerned with the origins and development of patterning. Principally, we would like to know what processes are important in initiating, maintaining and changing surface patterning, and whether these processes are a result of autogenic influences or are primarily related to allogenic forcing, particularly by climate change. Glaser (1998) summarises these issues as four main questions:

(i) What is the timing of pool formation? Are they present from the start of peat growth or do they develop later and is there a relationship between pool initiation and climate change?
(ii) What are the dynamics of the pools? Are they growing or shrinking or are they a stage in a cyclic process?
(iii) In both the above issues, how important are external drivers compared with internal processes?
(iv) What is the relationship between pools and ridges? Does the origin of one cause the formation of another?

All these questions have not yet been answered but they do provide a good framework for consideration of research on patterning in peatlands.

It is generally accepted that the development of patterning involves two principal types of process: physical disturbance and biotic changes. However, the exact nature and relative importance of these vary between location and mire type. Physical processes are particularly important in arctic peatlands, where permafrost is very clearly the biggest influence. However, in more temperate areas, pool development is more likely to be a result of biotic processes, only aided in exceptional circumstances by physical mechanisms. In permafrost peatlands, the formation of ground ice raises the ground level in some areas and collapse of ground ice causes water-filled depressions in others. Within collapse scar areas of permafrost peatlands, a cycle of infill and collapse may establish itself (see Figure 7.6). On non-permafrost peatlands, hollow and pool development is thought to be primarily a result of differences in accumulation rates and feedback mechanisms that amplify relatively small initial differences in peat growth rate. Processes in bogs and fens may be somewhat different and physical factors such as seasonal freeze/thaw activity and hydrological events may assist this process, but the biotic effects are thought to be the chief process underlying pool development (e.g. Seppälä and Koutaniemi, 1985).

Pool initiation generally occurs sometime after the start of peat growth, and often not until the mire has already developed some considerable depth of peat. In general, pool formation does not appear to be synchro-nous in the same area or even on the same site (Foster et al, 1988; Tallis and Livett, 1994; Karofeld, 1998), suggesting that it is primarily an autogenic process driven by local variability and change in hydrology and vegetation. However, a rise in the water table from climate change would presumably make the formation of pools more likely, especially where moisture supply is close to the threshold for pool formation. An intensive study in Estonia showed that pool initiation may be an autogenic process but pool widening was related to later periods of climatic change (Figure 7.13; Karofeld, 1998). Foster and Wright (1990) found older pools in the centre of a Swedish raised mire than at the edge, suggesting that pool initiation is triggered by an internal hydrological change as the mire expanded laterally and a larger area formed the wet centre of the mire.

The generally applicable processes of pool or hollow development are summarised by Foster et al (1983). It begins with the minor diversity in elevation shown by almost all plant communities. Some plants tend to form gentle hummocks either by growth form or accumulation of litter. An increase in surface wetness, either by locally reduced hydraulic conductivity or by increased supply of water (from climate change or increased focusing of runoff on a fen, for example), will result in localised ponding of water. It is also suggested that the development of a layer of peat is important in developing sheet flow of water rather than channel flow, presumably by spreading the water across the surface rather than concentrating it as an erosive force in one location (Foster et al, 1983). Once wet, it appears that accumulation rates in many mire plant communities are decreased, either from reduced growth rate of plants, a change to plants with higher decay rates or some other factor which inhibits peat accumulation on the bottom and edges of the incipient pool. The hollows can thus expand laterally if water supply remains adequate to maintain the water level. This basic process seems to adequately explain the formation of pools in patterned fens, where coalescence of the small initial hollows results in the long pools aligned across the slope (Figure 7.14). Figure 7.14 also shows how a drying of the surface can result in a reduction in pool area. While differential accumulation rates go some way to explaining how pools can begin to form, the active degradation of pool bottoms and sides is also an important process in many sites, as shown by both long-term observational data (Mets, 1982) and stratigraphic evidence (e.g. Foster et al, 1988).

Models of pool development such as these imply that peat accumulation in hollows is less than that

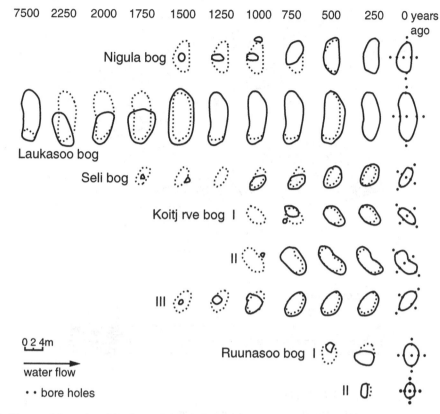

Figure 7.13 Temporal dynamics of the shape of eight hollows for every 250 radiocarbon years on five peatlands in Estonia. The shape of the hollows as they appear today is superimposed as a dashed line on the reconstructed outlines (solid lines). Redrawn from Karofeld (1998) by permission of Arnold Publishers.

of hummocks. Estimates from stratigraphic studies suggest that this is the case in some situations. For example, van der Molen and Hoekstra (1988) calculate rates of 62–73 g cm^{-2} yr^{-1} (1.03–1.59 mm yr^{-1}) for *S. imbricatum* hummocks compared to 20–28 g cm^{-2} yr^{-1} (0.36–0.47 mm yr^{-1}) for *Sphagnum cuspidatum* hollows. While this is sometimes the case, detailed evaluation of production and decomposition along the water table gradient reveals a more complex pattern (Belyea and Clymo, 1998). They show, by field measurements, that the relationship between the hummock–hollow transition and peat accumulation rate is non-linear (Figure 7.15). Here they suggest that the 'local rate of burial' (LRB, roughly equivalent to accumulation in the short term) varies in a more complex way and that although it is related to the depth to water table, some hollows or lawns will have the same LRB as some hummocks. What matters is the relative LRB in adjacent microforms as this will determine which

direction the hollow formation goes in. There are two balance points in the gradient from high to low water tables. The lower point (A) is unstable because it depends on which direction LRB in the hummock moves and how high LRB is compared to the adjacent hollow. If hummock LRB decreases a little and is less than that of adjacent hollows, the hollows 'catch up' with the hummocks and the hummock becomes a lawn or hollow as well. If hummock LRB increases, the hummock continues to rise still further above the water table until LRB slows after it has gone over the high point of LRB towards the second balance point (B). Once at B, the hummock tends to stay at the same LRB because this balance point is stable. Thus hollows and pools can expand and contract around the lower balance point, but once hummocks are formed as high as the upper balance point, they are unlikely to alter significantly as they are restrained by the growth rate of the hollows which will control the local rate of bog

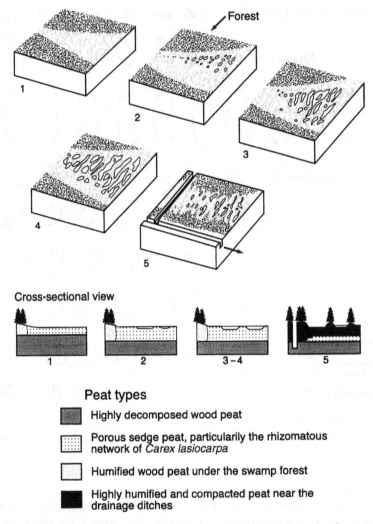

Figure 7.14 Developmental model for a patterned fen in Minnesota, USA. (1) A featureless fen water track within a swamp forest. (2) The water track expands as drainage converges into the zone of porous peat that is deposited by the hydrophilous sedges growing in the track. The water table rises in the track and ponds up at the surface because of the levelling of the slope and the convergence of the drainage. (3) and (4) The ponds expand parallel to the contour, restricting the sedge lawn into narrow sinuous ridges, because of reduced peat growth rates at the edges of the ponded areas. (5) A lowered water table from drainage allows trees and shrubs to reinvade the water track. Redrawn from Glaser (1987b). In other peatlands, for step 2, the water table rise could be caused by other autogenic and allogenic factors. Step 5 shows how pool extent can be reduced by a reduction in water supply, although it may not occur or could be due to other factors such as climate change on other peatlands.

growth. This model of the dynamics of hollows, lawns and hummocks would explain why stratigraphic studies often show alternation between lawn and hummock communities but rarely show a complete 'regeneration cycle' (e.g. Boatman, 1983). Furthermore, Belyea and Clymo (1998) show how the relationships between microforms and LRB can combine with water table changes to produce permanent deep pools such as those described from various sites around the world (e.g. Foster *et al*, 1988).

A third phase of pattern development that is not discussed as widely as the processes of initiation and development is the collapse and disintegration of pool systems. There are two potential problems with study-

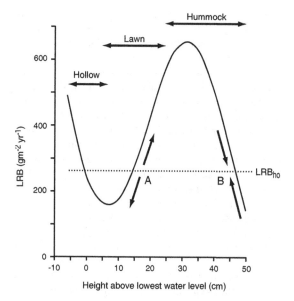

Figure 7.15 Changing local rate of burial (LRB) along the water table gradient in a raised bog. The dotted line is a specific LRB_{ho} (LRB in a hollow) that intersects the curve for the lawn and the hummock microforms as well, showing points where LRB is equal in all three microforms. The arrows show the direction a lawn or a hummock would move from the balance points A and B. The lower balance point (A) is unstable and the upper balance point (B) is stable. See text for further details. Redrawn from Belyea and Clymo (1998) by permission of Blackwell Science Ltd.

ing this process: (i) the stratigraphic evidence of past disintegration of pool systems may have been removed by erosion and (ii) rather few mires may have reached this stage of maturity. Three processes are probably important in the breakdown of patterning on blanket mires (Tallis, 1994) and these processes could also be acting elsewhere. First, headward erosion by streams within the mire drawing off water from pools. Second, drying of pools through climatic change. Third, anthropogenic impacts from grazing, burning and air pollution can damage the structure of pool–hummock systems and result in their disintegration. On patterned fens in Scotland Charman (1998) also found evidence of pool breakdown in observations of existing mires, which suggested that the gradual encroachment of pools on ridges could eventually result in the break-up of ridges due to water and sediment flow on sloping mires. The drainage of pools is also noted in Southern Hemisphere peatlands by Mark *et al* (1995), as a result of both surface and subsurface streams, although it is not clear to what extent this

would eventually lead to the break-up of the pool system.

7.10 PLANT-MEDIATED CHANGES

There is a wide range of interactions occurring at small scales within plant communities on mires, some of which have profound effects on changes within the system. While a number of these changes occur over seasonal or annual timescales, others are much longer lasting, ultimately resulting in new directions of development. We have already referred to the idea of 'ecosystem engineers' (Jones *et al*, 1994), with respect to beaver activity and peat initiation. *Sphagnum* mosses and many other species found on peatlands could also be said to be ecosystem engineers as they modify the habitat to such an extent that other species are affected in a major way. Another useful concept that has been applied to peatlands is that of the vegetation 'switch', where positive feedback between vegetation and environment results in sharp changes in vegetation along gradual environmental gradients (Wilson and Agnew, 1992). Several authors have used this idea explicitly to explain spatial patterns in peatland vegetation. Agnew *et al* (1993b) suggest that this type of mechanism operates on the margins of pakihi peatlands in South Island, New Zealand. Within the pakihis, there is a spatial transition from beech–podocarp forest on the mineral margins through to a shrub zone of *Leptospermum scoparium* and then to open *Empodisma minus* bog in the centre. The transition from forest to shrub zone is unusually sharp and is not explained by topography or substrate. It is suggested that tree seedling establishment is only possible on the dead fallen tree boles, as the ground level is too wet. Thus the forest maintains its own boundary but cannot establish itself within shrub or open bog due to a lack of suitable dry microsites (Figure 7.16). Feedback between trees and hummock-forming mosses in Rocky Mountain fens in northern Colorado, USA, is another example of such a process. Here, tree establishment is favoured only on low hummocks within an open fen. Once trees begin to grow on scattered hummocks, local water tables are further reduced due to increased interception and evapotranspiration. The growth of hummock-forming *Sphagnum* is also increased, with tree stem and shrub branches as a physical support. Increased hummock height and reduced water tables further accelerate tree growth in a positive feedback loop.

Sphagnum is the most obvious example of plant-mediated change in peatlands, although other major

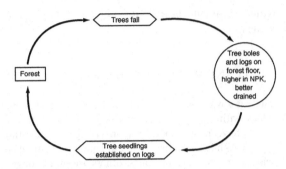

Figure 7.16 Schematic model of the 'one-sided' vegetation switch mechanism maintaining forest–mire ecotone in pakihi peatland in South Island, New Zealand. The dead tree boles provide drier, more nutrient-rich microsites than the mire surface and tree seedlings can establish only on these sites. Redrawn from Agnew *et al* (1993b) by permission of the International Association for Vegetation Science and Opulus Press.

peat-building taxa such as the restiad rushes in Southern Hemisphere peatlands display some similar attributes. The ability of *Sphagnum* to acidify peatland waters, its low decay rate and the fact that it forms deep peat deposits have already been referred to several times in this book (see especially Chapters 3 and 5). Van Breemen (1995) discusses these and other aspects of the impact of *Sphagnum* growth in reducing growth of vascular plants on peatlands (Figure 7.17). Here the model illustrates how increased peat growth leads to decreased trophic status which in turn leads to greater *Sphagnum* growth. Various factors, including increased acidity and low nutrient availability, lead to decreased vascular plant growth which in turn lead to increased light availability and decreased transpiration. Both of these effects provide positive feedbacks to *Sphagnum* growth. Certainly *Sphagnum* is a rapid invader of vulnerable habitats such as the moist, shaded habitats in boreal forests where taxa such as *S. girgensohnii* can colonise large areas of forest floor, perhaps as a precursor to more advanced peat development (Foster, 1984). However, the implication that *Sphagnum* acts purely as a negative factor in vascular plant growth is an oversimplification, although it is a major process in the establishment of a *Sphagnum*-dominated peatland.

In established peatlands where vascular plants coexist with *Sphagnum*, the moss may have rather different effects on higher plant growth. For example, the sundew, *Drosera rotundifolia*, extends its growth vertically to keep pace with the upward growth of *S. fuscum* in bog communities (Svensson, 1995). The

growth form of shrub taxa in an Alaskan peatland was also found to be strongly modified by *Sphagnum* growth (Luken and Billings, 1985). Some taxa (*Ledum palustre* and *Andromeda polifolia*) showed increased shoot density and structural biomass with *Sphagnum* growth, probably due to increased production of lateral shoots that eventually become individual stems. The plants thus respond positively (at least initially) to overgrowth by *Sphagnum*. Other taxa such as *Vaccinium vitis-idaea* are unable to respond in this way and would eventually be engulfed by the moss growth. Finely balanced communities of *S. fuscum* and *Picea mariana* (black spruce) in Canada can result in a cyclic succession between the two species (Payette, 1988). The *Picea* growth favours *Sphagnum* because of increased snow accumulation and therefore increased water supply. The *Sphagnum* favours *Picea* growth because of its suitability for layering of lower branches. The two species therefore coexist in a state of unstable equilibrium with regular flips between *Sphagnum* and *Picea* dominance at any particular location.

Sphagnum growth can also determine the potential for growth of other bryophytes. Liverworts, for example, are divisible into those having a 'compromise strategy' or those with an 'avoidance strategy' (Albinsson, 1997). The former are those that rely on habitats created by *Sphagnum* and other bryophytes, and coexist with them by occupying the niches such as the gaps between *Sphagnum* stems. The latter are the taxa that are effectively excluded by *Sphagnum* and must rely on areas of bare peat or plant litter for growth space. Because of the mechanisms summarised in Figure 7.17, *Sphagnum* is a resilient plant, but other plants can and do outcompete it on occasion, even where conditions remain favourable for its growth. Algae are one of the most noticeable successors to *Sphagnum* on bog surfaces. Algal mats, often mainly of *Zygogonium ericetorum*, can infest and ultimately slow down *Sphagnum* growth or even kill it off completely (Hulme, 1986; Karofeld and Toom, 1999).

Restiad species such as *Empodisma minus* and *Sporadanthus traversii* must promote similar positive feedback processes to establish the deep peats found in New Zealand. The mechanisms are less well known, but it certainly has a very similar base exchange capacity to *Sphagnum cristatum*, the main bryophyte competitor on the same mires (Agnew *et al*, 1993a). It has also been suggested that the water-holding capacity of the fine root system could be as efficient as that of *Sphagnum* (Campbell, 1964). However, more recent work suggests that it is not the capacity to hold water that enables *E. minus* to dominate the vegetation but

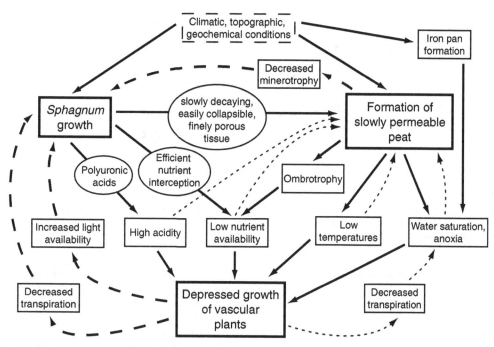

Figure 7.17 Pathways of peat formation (solid arrows) and feedbacks increasing the growth of *Sphagnum* (dashed arrows). The dotted arrows show feedbacks involving physico-chemical processes and suppression of decomposers. The oval outlines show the morphological, anatomical, physiological and organochemical properties of *Sphagnum*. Redrawn from van Breemen (1995) by permission of Elsevier Science.

rather its efficiency in capturing water through stem-flow (Agnew *et al*, 1993a), and the very low rates of evapotranspiration created by the mat of surface litter (Campbell and Williamson, 1997). The effects on the surroundings are similar to those of *Sphagnum* in Northern Hemisphere peatlands: the creation and maintenance of an acid, low-nutrient and very wet environment, which other plants find difficult to persist in.

Sphagnum is also involved in many of the small-scale dynamic interactions in permafrost and boreal peat-lands through its ability to insulate the ground and thus prolong the effect of winter cold into the growing season. This can be a major factor in the stability of peatlands affected by permafrost thaw. Camill and Clark (1998) found that *Sphagnum fuscum* cover was one of the most significant factors in controlling the stability of permafrost peatlands in northern Manitoba, Canada. Thaw rate increased with mean annual temperature but the effect of climate was strongly mediated by local factors, especially those of *Sphagnum* cover and azimuth of sites (Figure 7.18). Other taxa can also moderate the effect of temperature

in cold climate peatlands. Trees can reduce snow depth under the canopy of branches, which allows greater penetration of cold and may promote the formation of permafrost (Camill, 2000). Although clearly the cli-matic conditions determine whether permafrost forma-tion is possible, the location of trees determines precisely where it begins. In contrast, established forest vegetation may insulate the ground from winter cold, and only when trees are removed can features such as ice-wedges begin development (Payette *et al*, 1986). In forested areas, wind is reduced and drifting of winter snowfall is limited. Removal of trees results in snow-drift formation and the creation of snow-free sites with deeper cold penetration. In more temperate areas, forest patch dynamics, such as those created by wind damage and blowdown of trees, can also be important local influences on microtopographical and plant diver-sity (Ehrenfeld, 1995).

Finally, while recognising that there are many plant-mediated mechanisms of change in peatlands, there are often relatively stable communities. Changes in composition may only vary in a minor way with small-scale competitive interactions and species

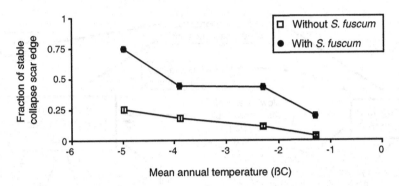

Figure 7.18   Comparison of the stability of collapse scar edges between sites with and without *Sphagnum fuscum* cover from northern Manitoba, Canada, along a mean annual temperature gradient. The difference between the two types of site is significant at $p < 0.0001$ and suggests that sites with *S. fuscum* are two to three times more stable than those without. Redrawn after Camill and Clark (1998) by permission of the University of Chicago Press.

replacement. Although such communities may show significant change in small-scale spatial patterns over time, the overall community composition remains essentially the same (Gibson and Kirkpatrick, 1992; Nordbakken, 2000).

7.11 PHYSICAL PROCESSES IN COLD CLIMATE PEATLANDS

Climatic conditions are clearly important in controlling peatland vegetation and landforms, and climate is unarguably an external factor. However, at the beginning of the chapter we made the distinction between *changes* in allogenic factors (such as climate) as influences on peat development, and allogenic factors as *background* conditions against which autogenically driven changes occur in the peatland. The nature of cold climate peatlands is strongly affected by processes related to low temperatures and seasonal fluctuations in temperature, but these can occur within a stable set of climate conditions and I will consider them here as autogenic influences on peatland change.

The growth and collapse of permafrost is one of the most notable influences on peatland landform development in boreal and arctic peatlands. Growth of palsa mires is largely a result of permafrost formation with mounds of ice growing under the surface. It has been known for some time that the initiating factor is most likely to be related to locally thin snow cover in winter, allowing greater cold penetration (e.g. Sjörs, 1959). Snow is most easily blown away from hummock microforms, which would form the most likely location for the embryonic permafrost mound to form. In summer

these locations also remain colder for longer because of the insulative effect of the hummock vegetation. The nature of vegetation cover (moss, lichen, forested, non-forested) will also have some effect (e.g. Railton and Sparling, 1973; see above). However, the precise mode of origin of these landforms is still debated and several different processes may be operating (Matthews *et al*, 1997).

Other high-latitude peatlands are also strongly affected by snow and ground frost. The effect of frost heave was thought to be an essential part of the formation of string–flark microtopography on aapa mires, but biotic factors are now regarded as more important (see above; Seppälä and Koutaniemi, 1985). The effects of spring melt are perhaps more important physical processes in string–flark fens than conditions during winter. The movement of large amounts of meltwater is sufficient to move loose litter and even to alter the position of entire strings, so that in the area of most rapid flow, the strings are bowed out in a downslope direction (Seppälä and Koutaniemi, 1985). Ground frost development is certainly present in a variety of other mire types although the effects on vegetation and mire morphology are less clear (Eurola, 1975). The relationship between hydrological conditions and ground frost is also of interest here. For example, Kingsbury and Moore (1987) found that melting from below was more important in spring than from above when the peat surface was dry. Groundwater flow is also likely to be important in prolonging ice-free conditions in flarks and in accelerating thaw in the spring. Conversely, increased water supply under colder conditions in the palsa mire zone can lead to an increase in height of palsa mounds due to increased ice formation, including pure ice (Thorhallsdottir, 1994).

7.12 SUMMARY

In this chapter we have explored the various ways in which change occurs in peatlands in the absence of major disturbances of climate, hydrology or fire regimes. The peatland system is a constantly growing, sometimes shrinking ecosystem, whether or not the external conditions alter. It is sometimes easy to forget these autogenic processes when we are interested in using peatlands purely as an archive for the detection of the external changes. However, the autogenic factors cannot be ignored, nor can they always be expected to be subordinate to allogenic changes. Indeed we can only realise the potential of the peatland archive when we understand the internal mechanisms more fully. The next chapter will explore the influence of allogenic factors on peatland development and summarise the ways in which the peat record has been used to reconstruct changes in past climate. Natural variability is also an important consideration in assessing anthropogenic damage to peatlands and the conservation and management issues that arise from this (Chapters 10 and 11).

8

Allogenic Change

8.1 INTRODUCTION

The 'external' conditions are major determinants of where and when peat growth can occur (see section 2.2 for example). One only has to look at where peat-lands occur to see there is a general association between cool and/or wet climates to observe this relationship. Other factors such as fire, sea level, tectonism and groundwater status are all important in different peat-land areas for establishing the right template against which peat formation takes place. However, in Chapter 7, we distinguished between (i) stable back-ground conditions, and (ii) changes in those conditions. Even given stable background conditions, we can see from the various processes of change described in Chapter 7 that over time, considerable variation in peatland systems can be expected. Here, we are moving on to consider the ways in which *changes* in allogenic factors affect peatland ecosystems. The conceptual framework of stability and change provided by Figure 7.1 is equally applicable to these processes.

While we have divided our discussion of long-term change on peatlands into two separate chapters for practical purposes, it is important to remember that 'external' and 'internal' changes are constantly inter-acting. It is only by putting both parts of the picture together that we can really understand the dynamics of the ecosystem in any balanced way. Relationships with climatic conditions and with climate change will form the core of the discussion here since moisture balance is the most critical factor in peatland development. We will also review some of the applications of the peatland–climate relationship to the reconstruction of past climate changes from the peat archive. This area of research has received much attention from Quaternary scientists, especially in the last decade or so, and while it has focused primarily on climate reconstruction, it has also yielded further clues on the impact of climate on peat growth.

8.2 CLIMATE

The influence of climate on the occurrence of peatlands is clear simply from looking at the distribution of major peatland areas in the world (see Chapter 1). Peatlands are predominantly ecosystems of either cold and/or wet climates and the type of peatland that develops is also strongly related to climate when patterns over large spatial scales are considered. The relationship between climate and peatland types established by Moore and Bellamy (1973) (Figure 8.1) reflects the main zonation of mires across Europe and similar relationships can be found in North America and elsewhere (Frenzel, 1983; Zurek, 1984). There are clear distinctions between the types of peatland system found in oceanic–continental and arctic–temperate–tropical gradients of climate. We therefore know that climate is one of the key factors influencing peatland ecology and functioning. What is much less clear is the extent to which climatic *change* modifies the peatland environment through time. The answer is of course not always straightforward, and it relies almost entirely on the palaeoecological record for evidence since changes in climate occur over timescales that are usually too long for direct observation of eco-logical change. As we have already found, inferences from the palaeoecological record are often open to interpretation, and multiple causes are often suspected for many changes observed in the record. Despite this, our understanding of the influence of climate on peat development is now quite good and there is growing agreement of the importance of climate in determining major changes in peatlands at broad spatial scales. As we have discussed the links between climate change and many other factors throughout this book, the aim here is to provide an overview of some critical over-riding considerations. This is followed up at the end of the chapter with a review of the ways in which peat records have been used to reconstruct past climate changes.

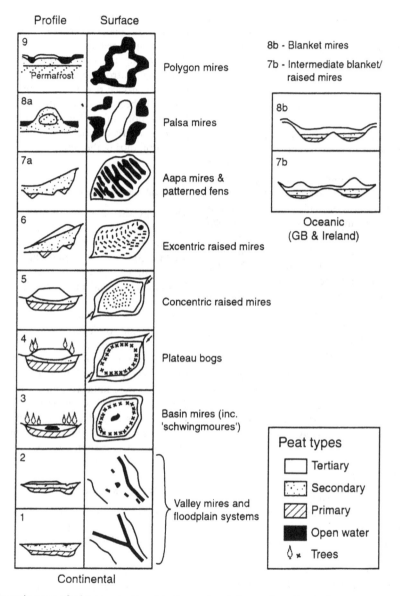

Figure 8.1   (a) The main types of mire complex found in Europe in relation to the main north (top) to south (bottom) gradient. The stratigraphic diagrams include reference to the peat types primary, secondary and tertiary, representing different stages of peat formation (see Chapter 4). The distribution of these peatland types is shown in Figure 1.11b. The extreme oceanic areas have the two additional mire types reflecting the secondary east–west climate gradient in peatland distribution. Redrawn from Moore and Bellamy (1973) by permission of Peter Moore.

The extent to which any particular peatland is affected by climate depends on a range of other factors, including:

(i)   *The location of the peatland with respect to major peat–climate thresholds.* For example, palsa mires

in the centre of palsa distribution are likely to be less sensitive to small changes in the thermal regime than those on the margins of palsa distribution.

(ii)   *Seasonal variation in climate.* In some areas, peatlands are sensitive to changes that occur only at a particular time of year. For example, many mires in temperate areas have excess water throughout the year except in the summer season. Therefore, changes in winter precipitation or temperature are unlikely to have a significant effect. In contrast, even small changes in the summer water balance may be enough to alter conditions for plant growth and peat decay.

(iii) *The type of peatland.* While all peatlands are clearly dependent to some extent on a favourable climatic regime, some are much more sensitive than others. For example, minerotrophic mires with surface runoff or groundwater as their main water supply are less dependent on direct atmospheric input of moisture than ombrotrophic peatlands.

(iv) *The strength and impact of autogenic and other local allogenic factors.* Climate may be subordinate to one or more other factors so that changes in climate have little effect on the site. More often, climatic change would serve to either accelerate or retard an effect of another factor, for example, in the case of fire and climate in retarding tree growth (see below).

The direct impacts of climatic change on peatlands are many but are divisible into a number of main categories: hydrological change, physical processes, peat accumulation and carbon storage, biotic impacts and influences on mire morphology.

8.2.1 Hydrological change

Hydrological change is probably the most important and universal response of peatlands to climate change. It is at the root of most of the other impacts of climate change, including all of those mentioned here. Ombrotrophic mires are likely to be the most strongly affected by such changes since they are solely dependent on atmospheric water to maintain a saturated surface. Because of this, raised and blanket mires have provided a focus for much of the work on the reconstruction of climate change from peatlands (see section 8.6 below). However, surface runoff and groundwater flow are also strongly linked to climate, although they are affected by many other factors as well. Some of these are discussed below (section 8.4).

8.2.2 Impacts on physical processes

Physical processes are important in the formation and maintenance of many geomorphic features of cold

climate peatlands in particular (see Chapter 7). Changes in temperature can affect these processes. For example, regional cooling in the Canadian subarctic during the later Holocene is thought to have caused permafrost aggradation, which then led to the peatland surface rising above the influence of groundwater and becoming ombrotrophic (Vardy *et al*, 1998). Colonisation of *Sphagnum* could then provide a positive feedback mechanism by providing greater insulation during the summer.

8.2.3 Influence on accumulation and carbon storage

The processes of peat accumulation and the various factors that affect the rate of peat growth were discussed in Chapter 5. Large-scale variability in peat accumulation rates is undoubtedly strongly affected by climate change (see section 5.6), due to alteration in temperature and moisture supply and their effects on productivity and decay. Clearly any effect on accumulation also impacts on carbon storage. Until around 6000 radiocarbon years BP, the ability of Canadian peatlands to act as a carbon dioxide sink was limited by a relatively dry climate. After this time, carbon accumulation increased rapidly as new peatlands were initiated and existing ones grew faster. Regional cooling in subarctic and arctic peatlands reduced this rate of carbon accumulation after about 3000 years ago, but the existing total peatland area is still a powerful carbon dioxide sink (Vitt *et al*, 2000).

8.2.4 Biotic changes

The biotic impacts of climate change are probably largely secondary. Rather few peatland plants are limited in their distribution by climatic constraints *per se*. Most are more strongly controlled by the local hydrological and nutrient conditions. Of course ultimately they may be susceptible to climate change, but this is only a result of any hydrological change rather than the direct impact of climate itself. Stratigraphic studies on *Racomitrium lanuginosum*, a bryophyte found in oceanic European mires, show periods of dry, warm climate are associated with increases in its abundance (Tallis, 1995b). The response is to the relative surface wetness of the mire rather than to the temperature or rainfall changes themselves. Some groups of plants may be particularly sensitive to changes in hydrology caused by climate change. The response of many bryophyte species to hydrological gradients is well known (see Chapter 3), a relationship that is exploited in palaeoclimatic reconstructions from peat deposits (see below). Plant communities can also be shown to be sensitive to change in water table conditions, with implications for future plant distribution under the enhanced greenhouse effect (Camill, 1999a). Taxa such as woody shrubs and trees more common in dryland communities also ought to be sensitive to shifts in hydrology. Various palaeoecological studies have looked at tree growth in response to climate change on peatlands. Periods of relatively dry conditions often favour spread of trees (e.g. Bridge *et al*, 1990), but research on changes in growth rate of trees on bogs through tree-ring studies suggests that climate is not the overriding factor in controlling interannual variations in tree growth on most peatlands (Linderholm, 1999).

A few taxa may be more sensitive to a direct climatic influence. For example, *Sphagnum lindbergii* occurs in colder mires in Europe, under rather similar hydrological conditions to other taxa in mires in more temperate areas (Daniels and Eddy, 1985). When broad-scale patterns of bryophyte distribution are assessed there is certainly a clear association between many taxa and climatic gradients of temperature and rainfall (e.g. Gignac, 1994). In cases such as these there may be direct impacts of temperature change, although these have not been documented empirically to date.

8.2.5 Alteration and limitation to peatland landform development

We have already noted the relationship between the prevailing climatic conditions and peatland typology (Figure 8.1). This would suggest that long-term climate change could bring about changes in the morphology of peatlands where significant climate thresholds are crossed. The amount and distribution of excess moisture through the year are an important influence and changes in the moisture balance can result in changes to the overall shape of the peat deposit. In raised mires, the shape of the surface is a function of the moisture surplus, with increasing moisture surplus allowing a higher dome to develop (Figure 8.2). It is therefore possible that mires have changed their overall shape in response to climate change in the past. However, this is almost impossible to discern from most palaeoecological records in the absence of very intensive coring, sampling and dating programmes to determine contemporaneous levels of peat across large areas of a mire.

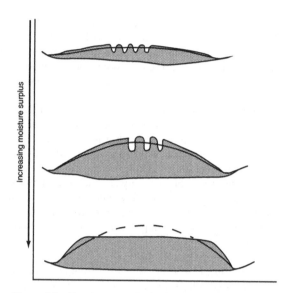

Figure 8.2   Development of raised bogs in relation to moisture surplus as suggested by Damman (1986). The heavy line indicates the critical level (the maximum possible convexity of the mound), which becomes increasingly convex with higher levels of moisture surplus. The bottom cross-section shows a plateau mire from Newfoundland, Canada, where the maximum convexity of the mire is not reached due to other factors that inhibit peat growth. Redrawn from Damman (1986) by permission of NRC Research Press.

8.3 FIRE

Natural fires occur throughout many parts of the world, including areas where peatlands are common. Fire also has a long history as a tool of landscape management by human populations and we know that in some cases this has had important implications for changes within peatlands. Of course, in general, peatlands are some of the wettest locations in the landscape and are therefore least susceptible to fire. Nevertheless, in dry summer seasons surface layers of peat and plant litter can provide a highly combustible mixture for fire to start in or spread to from surrounding upland vegetation. In the northern boreal forest zone, where peatland and forest are often intermixed, fire return periods are of the order of 60–100 years in the forests (Wein and MacLean, 1983). Not all such forest fires would be expected to spread to peatlands, but they are frequently at risk from fire in these environments. Given the potential impacts of a single severe burn, a high frequency of fires is not necessarily required to have a profound and long-lasting effect on peatland development.

Because of the long period between fires, studies of fire impact on peatlands generally fall into two categories: those on the impact and recovery from a single fire event and those using palaeoecological techniques to reconstruct fire history and impact over millennial timescales. The palaeoecological techniques used to reconstruct fire history were reviewed in Chapter 6. Using these techniques, data on local and regional fire history can be derived, although it is not always clear whether charcoal found within peat is derived from one or other of these sources. Only moderate to severe fires are usually detectable with any confidence in the peat record, as other smaller particles may be derived from further afield, such as from a fire in surrounding forest communities. The chief impacts of fire on peatlands are as follows:

(i) *Destruction of living biomass.* Fire may have differential effects on individual species and their subsequent survival and regeneration. Many peatland taxa regenerate vegetatively and should therefore be able to regenerate relatively quickly unless all living parts of the plant are killed. Soil organisms may also be killed or their populations reduced by heating of the peat (Wein, 1983).

(ii) *Destruction of plant litter and surface peat.* The loss of surface peat layers can have significant effects on peat accumulation rate and the carbon balance of the system.

(iii) *Release of nutrients.* The ash from burnt plant material and peat makes some nutrients more available immediately after a burn. In oligotrophic peats, where plant communities are limited by nutrient availability, even subtle changes in the nutrient balance can affect the competitive ability of some plants and therefore alter the plant community structure. In addition to releasing nutrients for plant growth, fire may also lead to a net loss of nutrients through gaseous losses (nitrogen and sulphur), and water- and airborne movement of ash (Moore, 1982a).

(iv) *Hydrological change.* The reduction in plant cover after a burn reduces the transpiration rate and thus may encourage an increase in surface water. In addition, the removal of the low bulk density surface material in established peatlands means that the burnt surface has a lower hydraulic conductivity and less capacity to store and retain water. In shallow organic soils and mineral soils bounding peatlands, the hydraulic conductivity may also be reduced as a result of pore blocking by fine charred particles (Mallik *et al*, 1984). This

may lead to establishment of new peatland areas or spread of existing peat soils.

8.3.1 Peat initiation

Various aspects of peat initiation were covered in Chapter 4, including the possible role of fire in the conversion of mineral soil to peat growth. If surface water increases as a result of reduced transpiration and decreased hydraulic conductivity, invasion of mosses and other peat-forming plants will be encouraged. The work by Mallik *et al* (1984) is usually quoted as support for this process, and increases in the abundance of charcoal at the base of blanket peat soils especially are taken as evidence that fire was important in early peat growth. However, the increase in wetness may be very short-lived and perhaps inadequate to establish peat development unless the soil is already close to the hydrological threshold for peat formation. Wein (1983) suggests that the intensity of burn may be a critical factor in the pathway of soil formation after a fire on mineral soils with shallow organic surface horizons. Where fires are low-intensity surface fires, only superficial burning of vegetation takes place. This reduces transpiration but leaves the shallow organic layer intact, allowing the increased wetness to further inhibit decay and promote the build-up of organic matter. In more intense fires, where the deeper burning removes much of the existing organic material, nutrients are released and the effect of an initial reduction in transpiration is less pronounced. The pedogenic trend towards deeper organic soil is reversed and processes of podsolisation and development of surface humic layer have to begin again. Certainly the fires associated with Mesolithic land use in Britain are likely to have been low intensity, although probably relatively frequent. One example of the conversion of woodland to blanket bog is provided by Caseldine and Hatton (1993). Here high-altitude woodland on Dartmoor was subject to enhanced burning for over 1000 years (7700–6300 radiocarbon years BP), associated with the hunter-gatherer economy of the area at the time. A long charcoal record shows that the burning episode between 7700 and 6300 BP is well defined, starting and finishing quite abruptly (Figure 8.3a). The base of a peat profile from nearby records the changes in plant communities associated with the burning during the inception of peat (Figure 8.3b).

8.3.2 Peat accumulation

The most obvious potential impact of fire on peatlands, especially of severe burning, is the removal of surface

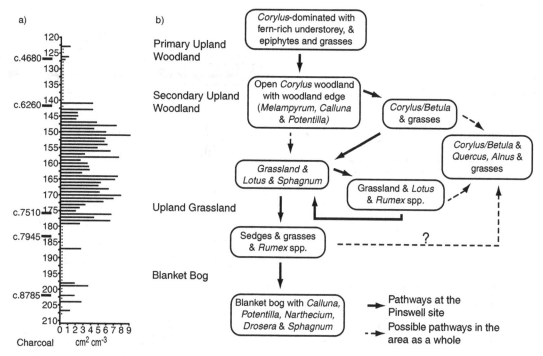

Figure 8.3 (a) Variations in the abundance of charcoal found in a deep peat record from Black Ridge Brook, Dartmoor, England, showing a well-defined phase of increased fire between about 7700 and 6300 radiocarbon years BP. (b) The sequence of vegetation change during the inception of peat at Pinswell, a nearby blanket mire site where burning was important in removing the early woodland and encouraging the spread of open ground taxa. Some taxa such as *Melampyrum* (Cow-wheat) may be particularly encouraged by fire, but most taxa are simply responding to the increased light levels and acidification caused by the destruction of the *Corylus* woodland. Although peat developed at Pinswell, there was some woodland regeneration in other areas (shown by the dashed arrows), in suitable microclimatic and topographic conditions. Redrawn from Caseldine and Hatton (1993) with kind permission of Kluwer Academic Publishers.

layers of peat, thus reducing average peat accumulation in the long term. Unless very large amounts of peat are removed, it is difficult to assess the total amount of peat lost in this way during a single fire, but palaeoecological data can provide information on longer-term changes in accumulation rates associated with repeated fires. For example, in western Canadian boreal peatlands, fire frequency is strongly negatively correlated with peat accumulation in terms of both depth of peat and mass of carbon (Figure 8.4). Kuhry (1994) calculated that an increase in fire frequency of between five and seven times would reduce the accumulation rate to zero. In this area, fire frequencies have been higher in the past than they are today: around 7000 radiocarbon years BP fires occurred approximately twice as often as they have done over the past 2500 years. Pitkänen *et al* (1999) also found that increased fire frequency reduced carbon accumulation rates and calculated that the average loss of carbon in an individual fire to be around 2.5

kg m^{-2} at a site in Finland. This equates to approximately 140 years of peat growth at this site, so one can see that even a relatively low number of fires could have a very significant impact on peat growth rates. Increased fire frequency may be one of the consequences of global warming in the twenty-first century for some regions, where increased summer drought is predicted. Gaseous carbon losses to the atmosphere would then be increased, although this could be slightly offset by the effect of the aerosol and particulate emissions from combustion (Pitkänen *et al*, 1999; see Chapter 9). In some peatlands, there is already zero net peat accumulation, and this may be at least partly a result of the fire regime (Jasinski *et al*, 1998).

8.3.3 Peat mound development

The microtopography of peatlands features in many parts of this book, and the occurrence of hummock–

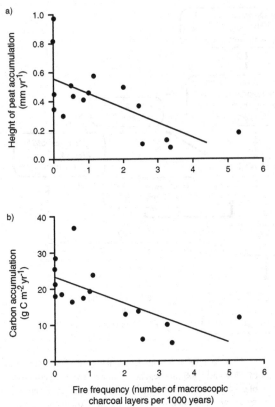

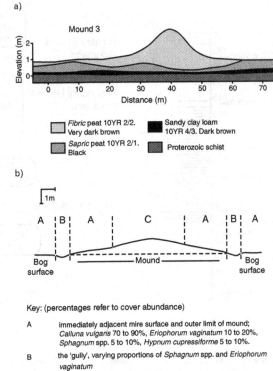

Figure 8.4 The relationship between fire frequency (expressed as the number of macroscopic charcoal layers per 1000 radiocarbon years) and peat accumulation (depth of peat and mass of carbon). Linear regressions between charcoal layers and (a) peat depth accumulation and (b) carbon accumulation. Similar relationships occurred when the data were divided into subcategories according to peat age classes and peat types. Redrawn from Kuhry (1994) by permission of Blackwell Science Ltd.

Figure 8.5 Peat mounds from (a) Tasmania and (b) northern Scotland. These peat mounds are much bigger than the 'hummocks' that commonly form part of the microtopography of mires. Both landforms occur in temperate blanket mires and both form mainly as a result of fire. See text for details. Redrawn from (a) Macphail et al, (1999), (b) Robinson (1987) by permission of the New Phytologist Trust.

hollow patterns is common. On some peatlands much larger hummocks are found, often referred to as 'peat mounds'. These are not associated with permafrost development as in the case of palsa mires, but are known from temperate blanket mires in both Northern and Southern Hemispheres (Figure 8.5). The mounds vary in size and shape but are broadly circular to oval, around 5–15 m in diameter and up to about 1.5 m in height. These dimensions appear to be about the same in both Scottish and Tasmanian peatlands, although diameters vary a lot, depending on the precise point of measurement. The mounds from Caithness, Scotland, were found to be surrounded by a circular depression (Robinson, 1987), but other sites

in the Western Isles and Shetland do not always have this feature (Goode and Lindsay, 1979). Various hypotheses on their formation existed, with suggestions that they could be archaeological structures, or result from relict or low-intensity periglacial action, differential peat growth or erosion. However, in both northern and southern peat mounds it has been shown that fire played a crucial role in their formation. In both cases fire seems to have been retarded at the locations of the mounds due to locally wetter conditions. In the Caithness peat mounds, the wetter conditions were caused by the combination of particular stratigraphy and local soil conditions (Robinson, 1987). In Tasmania, higher water tables at the mound locations are likely to be related to groundwater discharge. There may be other processes operating, including some erosion and enhanced peat growth due to increased nutrient loadings from birds and animals that use the mounds as lookout posts. However, fire in the surrounding peatland seems to be critical in both areas and provides a common basis for mound development.

8.3.4 Vegetation and post-fire regeneration

Vegetation is often altered significantly in the short term by fire. In less intense burns, only the surface dead plant litter may be removed and light burns such as this have often been used as a management tool (see Chapter 10). However, many wildfires are more intense and damage or kill some of the plants on the surface. The restoration of the original plant cover depends on how well these taxa can regenerate, and how well they compete with other taxa already present or recently invading the burnt peat surface. There are surprisingly few studies documenting vegetation recovery from fire on peatlands, perhaps because of the relatively slow recovery compared to some other ecosystems. One of the longest periods of monitoring specifically to track recovery from fire is that related by Clarkson (1997) for two peatlands in North Island, New Zealand, examining recovery from fires over approximately 12 years. Differences in plant recovery rates were primarily related to their physiognomy and to whether or not they were completely eliminated by the fires. Species with underground rhizomes resprouted rapidly whereas recovery from seed in other taxa was much slower. Recovery to vegetation to pre-fire condition took 12 years at one of the sites (Moanatuatua bog). Given a frequency interval of between 100 and 200 years for such sites, the vegetation suffers no long-term change, a finding also suggested by stratigraphic studies of Canadian peatlands (Kuhry, 1994). On oceanic peatlands, fires may be a less natural part of the system and catastrophic fires can occur. A severe fire largely destroyed the vegetation cover (including *Sphagnum*) on Glasson Moss, a raised mire in Cumbria, England in 1976 (Lindsay, 1977). The fire was a result of an exceptionally dry summer season together with artificial drainage of the periphery of the site, which exacerbated the dry conditions. It seems unlikely that such severe fires would occur naturally and recovery from this fire is still incomplete, more than 25 years later. Little is known of the influence of natural fires on tropical peatland vegetation. It has been shown that tropical swamp peatlands in northern Australia are strongly modified by fire, although this may not be representative of other tropical peat areas (Kershaw and Bohte, 1997).

Forest history and development can be affected in a major way by fire regimes. Trees growing on peatlands are no exception. Destruction of forests by fire may play a part in the initiation of peat growth (see above). In some regions, the timing of peat initiation appears to be intimately linked with the gradual fragmentation of forest and the reduction in tree regeneration over several millennia. On the high-altitude areas in the Charlevoix highlands area of southern Quebec, Canada, fire served to open up the landscape and provide suitable areas for peat formation as the climate became more favourable to waterlogging (Bussières *et al*, 1996). Fire and climate are also linked in reducing forest extent at the boreal forest–tundra boundary across northern continental areas. Here boreal forest can be pushed back by a combination of unfavourable climate and fire acting together when one of these factors alone may not have been sufficient to change the status quo. While established trees may tolerate a slight climatic downturn, regeneration following fire may be impossible. This combined effect of fire and climate is thought to operate in the Canadian north and in Siberia (Jasinski *et al*, 1998). In other areas, forest growth and regrowth occur in unstable equilibrium with climate and fire. Detailed studies of changes in tree rings from living and buried remains of black spruce in northern Canada reveal that regeneration following two fires around 350 and 10 calendar years BC was good and the forest maintained itself through regeneration quite well through these disturbances (Arseneault and Payette, 1997). However, a fire in AD 1568 had a catastrophic effect on the forest, reducing it to a 'krummholz' form (stunted spruce). The destruction of the forest canopy would result in a change in microclimate conditions that could prevent growth of fully formed upright spruce stems. In effect the tree line was being maintained in previous years by positive feedback from the forest itself.

8.4 HYDROLOGICAL FACTORS

Much of the variability in hydrological factors can be traced to other natural causes covered elsewhere in this book, such as climate and hydroseral succession, or to anthropogenic impacts such as drainage. However, in some cases, significant changes in groundwater and runoff inflow to peatlands can be caused by the impact of shifts in local fluvial patterns, the rise or fall of groundwater due to sea-level fluctuations or other external but non-climatic factors.

8.4.1 Anthropogenic impacts

One effect that we discussed in relation to peat initiation is the role of human impact in forest clearance and alteration of the hydrological balance. In the case of blanket mires, this can cause peat initiation (Moore, 1993). In other mires, it may result in a major change from fen to floating bog (Warner *et al*, 1989). Other changes in hydrological status causing vegetation change can be caused by human activity. Forest clearance by European settlers caused changes in mires in many parts of the New World over the past few hundred years. In addition to the development of floating kettle peatlands already described for Canada, increasing wetness of other peatlands occurred, including the development of apparently 'natural' forested swamp communities. In New Zealand, there is also growing evidence that European settlement caused increased surface wetness, promoting a greater extent of *Sphagnum* growth to establish the present vegetation of sites such as that at Glendhu in the South Island as recently as AD 1860 (McGlone and Wilmshurst, 1999). Clearly any such disruption is most likely to have a serious effect in fens receiving significant amounts of runoff from the slopes where deforestation has taken place. Some peatlands with well-established *Sphagnum*-dominated vegetation are also known to have been affected by forest clearance. For example, a bog in the Czech Republic became wetter at times of increased human impact at around 1400 and 900 calendar years BP (Speranza *et al*, 2000). However, although such peatlands have many of the characteristics of true ombrotrophic peatlands, they are usually found to receive at least some runoff from surrounding slopes or to be influenced by increased water levels in surrounding fen areas. This is likely to be especially true for smaller peatlands such as the one described in the Czech Republic.

8.4.2 Sea-level change

Sea-level change at the global scale is closely related to climate change and ice sheet history, but at the local scale it is also affected by isostatic changes in the height of the land surface independent of climate. Changes in sea level can have an effect on freshwater peatland development in a number of ways. A rise in sea level may be sufficient for an incursion of seawater on to the mire surface, causing deposition of saltmarsh peats or marine clays. Likewise a drop in sea level may initiate freshwater peat development over the top of coastal deposits. However, there are more subtle ways in which sea-level change can affect freshwater peat growth. In peatlands close to the coast, the regional groundwater table is often linked to sea level so that any changes in the sea level have an impact on the peatland hydrology without any direct contact between sea water and the peatland.

Frequently changes in coastal peatlands reflect influences from other factors as well as sea level. Cobweb Swamp is a coastal wetland in Belize which became wetter after about 5600 radiocarbon years BP in response to rising sea levels, was infilled by peat growth and then became aquatic again after 3400 BP, when Mayan populations cleared forest from the catchment (Jacob and Hallmark, 1996). Peat growth was subsequently renewed when conditions became drier, with reforestation of the surrounding slopes from about 500 years ago (Figure 8.6). Examples such as this underline the importance of considering multiple causes for similar stratigraphic changes in peatlands.

8.4.3 Changes in groundwater supply

We know that groundwater is an important component of the hydrological budget in many peatlands and any significant long-term alteration in either the supply or quality of groundwater is likely to have far-reaching effects. Often such changes will be linked to climate but this is not necessarily always the case. Rather few clear-cut examples of groundwater changes causing long-term alterations have been documented, but several recent studies show how important this effect can be. Groundwater discharge zones are known to be critical to the modern distribution of many of the peatlands in Minnesota, USA (Siegel, 1983), but the changing influence of groundwater flow over time is more difficult to assess.

With the use of palaeobotanical and mineral indicators, Glaser *et al* (1990, 1996) show that reversals in the direction of groundwater flow have altered peatland type and surface vegetation a number of times during the last 4000 years in Minnesota. Here, two topographically similar mounded areas of peatland currently have different vegetation: one has typical raised bog vegetation, the other is a rich fen with groundwater discharge at the surface (a 'spring-fen mound'). Both areas were initiated by mounding caused by groundwater discharge. Subsequently this flow reversed and the sites became groundwater recharge zones as the water table mounds built up adequate hydraulic head. After 1200 radiocarbon years BP, the recharge weakened and groundwater was discharged to the surface

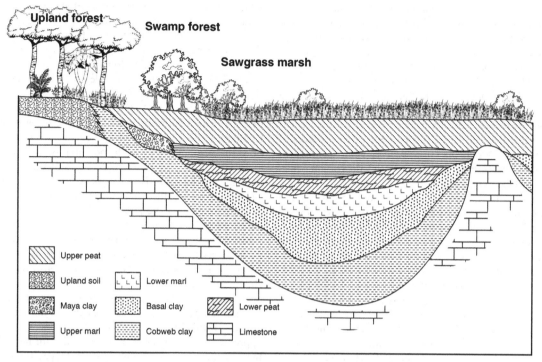

Figure 8.6   Schematic stratigraphy of Cobweb Swamp in coastal Belize. The site is currently forming peat (the upper peat layer) following previous aquatic phases (the marl layers), a peatland phase (the lower peat) and a period of major inwash (the Maya clay). The first phase of peat growth represents terrestrialisation of the saline lagoon created by sea-level rise. The second phase of peat growth followed the abandonment of the area by the Maya, allowing reforestation of the catchment and terrestrialisation of the basin by peat growth. Horizontal distance *c*. 0.5 km, vertical exaggeration *c*.× 80. Redrawn from Jacob and Hallmark (1996) with permission of The Geological Society of America.

at one site but only penetrated part-way up the profile of the other site. A change from groundwater discharge to recharge as a result of water table mound development may be important in the transition from fen to bog conditions at other sites (e.g. McNamara *et al*, 1992). Although this is essentially an autogenic process (see Chapter 7), it seems likely that the strength of groundwater discharge is an important factor at this point in peatland development. This is especially so where peat has developed over a relatively porous substrate and there is less resistance to flow between the peat body and groundwater than there would be over a clay substrate, for example. Modelling studies certainly suggest that vertical flow of groundwater in peatlands is strongly controlled by substrate permeability, which would support this view (Reeve *et al*, 1996).

Spring-fed mires are probably most sensitive to changes in groundwater regime. At one site in Yorkshire, England, interbedded peats and sands reflect changes in groundwater flow (Younger and

McHugh, 1995). Using the palaeoenvironmental data and a groundwater model, changes in water table profiles over a transect of approximately 5 km were made showing higher water tables at 5800 and 1300 radiocarbon years BP, but variable magnitude of change in different parts of the landscape. These fluctuations are a result of recharge from precipitation variations and in boundary conditions altered by sea-level change (Younger and McHugh, 1995).

8.5 VOLCANIC INFLUENCES

Peatlands occur throughout many of the active volcanic regions of the world, including Japan, Iceland and New Zealand. Ash deposition on the mire surface provides a convenient dating tool (see Chapter 6) but may also have impacts for the peatland system itself. Evidence for changes on peatlands brought about by ash falls is limited, however, since many studies on peat are focused on the pollen indicators of changes in the dryland forest history rather than the *in situ* vegetation. Following major eruptions such as the one at Taupo in New Zealand at *c*.1850 radiocarbon years BP, major fires could be expected in the surrounding landscape which could have spread to peatlands (Wilmshurst and McGlone, 1996). Burial of many low-growing plants would also be a serious short-term effect, as well as possible toxicity of the ash and volcanic gases. Despite these potential effects, bogs with sequences of volcanic ash falls do not appear to show any serious long-term consequences of ash deposition (e.g. Newnham *et al*, 1995).

There have been suggestions that even very long-distance tephra deposition could have an effect on vegetation but these remain controversial. For example, Blackford *et al* (1992) proposed the idea that the Hekla 4 eruption in Iceland could have triggered the decline of Scots pine (*Pinus sylvestris*) in northern Scotland, as it was growing at its climatic and edaphic limits on peatlands in this area. The amount of ash falling in the area produced only microscopic layers in the peat deposit but could have had a short-term effect on climate and pollution loads. Even if the pine decline was not linked to tephra fall, detailed analysis of microfossil changes immediately after the same volcanic event shows that in Ireland a mire surface became much wetter following tephra fall, suggesting a significant short-term climate change in response to the eruption could have occurred (Dwyer and Mitchell, 1997). However, it seems highly unlikely that long-term change would be effected by this level of impact.

8.6 CLIMATE RECONSTRUCTION FROM PEAT

The reconstruction of past climate changes is a major area of research in Quaternary science, in an attempt to understand natural variability of climate, the mechanisms that force global climate and the interactions between climate and the biosphere. Data are needed from a range of time periods and locations. Because of the links between changes in peatlands and climate, the peat archive has been thoroughly investigated as a possible source of palaeoclimatic data for a number of years. Because most peatlands are strongly controlled by hydrology, and hydrology is often strongly influenced by climate, reconstruction of past hydrological change in peatlands should yield proxy data on climatic change if all other potential causes of hydrological change can be ruled out. While this sounds like a simple idea, in practice it is often difficult to sift the climatic

information from the other sources of change in peatland hydrology. I have deliberately left discussion of climate reconstruction to the end of the two chapters that consider all the other natural autogenic and allogenic processes influencing long-term change in peatlands. This is so that the complexity of the task of establishing climatic records from peatlands can be properly appreciated. Despite this complexity, a range of climatic data has been derived from peat records and techniques and research approaches have been developed to overcome many of the potential problems.

In section 8.2, we summarised some of the factors that affect the sensitivity of peatlands to climate change. In essence, these are the same factors that influence the quality of the palaeoclimatic record in peatlands. Most peatlands will have been influenced to some extent by climate change throughout the course of their development, but we need to identify those that are particularly sensitive if we wish to reconstruct past climates with any measure of accuracy. There are two main areas of climate reconstruction from peat: those using data on the start of peat growth and those using changes within the peat profile. The second of these approaches is used more often than the former and there is a greater variety of methods and types of record available to tackle it.

8.6.1 Peat initiation and climate change

Despite the variety of mechanisms and causes of peat initiation (see Chapter 4), a number of studies have successfully used peat initiation data to derive palaeoclimate records. The main difficulties associated with doing this are: (i) to identify locations which would have been primarily dependent on climate change for the start of peat growth, and (ii) to accurately date the base of the peat profiles at these locations. Tackling the first of these problems means using only sites where peat growth proceeded via paludification, and avoiding areas where human impact, burning or hydrological changes unrelated to climate may have occurred. The second issue is more difficult and radiocarbon dating is the only viable dating tool, despite the potential difficulties of accurately dating basal peats. Both problems are eased by using large data sets, where anomalous basal ages can be identified, or have limited influence on the overall patterns of change. Because of the problems identified above, interpretation of the data has to be limited to identifying broad-scale spatial and temporal patterns in climate change. The other issue to be considered is that peat initiation will not show a linear relationship with climate over time because as more of the landscape becomes covered by peat, the remaining area is less prone to peat formation (Payette, 1984).

Some of the most productive work on peatland initiation and climate change is that in western Canada, where clear regional patterns of peat initiation can be identified. Initially a distinction between pre- and post-6000 BP conditions was identified, with earlier peat initiation north of latitude 54°30′ N and in the foothills of the Rocky Mountains (Zoltai and Vitt, 1990). This was due to more arid conditions in the early to mid-Holocene. Progressively wetter climates after 6000 radiocarbon years BP resulted in the present distribution of peatlands by around 2000–3500 years ago. Subsequently, more detailed patterns of change have been discerned and related to climate change modified by local topographic and edaphic factors (Halsey *et al*, 1998). Combining the influences of climate with the local factors allowed the production of modelled peat initiation times over a wide area of western Canada (Figure 8.7). Early peat initiation occurred in some zones in northern Alberta around 9000 to 8000 cal. BP after an initial deglacial lag unrelated to climate. Further peatland initiation was limited by warmer summers at this time but peat growth expanded east as a result of decreasing summer insolation after this. The spread of peatland was more pronounced in the west (Alberta) as a result of incursions of moist Pacific air. However, after 6000 cal. BP, peatlands spread south and east in response to the more general increasing climatic wetness identified from the earlier work. The pre- and post-6000 cal. BP patterns of peatland initiation have been shown to be consistent with other palaeoclimatic data and with climate models (Gajewski *et al*, 2000). The data have more recently been shown to reflect periodicity in climate detected from many other palaeoclimate records in ocean and ice core records, showing that global climate changes may be reflected in patterns of peatland initiation (Campbell *et al*, 2000).

It may be more difficult to use peat initiation data from some other areas to derive palaeoclimatic data. For example, in Europe, the problem of anthropogenic influence on peat formation may be greater (see Chapter 4). However, given careful selection of data sets to be used, some coherent patterns unlikely to be related to human impact can be found. More intensive

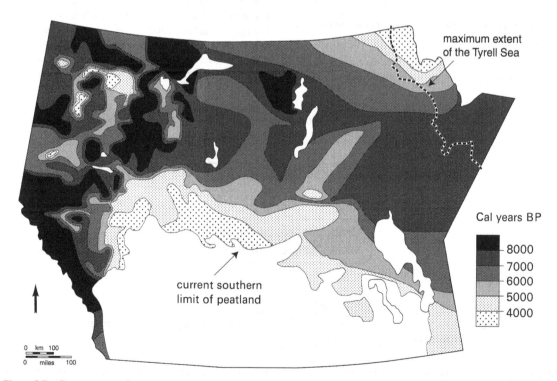

Figure 8.7 Contour map of peat initiation dates from a peatland initiation model developed for western Canada, showing the spread of peatlands in 1000-year increments. Redrawn from Halsey *et al* (1998) from an original kindly supplied by the authors, with kind permission from Kluwer Academic Publishers.

studies within a smaller area can help identify region-ally synchronous patterns of peatland initiation that can only be attributed to climate change. In southern Finland, two distinct phases of peat initiation and spread were identified between 8000–7300 and 4300–3000 calendar years BP. Although peat initiation was not restricted to these periods, a major acceleration of peat extent is notable during both periods which coin-cided with wetter climate, as indicated by lake levels in southern Sweden (Korhola, 1995). In contrast, human impact on vegetation and soils appears to have con-fused the pattern of blanket peat initiation in England and Wales, so that clear overall patterns are not so easily discerned (Tallis, 1991).

The use of peat initiation data to infer climatic change should be pursued cautiously. Large, high-quality data sets are needed to be able to identify climatic change securely. At the scale of individual sites and small areas, local factors are likely to have a more noticeable effect. Having said that, many studies of individual sites do find basal peat dates that fit within broader-scale patterns of peat initiation and climate change (e.g. Miller and Futyma, 1987; Korhola, 1996).

8.6.2 Signal and noise in climate change records in peatlands

Patterns of peat initiation are only ever likely to iden-tify very broad-scale, generalised, long-term climate

changes, mostly because of the other influences on the process of initial peat development. Peat profiles offer the possibility of continuous records of climate change through time from the reconstruction of hydrological conditions in sites sensitive to changes in atmospheric moisture supply. As we have already suggested, many peat profiles will contain a palaeoclimatic signal but often this is blurred by the 'noise' of other influences on the record. To derive the maximum amount of cli-matic information, we therefore need techniques to select sensitive sites and to filter out potential 'noise' in the signal. This has still not been adequately achieved, although significant progress has been made towards it. Here we restrict the discussion to ombro-trophic peatlands, as they ought to be most closely linked to climate. This area of palaeoclimate recon-struction from peat is particularly well reviewed by Blackford (1993, 2000).

Figure 8.8 shows a conceptual model of the relation-ships between climate, ombrotrophic peatland hydrol-ogy and other factors which influence hydrology – in this case they are the environmental noise we are trying to filter from the climate signal. We can use the rela-tionships suggested by the model to design effective research strategies for palaeoclimatic research. Autogenic factors cause most of the variability on small spatial scales within sites and sometimes at the whole site scale. Non-climatic allogenic factors also mostly act at these scales. Conversely, climatic change

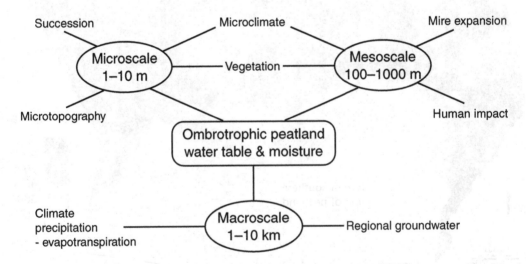

Figure 8.8   Conceptual model of the factors affecting ombrotrophic peatland surface wetness at three spatial scales. Microscale variability is principally affected by local autogenic factors. At larger spatial scales, allogenic factors are more likely to influence the record. Climate acts at the broadest spatial scale so that replicable changes across several sites can normally only be attributable to climate change.

affects very large areas, so that if it is the cause of past surface wetness changes it should be identifiable at a number of sites within the same region. Studies of multiple profiles within the same peatland are also useful to identify surface wetness changes at the mesoscale that are more likely to be linked to climate.

Multiple short cores from two sites in Cumbria, England, show that many of the major changes in past hydrology as reflected in plant macrofossils are very similar within the same site and between sites (Barber *et al*, 1998). This supports the notion that these changes are climatically driven. However, the variability in the detail between cores reflects the different microtopographical niches that have been sampled (Figure 8.9). Some locations are less sensitive to fluctuations in hydrology than others. For example, hummock tops are particularly well buffered against a rise or fall in water table, whereas *Sphagnum*-dominated lawn communities require only a slight shift in hydrology to result in a significant change in the plant community to a shallow pool or low hummock. The consequence of this is that variation between profiles on the same site may not necessarily mean that hydrological fluctuations recorded in only one core are due to non-climatic factors; they may simply be smaller magnitude climate changes which only cause a change in communities with particularly narrow ecological niches. One of the questions arising from data such as these is how does one decide which data to use in a palaeoclimatic reconstruction. Data from apparently insensitive locations can be excluded completely or several records can be averaged on the basis that shifts recorded at insensitive locations would have reflected higher magnitude climate change (e.g. Charman *et al*, 1999). Another approach is to ensure that a single

sensitive location is chosen for detailed analysis by careful reconnaissance in the field (Barber *et al*, 1998). Comparisons between the centre and edges of ombrotrophic mires may be a further way of identifying the most significant changes in surface wetness and climate. Hendon *et al* (2001) show that the edges of ombrotrophic sites are usually less sensitive locations but perhaps identify the more significant shifts in past surface wetness.

8.6.3 Records of change in the peat profile

Early work on peatland climatic records attempted to identify points in the peat stratigraphy where there had been a rejuvenation of peat growth. These were termed 'recurrence surfaces' and were thought to indicate increased surface wetness at particular points in time. More recently, continuous variations in indicators of surface wetness have been used to establish climatic curves through time, surely a more realistic portrayal of many trends in past climate. There is a wide variety of potential indicators of past climatic change in peatlands. Most of these are related to changes in the hydrological status of the peatland but a few are independent sensors of temperature, although the latter are still poorly developed. The techniques that are especially important are peat humification, plant macrofossils, testate amoebae and hydrogen and oxygen isotopes. All of these techniques were introduced in Chapter 6, but here we will expand as necessary on their use as palaeoclimate indicators.

Plant macrofossil analysis exploits the strong response of peatland plants and especially of bryophyte taxa to the surface wetness gradient (see Figure 3.19 for example). Changes in plant macrofossils over time,

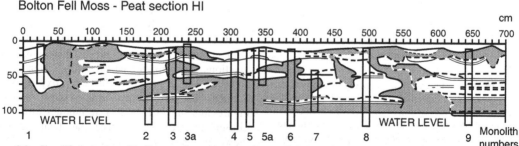

Figure 8.9   Simplified stratigraphic diagram from a peat section on Bolton Fell Moss, Cumbria, England, with the positions of nine separate sample monoliths. Shaded areas are highly humified peat and unshaded areas are lightly humified peat. The thick and thin lines represent major and minor stratigraphic boundaries respectively, with broken lines showing diffuse boundaries. The double lines represent pool muds. Only monoliths 2 and 9 record all of the possible changes to wetter conditions. Monoliths 3 and 6 are in less sensitive locations and do not record all these changes. Redrawn from Barber *et al* (1998) with permission. © John Wiley & Sons Limited.

therefore, can be taken to represent changes in surface wetness. There are many examples of this kind of work, especially in British raised mires, most notably by Barber (1981). More recent work by Keith Barber and his co-workers has derived indices of surface wetness changes for a number of sites in northern Britain. This is mostly achieved by summarising changes in plant macrofossil assemblages as a sample score on the first axis of an ordination analysis, where the plant taxa scores suggest that the axis is closely related to the changing water table gradient over time. One example of this approach is shown in Figure 8.10. Here, the distribution of species scores on the ordination axis 1 is strongly suggestive of a water table gradient, with taxa such as *Sphagnum Sect. Cuspidata*

and *Rhynchospora alba* at high values on the axis and others such as *Polytrichum* spp. and *Sphagnum Sect. Acutifolia* at low values. The plot of axis 1 scores (Figure 8.10b) can then be interpreted in terms of fluctuating surface wetness conditions, where very high values reflect a wet surface and low values reflect a dry surface. In this particular example, changes to much wetter conditions were noted at nine main points over the last 8000 calendar years (Hughes *et al*, 2000b). Bryophyte taxa from peatlands other than raised mires may also demonstrate a response to climatic wetness/dryness over time. For example in high-altitude blanket mires in the British Isles, *Racomitrium lanuginosum* has been used to good effect (Tallis, 1995b).

Humification of peat is one of the easiest techniques to apply and it has been more widely used than plant macrofossil analysis, with data from a variety of peatlands in Britain, Ireland, Denmark, Norway and New Zealand (e.g. Aaby, 1976; Nilssen and Vorren, 1991; Blackford and Chambers, 1995; Chambers *et al*, 1997; Anderson, 1998; McGlone and Wilmshurst, 1999). One of the key advantages of humification is the relative speed and ease of analysis so that high temporal resolution measurements are possible. It also appears to work well in quite highly decayed blanket peats as well as raised mire peats. A typical humification record is shown in Figure 8.11 over a 5500-year sequence from southern Scotland (Chambers *et al*, 1997). In this particular plot, humification is expressed as percentage humification, transformed from the raw data produced by the colorimetric analysis of the peat extract (Blackford and Chambers, 1993). Because of the continuous decay of peat over time, there is generally a systematic relationship between humification and peat depth that has to be eliminated from the data. This is shown as a separate curve in Figure 8.12.

Several additional techniques have been used for palaeoclimate reconstruction from peat. Testate amoebae analysis has only been applied to a relatively small number of profiles for reconstruction of past surface wetness changes. As described in Chapter 6, transfer functions to reconstruct past depth to water table can now be used to provide a further proxy for surface wetness changes. Several examples are shown in Figures 6.8 and 8.12. Other techniques also include isotopic analyses to reconstruct past temperatures, as shown in Figures 6.11 and 6.12.

8.6.4 Peat records and other records of past climate

Whatever technique is used to reconstruct a surface wetness record from an ombrotrophic peatland, there

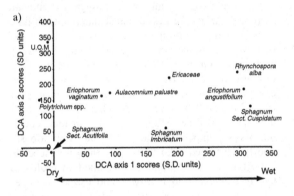

a)

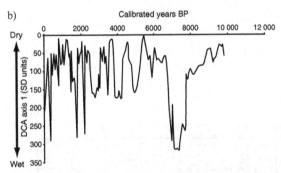

b)

Figure 8.10 Ordination analysis of plant macrofossil data from Walton Moss, Cumbria, England, used to summarise changes in surface wetness conditions over the Holocene. (a) Plot of taxa and scores on axes 1 and 2 from an ordination technique known as detrended correspondence analysis (DCA). The position of the different taxa along axis 1 suggests it is related to surface wetness. (b) Sample scores plotted through time, summarising changes in surface wetness. Arrows indicate the directions of the surface wetness gradient in both plots. See text for further details. Redrawn from Hughes *et al* (2000b) by permission of Arnold Publishers.

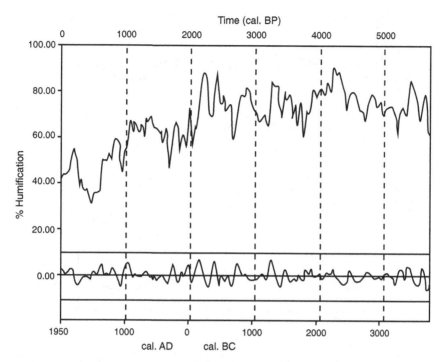

Figure 8.11 Plot of humification values for Talla Moss, southern Scotland. In this case, the data have been smoothed and the lower line shows fluctuations in values after the effect of continuous long-term decay has been eliminated from the data. Redrawn from Chambers *et al* (1997) by permission of Arnold Publishers.

remains a key question: what are the precise climatic parameters that the surface wetness changes are reflecting? Clearly surface wetness is a function of some combination of precipitation and evapotranspiration, but the actual relationship is not known. One approach is to compare records for more recent time periods with independent climatic data or with other proxy climatic data for which the influencing climate variables are better known. Broad similarities between past climatic data reconstructed from documentary records and peat surface wetness were suggested by Barber (1981) from his work on Bolton Fell Moss, England. More recently a 'climate response model' fitted changes in plant macrofossils from the same site to the historical records of summer wetness/dryness and winter mildness/severity over the period from cal. AD 1100 to 1800 developed by Lamb (1977). The model resulted in r^2 values of between 0.41 and 0.50, suggesting a good fit between climate and plant macrofossil variations (Barber *et al*, 1994). One graphical representation of the fit between surface wetness changes in northern England and Lamb's climate indices is shown in Figure 8.12. There is certainly mounting evidence that surface

wetness changes in ombrotrophic peatlands are strongly related to independently derived variations in climate. It is still not clear whether temperature or precipitation is the more important parameter and it may be different for peatlands in different areas of the world. The data from Britain suggest that temperature is perhaps more important than precipitation (Figure 8.12), whereas precipitation may be more important in continental peatlands. At present there are too few data of this kind to assess the possible spatial differences in these relationships.

Another approach is to avoid some of the problems of direct correlation with other climate records by performing time series analysis to identify periodicities in the surface wetness fluctuations that could be replicated at different sites, repeated in other proxies or fit with known fluctuations in solar activity, for example. This has only been applied in north-west European peatland records such as those of Aaby (1976), Barber *et al* (1994), Chambers *et al* (1997) and Hughes *et al* (2000b).

Other insights into the peat record can be gained from comparisons with independent proxy palaeoclimate data from the same areas. One of the problems

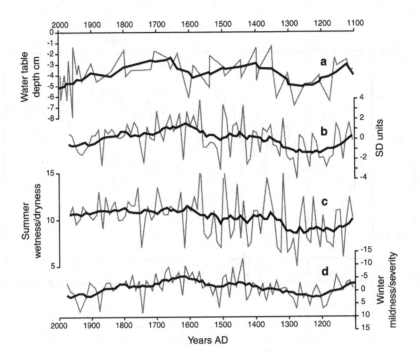

Figure 8.12   (a) Record of changes in mean annual water table depth produced by amalgamating records based on testate amoebae analysis of three separate cores from two sites in northern England. (b) Total of summer wetness/dryness and winter mildness/severity indices (Lamb, 1977) expressed as deviations from the AD 1100–1965 mean. (c) Raw summer wetness/dryness index and (d) raw winter mildness/severity index. The thick lines represent the raw data with 100-year running means as thicker lines. The match between the combined index (b) and the water table reconstruction (a) for the 100-year running means suggests a strong relationship between major trends in water table change and climate. The raw data in (c) and (d) suggest that the relationship is stronger with temperature (as expressed in (d)) than it is with precipitation (as expressed in (c)). The trend in water table change over the last three centuries is also consistent with this hypothesis, as long-term instrumental temperature data show a similar change over this time. Redrawn from Charman and Hendon (2000) with kind permission of Kluwer Academic Publishers.

here is that there are few proxies covering the same geographical areas over the same timespan as many of the peat records (the spatial and temporal coverage of peats is one of the main reasons for pursuing this line of research!). Baker *et al* (1999) and Charman *et al* (2001) report on comparisons between peat humification records and stalagmite luminescence in caves within the same catchment area in north-west Scotland. The records agree quite well, suggesting both are responding to the same climatic forcing. In the case of the stalagmites this is related to the amount and nature of dissolved organic acids in the cave waters, which in turn are linked to the hydrological conditions in the soils and peat overlying the cave system. Showing that the two types of record are linked conceptually and empirically gives much greater confidence in the use of either one in the same or other study areas.

8.6.5 Climate change records from non-ombrotrophic peatlands

The discussion above has been restricted mainly to ombrotrophic peatlands, which by definition should show strong links to climate, since they are wholly dependent on atmospheric precipitation for their moisture. However, there is growing evidence that other peatlands are also strongly influenced by climate changes during their development, and therefore records of these changes can also provide proxy climate data. Basin and valley peats appear to show many of the same major changes in surface wetness as those in ombrotrophic mires from the same regions of Britain (e.g. Anderson, 1998; Hendon *et al*, 2001). Even in the large floodplain fens of eastern England, it seems likely that many of the major shifts in wetness are linked to the same broad-scale climate fluctuations recorded further north (Wheeler, 1992; Wells and Wheeler, 1999). In arctic peatlands, permafrost development driven by climatic cooling through the late Holocene caused major changes in peatland landforms, a phenomenon reported from the North American and Eurasian arctic regions (e.g. Vardy *et al*, 1997; Blyakharchuk and Sulerzhitsky, 1999). At a broad scale, peat accumulation of more diverse peatland types may also be largely a function of changing climatic conditions, as suggested by studies in North America (Ovenden, 1990) and New Zealand (McGlone *et al*, 1997). We have not yet mentioned how pollen data are often used to trace the history of peatland development related to climate change. Although most often used to augment other data, pollen analysis alone can be highly informative. For example, both the development of individual peat deposits in the northern Andes and the general spread of peat moorland in the southern Andes have been traced through pollen analysis and related to climate change (Graf, 1981; Villagrán, 1988).

8.6.6 Future research

Much of the work we have discussed here has been carried out on peatlands in western Europe and parts of North America. However, it seems likely that similar peatland types in other areas of the world will also have a direct relationship with climate, and thus could be used to reconstruct climate changes elsewhere.

Adapting existing techniques and approaches to these new geographical areas will be a major challenge to peat palaeoclimate research over coming years. The other challenge is to derive reconstructions in terms of major climatic parameters such as temperature and precipitation, as this information is much more suitable for comparison with other data and for contributing to regional and global scale past climate reconstructions.

8.7 SUMMARY

The main aim of the last three chapters was to examine the many ways in which peatlands can change over time, taking a long-term perspective of natural changes, primarily through the use of palaeoecological data. We have seen how a variety of changes can occur through an equally wide range of both internal and external causes and we can now see what a complex system a peatland can be. However, because of the unique way in which peatlands preserve the evidence of their growth and change through time, we can look back into the past to understand the nature and processes of these changes. The focus so far has thus been on changes occurring on peatlands. We now turn to the flip side of the relationship – how do peatlands affect the rest of the environment in the immediate area and at the global scale?

9

Peatland–Environment Feedbacks

9.1 INTRODUCTION

Much of this book is concerned with changes within peatland systems brought about by a huge range of natural and anthropogenic influences, as well as the complex of internal dynamic processes of the growing peat body and its thin 'skin' of living biota. However, peatlands should not be seen solely as systems that *respond to* external influences in a unidirectional relationship with climatic, hydrological and land-use changes. They are also important *influences on* this external environment. For example, in Chapter 3 we saw how water enters but also leaves peatlands, and this can have important effects on the hydrological regime and water chemistry of a catchment. Likewise, in Chapter 5 we looked at peat accumulation and decay, a process that over time has removed significant quantities of atmospheric carbon and contributed to the global carbon cycle. The rate at which carbon sequestration in peat changes in the future may have significant effects on the anthropogenically enhanced greenhouse effect and global climate change. This chapter considers some of these aspects of peatland functioning in more detail, concentrating especially on the role of peatlands in the carbon cycle. An understanding of these processes is fundamental to a consideration of the impacts of our actions in modifying and exploiting peatland resources, an area that will be tackled in the final section of this book.

9.2 CATCHMENT HYDROLOGY

The exchange of water between wetlands and the surrounding areas is one of the more obvious processes by which peatlands can influence their external environments. They intercept precipitation and runoff as well as receiving groundwater from below (see Figure 3.1). Inevitably, they have some influence on the amount and timing of water flow elsewhere in the catchment, as well as impacting on aspects of water quality.

9.2.1 Water storage

The potential of peatlands as water storage 'reservoirs' in the landscape is one aspect that at first sight seems likely to be important. Certainly peatlands do store very large volumes of water in the catotelm, as the peat may be up to 95% water. Thus, 1 ha of peatland with an average catotelm depth of 2 m could store approximately 19 000 m^3 water plus any storage in the acrotelm. However, only a very small proportion of the stored water is normally involved in the seasonal exchange between peat and non-peat landscape (Eggelsmann et al, 1993), and the net gain or loss of water may also be limited. In the past, some arguments for the conservation of peatland resources have been made on the basis that they perform a valuable role in moderating fluctuations in runoff and flooding. However, this seems unlikely to be the case because of their generally limited capacity for temporary storage of water which could be released slowly following any storm event (Ingram, 1983; Eggelsmann et al, 1993). There is an important distinction between two main hydrological categories of peatland that should be made clear here (Burt, 1995). First, there are those that act as 'sources' of runoff, which include all of the ombrotrophic mires and many of the fen systems such as sloping fens, patterned fens and valley fens. Water drains from these systems more rapidly due to slope and topography. In contrast, there are some systems (e.g. floodplain fens, basin fens) that act as 'sinks' for runoff by virtue of occupying the lowest locations in the landscape. In the case of the latter types, water storage may not be limited by the factors discussed below.

Most of the potential for temporary storage of water exists in the acrotelm, where variations in storage are indicated by the changing water table. This thin skin can only contain a relatively small amount of water and, once it is saturated, additional water is shed quickly from the surface. Some slow release of water does occur during dry periods, but much of the fall in water table is due to evapotranspiration losses as shown by the diurnal changes in water tables displayed by many mires during such conditions (e.g. Figure 3.11). The rate of runoff from the acrotelm after precipitation will depend upon the prior status of the water table as well as the permeability and depth of this upper layer. One way of assessing the potential for water storage in the acrotelm is to look at the distribution of air, solid and water space in typical profiles. Figure 9.1 shows the differences between German raised mires dominated by dwarf shrub (*Calluna vulgaris*), grassland and forest vegetation. The grassland and forest vegetation types are grossly modified from the natural vegetation and the *Calluna* vegetation is near-natural. Comparisons such as these demonstrate that natural mire vegetation may not be particularly good at temporary storage of large volumes of water, whereas mires that have deeper active surface zones for rooting and aeration have a potentially greater water-holding capacity. It has also been suggested that mineral soils probably have a storage capacity that is greater still, and therefore the arguments for mire preservation on grounds of their water storage potential are misguided (Eggelsmann *et al*, 1993).

9.2.2 Runoff

Despite the limited temporary storage by mires, the amount and temporal distribution of runoff from peatlands are still of interest in management of the surrounding landscape. A better understanding of the processes and controls on runoff is desirable, especially as they also strongly influence water quality changes resulting from the presence of peatlands (see below). Some aspects of this were addressed in Chapter 3. One of the problems in understanding the impact of peatlands on runoff characteristics is to separate the effects of the peatland from that of other mineral soils within the same catchment. Smit *et al* (1999) used a method of area separation to do this so that the difference between mineral and peat elements of the catchment could be assessed (Figure 9.2). The results show that the start of the peatland runoff in the stream following rainfall events was generally delayed by around three to six hours, but was effectively over much sooner than the runoff from mineral areas of the catchment (Figure 9.2a). However, where a long dry spell preceded the rain, runoff from the mire was delayed by around 22 hours compared to the mineral soil. In this case, the peatland had greater temporary storage capacity because of the water table drawdown during the preceding dry spell. Only when water tables had risen again could runoff from the mire take

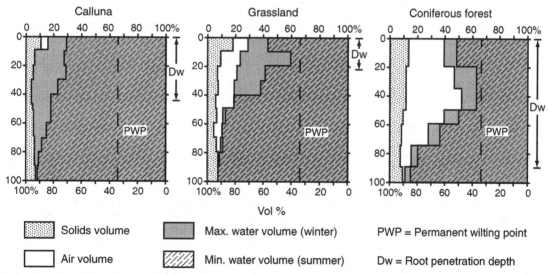

Figure 9.1   The volume of water, air and solids in the upper 1 m of three raised mires dominated by *Calluna vulgaris* (a dwarf shrub), grassland and forest in Germany. The observed temporary water storage is the difference between the maximum winter volume and the minimum summer volume. One estimate of the potential additional water storage possible would be the winter air space, which is currently not filled by water in either the grassland or forested bogs. There is no additional capacity in the *Calluna* bog. Redrawn and modified from Eggelsmann *et al* (1993) with permission. © John Wiley & Sons Limited.

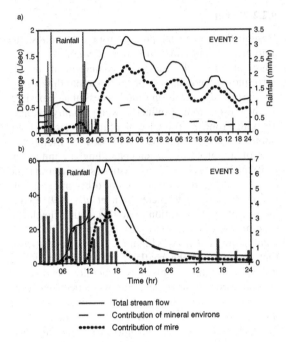

a)

b)

————— Total stream flow

— —   Contribution of mineral environs

••••••• Contribution of mire

Figure 9.2 Time course of rainfall and discharge from an area around Dun Moss, Scotland during two periods in 1994. (a) A period in June 1994, that was preceded by 50 dry days where the water table had fallen to 250 mm below the surface. (b) During October 1994, when the mire was wetter in the immediately preceding period. In both cases the contribution to runoff from the mire is delayed but this is more extreme in (a) than in (b). In (b) the runoff from the mineral ground continues for longer than that from the mire. Redrawn and modified from Smit *et al* (1999) by permission of Elsevier Science.

place. The effect of prior water table status is also seen in other studies linking surface hydrology to runoff (e.g. Figure 3.13; Evans *et al*, 1999).

However, despite the influence of peatland on the timing of runoff, at the catchment scale discharge characteristics are more likely to be mainly a function of the importance of groundwater flow rather than the amount and distribution of peatlands. Burt (1995) shows that the proportion of baseflow is much higher in a groundwater discharge-dominated catchment than in others dominated by runoff. Peatland catchments that sustain a high level of baseflow are usually connected to a larger hydrological system via groundwater so that observed patterns are not solely due to the function of the precipitation–peatland–runoff pathway (e.g. Roulet, 1990). Disturbance in the catchment may also affect the runoff from peatlands, especially through drainage. Despite the potential increase in storage

capacity in drained peats noted above, some studies have shown increased storm flow, perhaps because of a high density of drainage ditches (Nicholson *et al*, 1989; Burt, 1995).

9.3 WATER QUALITY

The effects of peatlands on water quality in the catchment include increases in suspended sediment, acidity and dissolved carbon; changes in nutrient status and water colour; and in some circumstances transfer of other pollutants such as methyl mercury and caesium. Many of these effects are important only in quite specific circumstances while others are much more widespread. In recent years there has been much greater emphasis on the influence of peatlands on water quality because of concerns over excess nutrients in watercourses and the export of carbon from terrestrial systems in relation to the carbon cycle and greenhouse gases.

9.3.1 Suspended sediment

Measurements of suspended sediment in discharge from peatland areas are limited, mostly to those made in areas where peatland erosion is a significant problem, such as the blanket mire regions in northern Britain. Unless there are areas of exposed, unvegetated peat on a mire, it seems unlikely that suspended sediment loads will be significant. While unplanned erosion is the major cause of peat exposure to the elements, intentional disturbance of peatlands from drainage for agriculture, afforestation or peat extraction may also contribute to this. The processes of erosion were discussed in section 7.7, and sediment loads in streams and rivers supplied by eroding peats can be very high. Figures of up to 2800 mg l^{-1} for peak concentration during individual events in the southern Pennines have been recorded (Labadz *et al*, 1991), and the total amount of peat removed per unit area is also surprisingly large in many cases. Often it is difficult to assess what proportion of total suspended sediment is derived from the peat itself rather than mineral soils, and a best guess of this can only be made by looking at the total amount of organic sediment. In excess of 30 t km^{-2} yr^{-1} of organic sediment appears to be a typical figure for eroding upland peatlands in Britain (Francis, 1990; Labadz *et al*, 1991). Sediment supply and transport seem to be highest in the late summer to early winter period (Francis, 1990; Heathwaite *et al*, 1993a). The timing of this suggests that both extensive desiccation during late summer and early frost action during the winter are critical in making sediment available for transport. The specific effects of suspended sediment supplied by peatlands are not known in any detail, but there are known to be severe effects of high overall sediment loads in the affected areas. Engineering problems result from rapid sedimentation in reservoirs for water supply and hydroelectric power, and salmon and trout fisheries are also adversely affected.

9.3.2 Dissolved organic carbon and water colour

In addition to suspended sediment from peat, much of which is organic particulate matter, there are several forms of dissolved organic carbon (DOC) present in peatland drainage waters. This component can be important in relation to the overall peatland carbon cycle (see Chapter 5 and below) and must therefore be considered in peatland carbon budgets and their relationship with the global carbon cycle. However, there are also other direct effects of increased levels of DOC, including changes in water colour and the problems this represents for the water supply industry, and increased acidity in runoff due to the higher levels of organic acids. Increasingly brown water has been a problem in Britain over the last 30 years, probably in response to increased inputs from peatlands following erosion, and exposure and remobilisation of organic material in reservoir deposits during periods of drought (Anon, 1986).

DOC in peatland waters results from the degradative pathways which break down the plant material (Hatcher and Spiker, 1988; see Chapter 5). DOC in peatland waters is primarily composed of the organic acids generally characterised as 'fulvic' acids or 'humic' acids, defined principally by low or high molecular weights respectively. Further breakdown could result in even shorter-chain molecules such as polysaccharides and these are also present in peatland waters. Generally these are assessed as the total DOC concentration and are not separated in most measures of DOC export from peatlands. Occasionally total organic carbon (TOC) is used as an approximation of DOC where particulate organic carbon (POC) is thought to be present in negligible amounts. An additional source of dissolved carbon is that held as dissolved carbon dioxide and methane. However, rather few estimates and measures of export of these dissolved gases have been made, aside from that of Waddington and Roulet (1997). DOC accumulates in peat pore waters as a result of production, consumption and transport. It is flushed out by water movement, and concentrations of DOC are often greatest following periods of dry conditions when DOC has had time to accumulate. However, while concentrations of DOC are higher during low flow periods, the total amounts of DOC exported are likely to be higher during storm flows. Schiff *et al* (1998) find that approximately 50% of the total DOC export occurs during the upper 10% of the flow values and suggest that many sampling programmes thus underestimate the total DOC export as they miss at least some of the high flow events. Studies of radiocarbon ages of DOC exported from peatlands in runoff suggest that almost all DOC is produced in the uppermost layers of peat, although groundwater is more likely to contain older carbon (Figure 9.3).

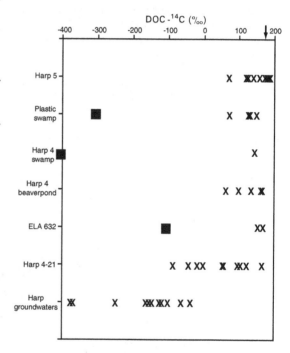

Figure 9.3 ^{14}C activity in outlet streams from six wetlands in Ontario, Canada. The groundwater measurements are from Harp 4–21 wetland. The ^{14}C activity in leaves from growing plants at Harp 4–21 is marked by the arrow. The ^{14}C activities in the streams reflect the decomposition of material fixed from atmospheric CO_2 within the last 45 years. For comparison, the radiocarbon activities of peat samples from 50 cm depth in three of the sites are shown as filled spaces. These have much lower activity than the water samples, showing that the DOC is all derived from the uppermost peat and vegetation. Redrawn from Schiff *et al* (1998) with kind permission from Kluwer Academic Publishers.

The presence of DOC in many peatland waters is obvious from the brown colour of the water, but lower concentrations of DOC are not so easy to observe visually. Measurements of water colour and DOC show that the two are very closely correlated (Kortelainen, 1993). At a catchment scale, the importance of peatlands on stream and lake water DOC is now well known from studies in a number of locations. In Britain, measurements on 11 subcatchments within the larger catchment of the River Dee in Scotland showed that the areal extent of peatlands explained between 59 and 72% of the variance in annual DOC flux estimates (Hope *et al*, 1997). In catchments where lakes are present, concentrations of DOC are likely to be less variable over time and more spatially extensive studies are possible. In Finland, a study covering almost 1000 lakes showed that peatland extent was an important influence on DOC concentrations in lakes in the autumn period of 1987 (Kortelainen, 1993). Within the northern area, where there has been less alteration

of the landscape due to changing land use, this relationship was particularly strong (Figure 9.4).

Other factors can complicate this simple relationship. For example, in Sweden, a general increase in DOC was recorded in a transect from the interior to the coast (Ivarsson and Jansson, 1994). This appears to reflect increasing temperatures and therefore increased biological activity along this altitudinal gradient. However, there is a greater extent of peatlands in inland areas, contrary to the relationship suggested above. Further examination of the data suggests that the character and distribution of the peatland in relation to watercourses are critical. The thinner peats close to the watercourses probably produce as much, if not more, DOC than the deeper more extensive fens that occur inland (Ivarsson and Jansson, 1994). In addition, the nature of other soils in the area and influences of drainage and forestry activities are likely to be important. Exports in the River Kiiminkijoki draining into the Finnish part of the Gulf of Bothnia were also raised

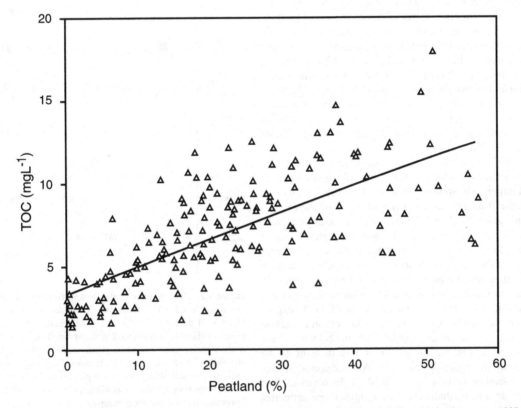

Figure 9.4   The relationship between total organic carbon (TOC) as measured in 183 northern Finnish lakes in autumn 1987 and the extent of peatland in the catchment (measured as % of land area). The proportion of peatland in the catchment explains 49% of the total variance in TOC. Redrawn from Kortelainen (1993) by permission of NRC Research Press.

by increased drainage and peat mining activity in the area (Heikkinen, 1994).

9.3.3 Acidity

As we discussed in Chapter 3, many peatlands are inherently acid for a variety of reasons related to water supply, influences of cation exchange processes in plants and release of acid compounds. In addition, some peatlands have a significant mineral acid load from precipitation, which can further decrease pH levels (Skiba *et al*, 1989). Thus there is an expectation that runoff from many peatlands will tend to acidify watercourses and lakes further down the catchment. Increased acidity can itself be of concern because of the biological effects on aquatic ecosystems. Secondary effects such as mobilisation of metals at low pH may also have an impact on water quality (Gorham *et al*, 1984), particularly the release of aluminium to watercourses (Muscutt *et al*, 1993). The amount of *Sphagnum*-dominated peatland present in lake catchments in north-eastern Alberta, Canada, has been found to be directly related to the acidity of the lakes. The peatland type may also be important here, with poor fens thought to have the greatest effect on lake acidity by virtue of their combination of *Sphagnum* cover and greater throughflow than many bogs (Halsey *et al*, 1997).

The potential for increased acidity of runoff is a function of both organic and inorganic chemical relationships. While runoff from many ombrotrophic and oligotrophic peatlands is expected to have a low pH, there is a growing appreciation that this may vary through time, causing temporary disruption to stream and lake ecosystems downstream. Sulphur dynamics are important in these changes. Sulphate is received at the surface in many peatland areas but it is also produced in the surface layers of peat, especially during periods of water table drawdown when sulphur is oxidised by unusually dry conditions. In a series of catchments in Ontario, Canada, levels of sulphate were up to five times higher in dry summers than in wet ones, with greater effects in catchments with shallow till coverage and more extensive wetlands. During wet summers, little variation between catchments was observed, suggesting that the impact of water table change was most marked in peatland and shallow till soils (Devito *et al*, 1999). Drainage may be able to produce similar effects of increased sulphate load following water table drawdown in freshly drained peat.

In saturated peatlands, the organic acid load represented by DOC can be the dominant factor determining acidity in associated streams and lakes. Kortelainen and Saukkonen (1995) found that forested catchments in Finland contributed between 25 and 85 keq km^{-2} yr^{-1}, an amount greater than that from acid deposition in any area of the country. Particularly important in boreal areas is the release of organic acids at the time of spring snowmelt when very large volumes of water are released in a short period of time. Occasional events such as fire and drought can temporarily enhance the acidity of runoff from peatlands. In the Experimental Lakes Area of north-west Ontario, a forest fire caused an elevated release of both cations and anions from the soils but the balance between the two increased acidity in the stream (Bayley *et al*, 1992b). Although these observations were not in a peat-dominated catchment, similar effects could occur with burning of deeper peat soils.

9.3.4 Nutrients and other solutes

Peatlands can either contribute additional nutrients to a catchment or they can act as a net sink. To a large extent this depends on the extent of damage and the degree to which the peat is actively accumulating. One of the well-known applications of bioengineering in wetlands is the highly successful and widespread use of purpose-built marshes (often reed beds – *Phragmites australis*) to absorb excess nutrients in water treatment. While this does occur in many wetlands, the documentation of such effects in natural peatlands is limited. One example of the potential for uptake of excess nutrients was carried out on peat soils in the Florida Everglades (Jones and Amador, 1992), where it was shown that the uptake of total phosphorus could exceed that previously estimated from consideration of phosphate alone due to a variety of different abiotic mechanisms. In oligotrophic bogs, although the absolute quantities are much less, there is also probably a net gain of nitrogen and phosphorus over time, as shown by various studies in different locations in the Northern Hemisphere (see review by Verry and Timmons, 1982).

Although it is especially common for fen systems to act as nutrient filters, many peatlands contribute additional nutrients to adjacent watercourses. This is often the case in disturbed peatland, especially where drainage has occurred (including drainage for forestry), but also with occasional events such as fire or major erosion. Nutrient release resulting from drainage of fen peats in England was examined by Heathwaite (1990, 1991) using laboratory experiments and field monitoring. Laboratory incubations of different peats under various waterlogging regimes showed that the prior status of the peat (drained or undrained) as well as the degree of waterlogging had a major impact on nutrient release. Previously undrained peats released the largest amounts of nitrogen, sulphur, calcium and magnesium, although only the last three were statistically significant. In the field monitoring, the effect of pump-assisted drainage was assessed and found to promote nutrient release to the drainage ditches and out from the peat. A cycle of nutrient release followed water table drawdown from pumping (Figure 9.5), which produced particularly high levels of ammonia N and hydrogen sulphide S in spring, sometimes linked to fish kills (Heathwaite, 1991). In this case, the release of nutrients from drainage was exacerbated by the fluctuating water tables induced from pumping of the drainage ditches.

Fire is a potentially potent agent of nutrient release from mineral and organic soils. Measurements of nitro-gen and phosphorus levels in streams draining burned catchments show elevated levels of various forms of these elements immediately after and for up to two years following a fire (Figure 9.6). For wetland-dominated catchments (such as the NE basin in Figure 9.6), there seems to be a particularly strong increase in phosphorus following fire, and it is certainly greater than in catchments where wetland is less important. Total nitrogen and total dissolved nitrogen also increased following fires, although increases in nitrate in particular were greatest in catchments dominated by mineral rather than peatland soils (Bayley *et al*, 1992a). Variations in response between basins and between fires also depend on the intensity of the fire and the interval between fires, as well as soil and vegetation characteristics of the burned area. The second fire monitored by Bayley *et al* (1992a) showed lower responses than the first, perhaps because it followed only six years after the first and the systems had not had adequate time to accumulate nutrient loads equivalent to those preceding the first fire. It is also apparent from this study that nutrient levels in the streams vary with other

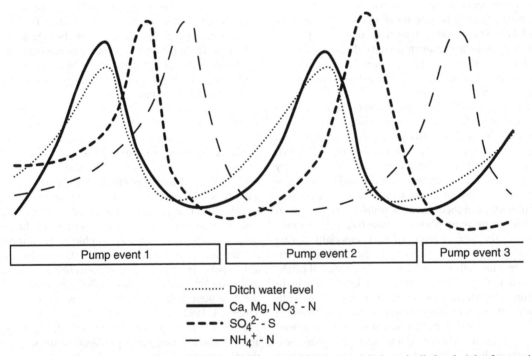

Figure 9.5 Schematic pattern of nutrient and solute concentrations during pumped drainage in ditches draining fen peat in the Somerset Levels, England. Pumping of the drains results in a fluctuating water table. Concentrations of Ca, Mg and NO_3^- are approximately in phase with the water table changes, while SO_4^{2-} concentrations are greatest immediately following water table drawdown and NH_4^+ shows increased concentration after this. Redrawn from Heathwaite (1991) with kind permission from Kluwer Academic Publishers.

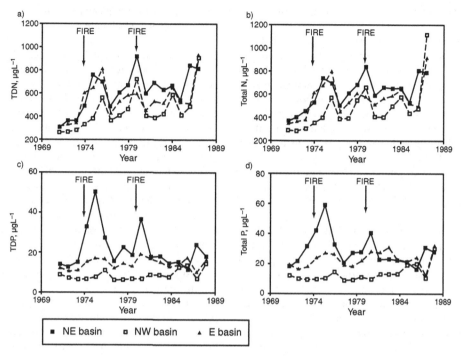

Figure 9.6 Changes in streamwater nitrogen and phosphorus before and after two fires in 1974 and 1980 in the Experimental Lakes Area, Ontario, Canada. Three basins were monitored, each with different extent of peatland. The NW basin only has a small area of peatland, the E basin is mostly upland with a larger area of peatland and the NE basin has the largest extent of peatland of the three. The 1974 fire affected only the NE and E basins and the 1980 fire affected all three basins. Changes in (a) total dissolved nitrogen, (b) total nitrogen, (c) total dissolved phosphorus, (d) total phosphorus. The greatest increases were recorded for phosphorus levels in the basin with the greatest extent of peatland (NE). Redrawn from Bayley *et al* (1992a), reproduced with the permission of the Fisheries and Oceans Canada, Minister of Public Works and Government Services, Canada, 2002.

factors, especially drought, and that this variability may be difficult to separate from changes caused by fire. For example, there is a general upward drift of values in some parameters, probably related to greater droughtiness in the period 1983–1988.

9.3.5 Other pollutants

Other geochemical effects of peatlands on catchment water quality also occur, but many of these depend upon specific combinations of geology, vegetation, climate and peatland type. One focus of recent attention in North American peatlands has been the dynamics of methyl mercury (MeHg), where levels of this chemical in lakes have been of concern. Branfireun *et al* (1996) showed that peatlands are a significant source of MeHg, and modelling suggests that recent rises in MeHg contamination in lakes is not necessarily associated with increased atmospheric deposition. Instead,

the release of stored mercury in catchment soils and sediments may have been promoted by deposition of some other industrially derived pollutant such as sulphate.

The transport and deposition of other pollutants of anthropogenic origin are also affected by peatlands. One example is that of radiocaesium released following the Chernobyl nuclear power station accident. There has been much attention given to the fate of caesium after deposition in many areas of western Europe. Particularly high levels of contamination have been found in western Britain, where peatland occurrence is also frequent. Comparisons of drainage waters from peat-dominated and mineral soil areas suggest that peatlands provide a much greater source of caesium than dry soils. Caesium is apparently easily replaced by calcium on simple ion exchange sites in fibrous peats so that it could be relatively easily released to solution and flushed out into the catchment lake (Hilton and Spezzano, 1994).

9.4 PEATLANDS AND GLOBAL CLIMATE

We briefly reviewed the relevance of peatlands to the global carbon cycle in section 5.2 and it is immediately apparent that, because peat represents about a third to a half of the global soil carbon pool, or almost the same amount as the global atmospheric carbon pool, peatlands must play a crucial role in the global carbon cycle. There is now reasonable consensus that concentrations of atmospheric greenhouse gases (CO_2 in particular) are intimately linked with the global climate system and especially with temperature change through time (Houghton *et al*, 1996). Therefore the size and functioning of any large carbon pool are of importance to past and future climate change.

In the long term (10 000–20 000 years), carbon accumulation in peat has removed much of the CO_2 that would otherwise be held within the atmosphere. If this carbon had not been absorbed by peatlands, atmospheric CO_2 would inevitably be higher, whatever ecosystem replaced them in this hypothetical course of global environmental evolution. Over shorter time-scales, the exchange of carbon between peatlands and the environment is of particular interest. Since the carbon pool is large we need to be able to understand and predict its fate over the next decades and centuries. If degradation of the peat occurs, carbon will be released to watercourses and the atmosphere and provide a significant additional contribution to atmospheric CO_2. Conversely, if peat growth continues or increases, peatlands may provide a large sink for atmospheric CO_2. While the basic balance between intake and export of carbon is of fundamental importance, peatlands exchange other greenhouse gases besides CO_2. Methane (CH_4) and nitrous oxide (N_2O) are also important greenhouse gases that are produced by peatlands. Other gases, such as non-methane hydrocarbons (NMHCs), dimethyl sulphide (DMS) and even the man-made chlorofluorocarbons (CFCs), also interact with peatlands and all play a role in the radiative balance of the atmosphere. Consideration of these other gases complicates the picture because their influence on radiative forcing is not the same as that of CO_2. Perhaps most important is the role of methane, as it is approximately 20 times more effective as a greenhouse gas than CO_2, yet its atmospheric lifetime is much shorter. Thus its effect on climate is rather different. Release of methane produces a short but intensive greenhouse forcing; release of CO_2 produces a weaker but more sustained effect. Thus the balance between different gases released (especially CO_2 and methane) is equally important to the overall balance of elemental carbon.

9.4.1 Glacial to interglacial variations in the global carbon budget and climate

Over very long (>100 000 years) periods, the global climate has fluctuated between glacial and interglacial periods of cold and warm climates respectively. Associated with the climate change there has been a change in atmospheric CO_2 and methane concentrations, with much higher concentrations of both gases during warm interglacial periods. The transition from glacial to interglacial periods and vice versa must have involved major reorganisations of the global carbon cycle, probably affecting the size and distribution of both ocean and terrestrial pools of carbon. The role of peatland growth and decline is not known with any certainty but several possibilities have been proposed, some placing peatland dynamics at the very centre of the climate change process. During the transition from the last glacial to the present interglacial and especially during the early Holocene, there would have been a shift in the areas favourable for peat growth with new peat beginning to form in areas previously too cold or occupied by ice sheets. Conversely, other regions would be becoming less suitable for peat growth or in coastal settings, being submerged by the rise in sea level that took place at the same time (Faure *et al*, 1996), although other evidence suggests that the amount of peatland carbon at this time may have been negligible (Adams *et al*, 1990). The net impact of peatland carbon balance during these kinds of transition has been debated but it seems likely that total soil carbon biomass (including peatlands) has increased by approximately 770 GtC between glacial and Holocene conditions (Faure *et al*, 1996). One estimate of the change in terrestrial carbon storage is shown in Figure 9.7, although the estimate for the total peatland pool is rather low by comparison with later work (Gorham, 1991; Immirzi *et al*, 1992).

The growth of peatlands can be seen as a kind of negative feedback to climate change: warmer global climates increase the spread of peat and the growth of peat reduces the quantity of CO_2 in the atmosphere, thus limiting further increases in temperature. It is not a true feedback as it is the increase in temperature rather than in CO_2 that causes the initial global increase in peat growth, but the effect on climate is the same. However, others have suggested that peatlands play a primary role in actually forcing climate change, particularly in the cooling down phase during the transition from interglacial to glacial conditions (Franzen, 1994; Franzen *et al*, 1996; Klinger *et al*, 1996). While it is generally thought that small changes in solar insolation

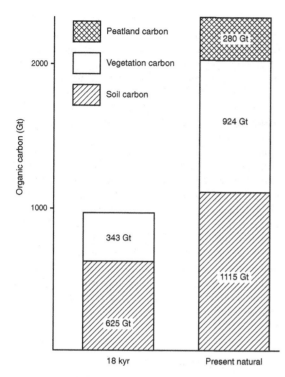

Figure 9.7 Comparison of estimated carbon pools in soils, vegetation and peatlands during the last glacial maximum (18 000 radiocarbon years ago) and the present (natural) state. Redrawn from Adams *et al* (1990) by permission of *Nature*. Note that the estimate of the total peatland carbon pool is probably underestimated here compared to later revised estimates.

resulting from variations in the earth's orbit are responsible for initiating transitions from interglacial to glacial conditions, the absolute changes in solar radiation need additional mechanisms to produce the observed changes in climate. The oceans are almost certain to hold one of the keys to this but peat growth may also play an important role. Franzen (1994) suggested that cooling during interglacial periods is caused by the gradual sequestration of carbon in peatlands. As peatlands expand and deepen, atmospheric CO_2 levels decline, causing a decreased greenhouse effect and global cooling to occur. During glacial periods, new locations suitable for peat development are produced by the action of ice sheets, effectively setting up the system to start again once the ice sheets have receded. However, the change back from glacial to interglacial conditions would have to involve more complex factors, including orbital forcing. The gradual breakdown of peat from beneath the ice sheets could still be one

factor that would enhance global CO_2 levels and provide the raw material for plant photosynthesis as well as enhancing greenhouse warming (Franzen *et al*, 1996).

Not all peatland and climate scientists would accept this theory and have argued that the hypothesis is inconsistent with observed CO_2 concentrations over the past few thousand years, as well as that CO_2 changes may actually lag temperature changes rather than precede them. They also suggest that it is not reasonable to ignore other terrestrial carbon stores and the changes in ocean carbon cycling that are also likely to have taken place. Finally they say that the necessary area and depth of peat growth are an overestimate of what would be possible, even if northern peatlands were in a completely natural state (Rodhe and Malmer, 1997). While most would accept that peatland carbon accumulation has been and still is an important part of the global carbon cycle, there are still some serious objections to the idea that peatlands play the central forcing role to glacial–interglacial cycles. However, it is possible that shorter-term changes in climate are partly related to peatland dynamics. An approximately 1400-year periodicity in CO_2 fluctuations over the Holocene is reflected in the same length of cycle in peatland initiation dates in western Canada. Campbell *et al* (2000) have suggested that peat growth phases moderated the trends shown in the global CO_2 record and thus have a detectable effect on the evolution of global climate during the Holocene.

9.4.2 Long-term variations in carbon accumulation and gas production

Within the current interglacial period, carbon has been sequestered in peat deposits at varying rates. In Chapter 5, some characteristics and processes of this long-term accumulation were examined in relation to the total mass of peat. Since peat is approximately 50% carbon (dry weight), many of these principles apply to carbon accumulation as well as total peat accumulation. However, in terms of the global carbon cycle, the accumulation of carbon at individual sites is of less significance than the totals across much larger geographical areas. In addition, there has now been some research attempting to reconstruct patterns of carbon gas emissions from peatlands over time.

In acknowledgement of the problem of accounting for continuous long-term peat decay (see Chapter 5; Clymo, 1984a), discussion of carbon accumulation in peat refers to 'apparent' rate of carbon accumulation (LARCA or LORCA) and the actual rate of carbon accumulation (ARCA). ARCA takes account of the long-term decay of peat whereas LARCA does not. Large data sets have been assembled for many of the northern peatland areas. Much of this work originates in Finland, where early data suggested younger mires have a much faster rate of LARCA than older mires, but that there is considerable variability dependent upon peatland type, geographical location (related to climatic conditions) and fire history (Korhola et al, 1995). Still larger data sets from wider regions show that long-term carbon accumulation is driven by prevailing climatic conditions, with the rate of addition of carbon determined by degree-days above zero, and decay rate is related to mean annual temperature (Clymo et al, 1998). These data allow an estimation of the average carbon sequestering capacity of peatlands to be made. In this case, the current rate for all northern peatlands is estimated as 0.07 GtC yr^{-1}. Other regional studies on long-term accumulation find similar rates on a unit area basis (e.g. Vitt et al, 2000). These approaches are valuable because they are an alternative method to more detailed considerations of specific peatland carbon budgets measured over very short ($<$ decade would be a long study) periods. Korhola et al (1996) extend the application of palaeoenvironmental data to estimating past changes in carbon storage and methane emissions for a peatland in south-west Finland. Here detailed radiocarbon and bulk density data were used to reconstruct the three-dimensional development of the mire and its main habitats. Using existing data on methane emissions from similar habitats allowed the reconstruction of changing methane emissions through time.

9.5 CARBON BUDGETS AND GAS EXCHANGE

The main processes of carbon import, export and transformation are summarised in Figure 5.2. Carbon enters the system mainly as a gas (CO_2) by photosynthesis and is fixed in plant tissues until decay of dead plant material occurs. On death and decay of biomass, the carbon is transformed into various components, including gaseous and dissolved methane and CO_2. Other compounds form mobile organic acids and move with water flow within and from the peatland. Some carbon is retained in the peat and smaller fragments may be exported as particulate organic carbon in runoff (for example during erosive events). Behind these relatively simple flows and changes in the carbon pool, there are complex processes and interactions that until relatively

recently were very poorly understood. While there are still many questions to be answered, there has been a series of major research efforts to understand processes of carbon cycling in peatlands during the 1990s that have improved our knowledge immeasurably. One of the largest of these efforts was the Finnish Research Programme on Climate Change (SILMU), and a major focus was on peat and peatlands (Kanninen and Anttila, 1992; Laiho et al, 1995). In North America, the Northern Wetlands Study (NOWES) and the Boreal Ecosystem Atmosphere Study were two of the biggest projects to concern themselves with gaseous emissions from peatlands (Glooschenko et al, 1994; Sellers et al, 1994). Beyond these large integrated projects there have been a multitude of smaller ones and the literature on carbon cycling in peatlands is now extremely large, especially concerning the amounts and controls on gas emission from peatlands. Here we will look at a series of the most important components of the carbon cycle in peatlands relevant to climate change feedbacks.

9.5.1 Gas production

Gas production in peat is the end result of a series of decay processes beginning with the recently dead plant material. We reviewed some of these processes in section 5.4 in relation to decay and peat accumulation. Here we will look at the pathways of production and transformation of gases in a little more detail, with an emphasis on understanding how gases are produced, the controlling factors on the rate of production and their subsequent movement through and from the peat profile.

The production of gases in the peat occurs in four distinct zones, defined by the position and fluctuations in the water table. These zones are approximately coincident with the functional zones identified in Figure 5.13. In the euphotic zone, the main process is the respiration of growing plants and the production of CO_2. Below this, in the aerobic zone above the water table, CO_2 production is most intense, with high availability of food sources for the decay organisms. In the zone around the water table, conditions fluctuate between aerobic and anaerobic so that microorganisms have to be capable of tolerating these two extremes. Aerobic decay still takes place here but anaerobic activity is also possible at times and this results in different pathways to production of CO_2 as well as methanogenesis. Finally, in the lowest zone, where conditions are permanently anaerobic, methane and CO_2 are produced from the continual but slow decay of the remaining organic matter.

9.5.2 Methane

In terms of the global methane budget, peatlands and other natural wetlands are an important source, roughly equivalent to the total anthropogenic emissions from livestock and rice paddies (Table 9.1). Figure 9.8 shows the main processes involved in methane production, transformation and transport in peatlands. Methane is produced by methanogenic bacteria and is restricted to the waterlogged zone. There is a variety of specific types of bacteria involved, some producing methane via different pathways than others (Games and Hayes, 1976). The dominant pathway is probably by fermentation of acetates which could be derived from *in situ* peat but may also have been transported as part of the DOC in pore waters or from root exudation (Charman *et al*, 1994). Other

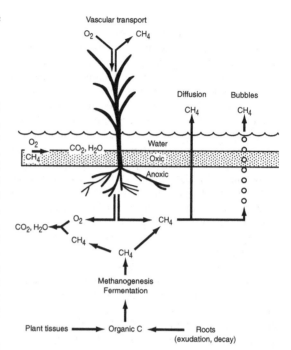

Figure 9.8 Main aspects of the methane cycle in peatlands, showing production, transformation and transport pathways. Redrawn from Conrad (1989) and Bubier and Moore (1994) with permission. © John Wiley & Sons Limited.

Table 9.1 Annual methane emissions from major natural and anthropogenic sources as estimated by Tyler (1991). The sources are divided into those that contribute mainly modern carbon (radiocarbon live) and those that are derived from fossil carbon (radiocarbon dead). In the case of peatlands, the carbon is a mix of very recent and moderately old carbon, although the mean radiocarbon activity suggests more modern carbon than fossil (see text for details).

Source of CH_4	Range of emissions (Tg CH_4 yr^{-1})
Radiocarbon live	
Natural wetlands	120–200
Rice paddies	70–170
Livestock	80–100
Termites	25–150
Solid wastes	5–70
Biomass burning	10–40
Oceans	1–20
Tundra	1–5
Subtotal	312–755
Radiocarbon dead	
Coal mining	10–35
Venting and flaring	15–30
Industrial losses	5–25
Pipeline losses	10–20
Methane hydrates	2–4
Volcanoes	0.5
Automobiles	0.5
Subtotal	43–115
Total	355–870

methanogens use hydrogen and CO_2 as substrates for methane production, both of which are likely to be available in the anaerobic zone. The rate of production of methane in the waterlogged zone is dependent on temperature, peat composition and nutrient and pH conditions. The effect of temperature has been shown most effectively by laboratory incubations of peat samples where very large increases in production of methane result from raising peat temperatures (e.g. Moore *et al*, 1994). Temperature effects can also be observed in the field (e.g. Dise *et al*, 1993), but these are often complicated by the effects of changing water tables, which have a strong influence on temperature of the peat and also exert an independent effect on methane production through changes in waterlogging. Peat composition determines the suitability of the substrate for methanogenesis (Nilsson and Bohlin, 1993), but it also affects the structure of the peat and therefore the movement and storage of dissolved and gaseous methane (Buttler *et al*, 1994). Acidity may limit methane production in some sites; Valentine *et al* (1994) found that a combination of low pH and substrate quality was likely to have been the underlying cause of low methane emissions in some bogs in the NOWES study area. It has also been suggested that the supply of carbonate through groundwater can be an important stimulant to production of both methane and CO_2 (Lamers *et al*, 1999). Although the detailed factors and interactions with methanogenesis are complicated, at the ecosystem scale methane production is strongly related to net primary productivity (NPP). This may be explained by the increased inputs from roots, recent litter and biomass turnover as well as increased potential for transport through plant tissues in systems with higher NPP (Whiting and Chanton, 1993). Certainly cutting of surface vegetation has been shown to reduce methane production, supporting the idea that plants play a crucial role in generating methane in the surface layers of peatlands (Waddington *et al*, 1996).

Isotopic signatures of carbon gases help elucidate the sources and pathways of production. In particular, radiocarbon provides information on the age (and therefore the location) of the substrates being used by methanogenic bacteria, and ^{13}C data suggest the main metabolic pathways of methane and CO_2 production. Methane emissions from peatlands are generally depleted in ^{14}C, which shows that some of the gas is being produced from older carbon sources, as expected (Wahlen *et al*, 1989). However, the activity is much higher than it would be if the dominant source of methane were peat older than the last few hundred years. Measurements of ^{14}C activity in methane from peatlands in Minnesota suggest that almost all of the methane production occurs within the upper waterlogged layers of peat, derived from recently fixed organic carbon, probably from root exudates and recently produced DOC from plant litter (Chanton *et al*, 1995). These observations tie in very well with the observations on plants and methane production (above). Even in gases recovered from 2–3 m depth in the peat, the radiocarbon ages are much younger than those in the adjacent peat, so that methane must either be transported downwards after it has been produced, or there must be a younger source of carbon for the methanogenic bacteria to use.

Once produced, methane can be transported by diffusion through the peat or in some circumstances as gas bubbles or through plants (Bubier and Moore, 1994). The net flux from the peatland can be seriously reduced as the methane has to pass through parts of the peat where it can be oxidised to CO_2. The principal zone in which this occurs is the aerobic zone immediately above the water table. Unless water tables are above or at the surface, much of the methane is consumed here. Some

can also be consumed below the water table, especially in the rooting zone of plants where oxygen is available for use by methanotrophic bacteria. Transport through plants is one way in which methane may move directly from the anaerobic zone to the atmosphere. This is variable between plant taxa, although it can be very important in some communities.

The strong effect of water tables on methane emissions from peatlands, due to the increased opportunity for oxidation where water tables are low, is well known, and has been shown for many different peatland types in many different locations. One example is shown in Figure 9.9, where the mean position of the water table explains about 65% of the variation in methane flux across three different microhabitat types. The relationship is similar for the wetter habitats (lawn and hollow areas), but less strong for the hummocks, where methane emissions were generally lower, despite warmer temperatures within the hummocks. This suggests that methane production rates were lower within hummocks or perhaps that consumption rates were higher, even where seasonal water tables were similar to lawn habitats (Bubier *et al*, 1993).

9.5.3 Carbon dioxide

Carbon dioxide production occurs in a greater variety of locations than methane in peatlands. Aerobic decay of organic matter in the surface layers is a major source, but it is augmented by anaerobic decay in waterlogged peats and by oxidation of methane transported from the waterlogged peats (Figure 9.8). It is also produced by plants through respiration at the surface and through respiration of roots within the peat. Many of the controls on CO_2 production are similar to those for methane; temperature and substrate have similar effects, although the influence of the water table acts in the opposite direction as higher water tables depress CO_2 production. The importance of some processes is only recently being appreciated. For example, some of the CO_2 produced from soil or plant respiration may be refixed by *Sphagnum* and other plants before it can escape to the atmosphere (Turetsky and Wieder, 1999).

Of fundamental importance are rates of photosynthesis and respiration, which combine to give a value for net ecosystem exchange (NEE) of CO_2. Clearly if the NEE is positive then the system is acting as a sink for CO_2 but if it is negative then the peatland is a source. Some of the earlier attempts to measure the net exchange of CO_2 used microcosms of surface peat and vegetation transplanted to the laboratory to assess

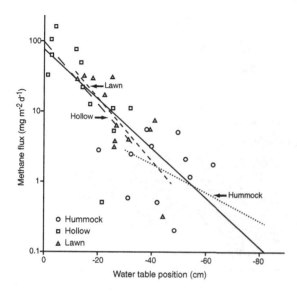

Figure 9.9 Relationship between the mean seasonal CH$_4$ flux and the mean seasonal water table position for different microtopographic locations at 12 peatland sites in northern Ontario, Canada. Note the logarithmic scale on the methane flux axis. The solid line is the regression line for the entire data set, with separate regression lines shown for the microtopographic subsets. The strong control of water table on methane flux is apparent, especially in the lawns and hollows, where methane fluxes are highest. Redrawn from Bubier *et al* (1993) by permission of NRC Research Press.

the response of different plants and of the peat itself to changing temperature and irradiance (Silvola and Heikkinen, 1978). This was followed by field data collection and simulation to estimate the total CO$_2$ balance for a *Sphagnum fuscum–Empetrum nigrum* community over the course of a year (Silvola and Hanski, 1979). Figure 9.10 shows some of the results of this work as the net gas exchange of the two main plant species and the peat, summed to provide a total CO$_2$ balance for the growing season. In this case, long daylength and moderate temperatures allow high growth rates in the early summer, but as the season progresses, respiration increases and photosynthesis falls so that the carbon balance is negative for some periods from late July onwards.

More recently, field monitoring of CO$_2$ fluxes has been carried out using chambers or at a larger scale using towers to measure CO$_2$ from a wider area. Some of these field data suggest that the balance between positive and negative NEE for CO$_2$ is rather finely poised, with variations between positive and negative both within years and between years

(Shurpali *et al*, 1995). Figure 9.11 shows data from a peatland in Minnesota during 1991 and 1992. In 1991, high temperatures and a relatively low water table kept photosynthesis low and emissions of CO$_2$ relatively high, making the site a net source of CO$_2$. The following year was cooler and moisture levels were non-limiting for plant growth so that photosynthesis was higher and soil respiration was lower, making the system a net sink for CO$_2$. Clearly, if the CO$_2$ balance is close to the threshold for continued net accumulation, it is important to understand the factors that could cause changes in the system and to quantify them adequately.

In order to maintain optimal photosynthetic rates, daylength and irradiance are clearly important (see above), while temperature becomes the main factor determining respiration rates, especially of the surface peat. However, other factors can exert a strong influence as well. One factor that is surprisingly important in such a saturated ecosystem is moisture, particularly for bryophytes including *Sphagnum*, that have little control on water losses. Silvola and Aaltonen (1984) showed that photosynthesis declines quickly once moisture drops below optimal levels in *Sphagnum*, although excess moisture had a much less marked effect. Different taxa have different optimal hydrological conditions for photosynthesis, one of the factors that helps explain the differing competitive abilities of *Sphagnum* species (Wallen *et al*, 1988).

9.5.4 Other gases

While exchange of carbon, and especially the production of carbon gases, is paramount in the feedbacks between peatlands and the climate system, a number of other gases are produced and consumed by peatlands. In contrast to methane and CO$_2$ these gases have received rather little attention. Nitrous oxide (N$_2$O) is an important greenhouse gas and is both produced and consumed by soils, including peatlands. Fluxes of nitrous oxide from undamaged peatlands may not be all that great (Martikainen *et al*, 1993), although episodic emissions which are hard to measure accurately may increase the estimates of this flux considerably (Schiller and Hastie, 1994). More importantly, drying out of peat either by drainage or through climate change may bring about increases in fluxes. However, these are only likely to be large in more nutrient-rich peats and even these may not be significant in terms of global nitrous oxide exchange (Martikainen *et al*, 1993). Nitric oxide (NO) can also be produced from peat soils, but again only where there

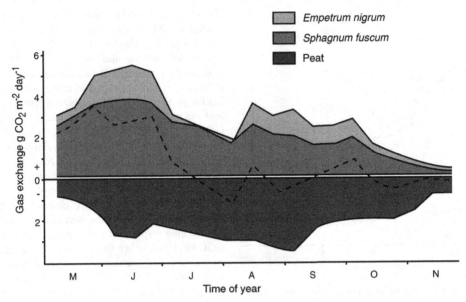

Figure 9.10   Net gas exchange for a *Sphagnum fuscum–Empetrum nigrum* community derived from laboratory experiments, field data and simulation modelling. Each of the curves shows the net exchange from the plants or the peat, and the total exchange of the system is indicated by the broken line. Each data point represents average values for 10 days. *Sphagnum fuscum* has a strong positive balance throughout the season but *Empetrum nigrum* shows relatively slow growth and in early August, it is estimated that respiration exceeded photosynthesis. Redrawn from Silvola and Hanski (1979) by permission of Springer-Verlag.

has been drainage or afforestation (usually with associated drainage activity), and especially where fertilisation of the peat has taken place (Lang *et al*, 1995). Sulphur gases are another group that is still relatively understudied. It is suggested that peatlands produce dimethyl sulphide (DMS) at a rate of between 4 and 400 nmol m^{-2} h^{-1}, whereas uptake of carbonyl sulphide (OCS) occurs making peatlands a net sink for the latter gas (Demello and Hines, 1994). Even artificial compounds such as the chlorofluorocarbons (CFCs) interact with peatlands and it has been suggested that peat will act as a sink to assist removal of these gases from the atmosphere (Bauer and Yavitt, 1996).

9.6 IMPACTS OF MANAGEMENT AND CLIMATE CHANGE ON CARBON CYCLING

Gas production and carbon cycling are inevitably affected by various external factors. Particularly important are any changes to the hydrological and temperature regime. The impacts of these changes will affect the

ways in which peatlands limit or reinforce the anthropogenic greenhouse effect.

9.6.1 Drainage and flooding

Hydrological change is critical to gas exchange. Drainage of peatlands has been taking place for centuries and new areas are still drained for forestry, peat extraction and agriculture (see Chapter 10). With water table drawdown from drainage, a series of effects is usually observed. First, increased respiration in the peat and greater release of CO_2 can be expected, with a concomitant decrease in methane release due to the more limited anaerobic conditions present. The increased depth of the aerobic zone also means that any methane produced at depth is oxidised before it can be emitted. Second, an increase in release of nitrous oxide as a result of mineralisation of nitrogen. Third, there may be some increase in standing biomass, depending on the change in vegetation and whether new species (e.g. trees in afforestation schemes) are introduced. Comparisons of undamaged and altered sites are especially important to determine the magnitude of these effects and their net effect on radiative

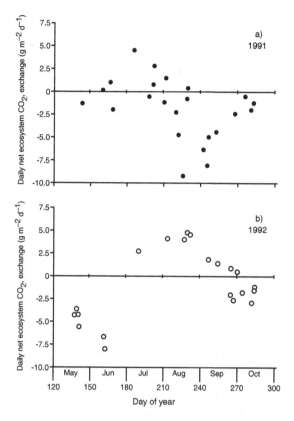

Figure 9.11 Net ecosystem CO_2 exchange for a peatland in Minnesota during the May to October periods in 1991 and 1992. There is considerable variability within the year and especially between years in the high summer season. See text for details. Redrawn from Shurpali *et al* (1995) by permission of the American Geophysical Union.

net effect on climate was calculated in terms of global warming potential (GWP, expressed as equivalent mass of CO_2) as representing an increase of 2.6×10^4 kg CO_2 ha^{-1} yr^{-1}. Clearly changes such as this over larger areas would have a significant effect on the greenhouse gas budget of peatlands.

Just as drainage alters the gas exchange balance of natural peatlands, management to restore damaged sites can also have significant effects on CO_2 and methane emissions. For example, restoration of Dutch peatlands, previously drained for pasture and growth of hay crops, usually involves rewetting by ditch blocking. Although this reduces CO_2 emissions by around 14%, it has the effect of increasing methane emissions by at least three times (Best and Jacobs, 1997), at least in the short term. Therefore, although it might seem sensible to restore peatlands to reduce emissions of greenhouse gases, it may not always be the best option if this is the sole purpose of the work. Of course there are usually many other good environmental arguments for peatland conservation – these may be a better justification than that of reducing global warming. In the long term, the net change may be rather different since a drained peatland converted to grassland will eventually emit almost all of the stored carbon to the atmosphere as CO_2, following slow wastage of the peat mass. Other increases in surface wetness are also known to increase methane emissions considerably, especially where open water areas are created over terrestrial vegetation such as in beaver ponds (Yavitt *et al*, 1990, 1992). Flooding of peatland forest due to reservoir construction may also have a similar effect (McKenzie *et al*, 1998).

9.6.2 Future climate change – 'global warming'

Because hydrology is of such importance in the carbon cycle and especially in gas production in peatlands, there has been much speculation on the reaction of peatlands to future global warming. It is normally assumed that, at least for the very large areas of boreal peatlands, increased temperature will lead to decreased water tables as well as increased peat temperatures. In some ways this would result in changes very similar to those known to occur during peatland drainage. However, there is considerable debate about what the net effects of future climate change might be and whether peatlands would be a negative or positive feedback to the anthropogenic greenhouse effect. Qualitative responses to climate change are listed in Table 9.2, together with an assessment of the estimated

forcing in the climate system. In peatlands drained for forestry in Finland, CO_2 exchanges changed from a sink of around 25 g m^{-2} yr^{-1} to being a source of approximately 250 g m^{-2} yr^{-1}, as a result of increased peat respiration and changes in surface vegetation, including the trees (Silvola, 1986).

Nykanen *et al* (1995) compared emissions of CO_2, methane and nitrous oxide on a virgin fen with those from a similar site converted to grassland in Finland. Methane emissions were almost zero on the grassland site with around 200–300 kg ha^{-1} yr^{-1} emitted from the virgin site. In contrast, nitrous oxide emissions were much higher on the drained site (8–9 kg ha^{-1} yr^{-1}) and CO_2 also increased by three to four times. Added to this, methane emissions from livestock on the grassland site further compensated for part of the decrease in methane resulting from conversion to grassland. The

Table 9.2 Predicted responses of northern peatlands to future climatic change, with an assessment of the confidence in each of these (Moore *et al*, 1998).

Hydrologic gradient Peatland type	Dry Hummock/plateau-palsa	Moist Lawn/swamp	Wet Floating mat/pool
Environmental response to climatic and atmospheric forcing			
Available CO_2	Doubled[+++]	Doubled[+++]	Doubled[+++]
Peat temperature	Warmer[+++]	Warmer[+++]	Warmer[+++]
Water table position	Lower (hummock)[++] Higher (plateau-palsa)	Lower[++]	No change (mat)[++] Lower (pool)[++]
Ecosystem response			
CO_2 exchange	Smaller sink or possible source (hummock)[+] Larger sink (plateau-palsa)[+]	Smaller sink[+]	Larger sink[+]
CH_4 emission	Smaller (hummock)[+++] Larger (plateau-palsa)[++]	Smaller [+++]	Larger[++]
DOC export	Smaller (hummock)[+] Larger (plateau-palsa)[+]	Smaller [++]	Smaller [++]
C storage	Possibly net loss (hummock)[+] Larger (plateau-palsa)[+]	Smaller [+]	Larger [+]

[+++] Reasonably confident, [++] moderately confident, [+] least confident.

confidence in these changes (Moore *et al*, 1998). Responses (and confidence in these) vary considerably between peatland types at various spatial scales. Some aspects of change are becoming clearer, yet there is still considerable progress to be made in understanding individual processes and the way in which these interact to yield net responses in gas and carbon exchanges.

Besides water table drawdown, the impacts of increased atmospheric CO_2 are not easy to predict. Atmospheric concentrations of CO_2 are predicted to be around double the current values by the end of the twenty-first century (Houghton *et al*, 1996). It has been shown that *Sphagnum* mosses are capable of increased productivity at raised CO_2 concentrations (Figure 9.12) and that this effect is aided by increased temperatures, although the longer-term impacts of such changes are difficult to predict (Silvola, 1985). Methane production may also be affected by changes in atmospheric CO_2 concentrations caused indirectly by increased plant growth. More vigorous vascular plants and especially greater root production would enhance the supply of more easily degradable substrates to methanogenic bacteria and would provide greater transport of methane through plant tissues. Initially it was thought that methane production from mires could be greatly increased by such a mechanism (Hutchin *et al*, 1995). Subsequent work has shown that any response is likely to be highly temperature dependent to the extent that actual impacts in most northern peatlands would be minimal (Saarnio *et al*, 1998). Certainly the potential increase in total methane production from raised tem-

peratures is likely to be far exceeded by the decrease due to predicted drops in water tables (Roulet *et al*, 1992). Temperature is also important for CO_2 changes because of its positive relationship with respiration. Combining changes in surface wetness and temperature for CO_2 exchange, Waddington *et al* (1998) found that the CO_2 sink in fens could be enhanced in future, but that it could switch to a source in bogs.

Results of some of the work on gas fluxes and responses to various specific changes in the environmental conditions on mires (temperature, precipitation, water tables, atmospheric CO_2, etc.) under future global warming scenarios are therefore still conflicting in a number of respects. Perhaps because of the difficulties involved in understanding each component of the carbon cycle and gas exchange adequately, there have been few attempts to predict the overall response of total peatland feedbacks to climate change in any detail. Laine *et al* (1996) use an extensive set of data on gas fluxes from undrained and drained peatland sites in Finland to predict overall changes in radiative forcing from peatland gas fluxes given future warming. Figure 9.13 shows the results of this work in schematic form, showing overall decreases in methane with smaller increases in CO_2 and nitrous oxide fluxes. Taking into account the radiative forcing of each of these gases and the changes over time, the total effect on the greenhouse effect is predicted to be a reduction in radiative forcing of about 0.1 W m^{-2} over a 100-year time period (Figure 9.13). This result is partly dependent on the migration of trees to currently open peatlands, increas-

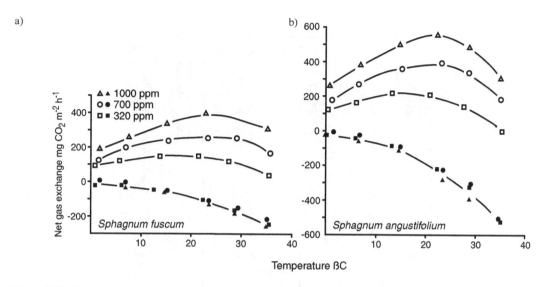

Figure 9.12 Response of *Sphagnum* to increased CO_2 and temperature. (a) *S. fuscum* and (b) *S. angustifolium*. Open symbols represent light conditions (500 μmol m^{-2} s^{-1} irradiation) and closed symbols are from dark conditions. Higher CO_2 concentrations lead to higher positive rates of gas exchange during daylight. Increased temperatures also lead to higher growth during the day but increased respiration at night. Redrawn from Silvola (1985) by permission of *Lindbergia*.

ing carbon stored in living biomass, but the negative feedback may not be reversed even without this change (Laine *et al*, 1996). Of course, reactions to warming in other peatland areas may not be the same, even if the proposal that drainage is a reasonable proxy for future climate change scenarios on peatlands is accepted (it does not for example allow for the direct effects of temperature and atmospheric CO_2 changes).

For example, in permafrost peatlands, warming may be associated with increased wetness through ice thaw rather than drying, providing increased methane emissions (Christensen and Cox, 1995). Also, in areas far north of the current tree line, it may take some time for forest to establish and increase the capacity for CO_2 drawdown. Some data suggest that tundra soils have already switched from being a net sink to a net source of CO_2, following recent warming of the Arctic (Oechel *et al*, 1993; Webb and Overpeck, 1993). Permafrost also currently stores significant amounts of methane in ice in both peat and mineral soils. Thawing of this ice is likely to release some of this methane, amounting to up to 3×10^4 t C yr^{-1}, possibly within the next 30 years (Kvenvolden and Lorenson, 1993). Other effects of anthropogenic activities on CO_2 and methane dynamics may also need to be considered. For example, 'acid rain' pollution is thought to enhance methane production through the provision of an additional source of nitrogen (Aerts and de Caluwe, 1999). It

may also increase anaerobic CO_2 production by promoting higher rates of sulphate reduction (Wieder *et al*, 1990).

Additionally, some potentially large components of the carbon budget are often not considered. Few studies include losses of carbon from DOC export when considering impacts of climate change on the carbon cycle, yet we know these exports can be considerable (see above) and they are likely to be responsive to changes in temperature and hydrology. Accumulation of DOC in pore water has been reported from peatlands in most areas (e.g. Yavitt, 2000), and it seems likely that at least some of this will be flushed from the system. For example, Waddington and Roulet (1997) found that the mass flux of DOC from a Swedish peatland was approximately 20% of the annual CO_2 fixation. Warmer conditions might be expected to accelerate the supply of DOC through increased microbial decomposition at higher temperatures and therefore exports would also increase. This effect has not yet been studied in any detail and its magnitude is unknown.

Of course peatlands are only one component of the landscape, and the potential change in the peatland carbon balance is only one of many to be considered in the big picture of future climatic change. National inventories of carbon stocks and land-use changes will provide part of the basis for negotiations on carbon

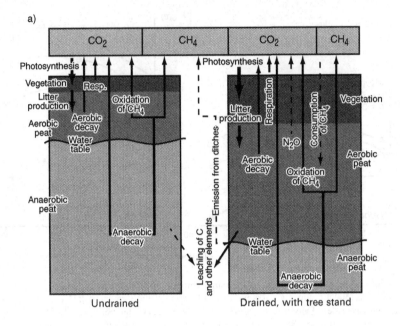

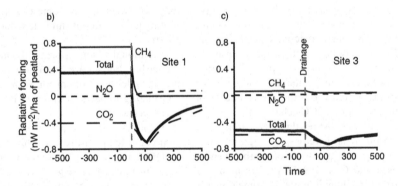

Figure 9.13 (a) Schematic changes in net greenhouse gas exchange under undrained and drained conditions in boreal peatlands. The drained condition is thought to represent possible peatland conditions under future global warming scenarios. The line widths represent approximate flow rates and the dotted lines indicate potential fluxes. (b) and (c) These show changes in radiative forcing following drainage separated into individual greenhouse gases and as the total sum. The effect of drainage is thought to be an approximate simulation of changes resulting from future global warming. (b) A site where higher CH_4 emissions gave a positive net radiative forcing changed to a negative feedback following drainage. (c) A site with a negative net radiative forcing further enhanced following drainage. Redrawn from Laine *et al* (1996) by permission of the Royal Swedish Academy of Sciences.

emissions limitations under international agreements such as the Kyoto Protocol. At the national level, where action to meet targets will be focused, some governments will need to consider the role of peatlands carefully. Certainly in a number of northern countries that currently emit relatively large amounts of CO_2 (on a per capita basis), peatlands have the potential to pro-

vide significant additional sinks or sources of greenhouse gases. For example, Britain has lost much of its actively growing peatland and may wish to reinvigorate peat growth since peatland is one of the main existing land uses where carbon accumulation may continue to grow (Heathwaite, 1992; Adger *et al*, 1992; Pearce, 1994). Forest growth is currently an important carbon sink in several countries such as Sweden (Eriksson, 1991), yet this sink is unlikely to continue at the same rate for longer than a few decades because there is a limit to the amount of additional growth possible. Peat and lake sediments are therefore the only effective terrestrial carbon sink in the longer term (>100 years) (Eriksson, 1991).

9.7 SUMMARY

Peatlands are important elements in the landscapes in which they occur because of the interchange of water, nutrients and other materials that pass in and out of them. They are also significant in the global environment because of their role in the carbon cycle. They thus play a vital role in modifying the local environmental and global climatic conditions. Understanding of many of these processes and interactions is limited and some of the observed relationships may only be applicable to particular peatland types or geographical locations. However, progress has been made towards a more universal understanding of the role of peatlands and their feedbacks to major environmental changes such as future climate change. In the study of peatland ecology and functioning we should not perceive peatlands as passive receptors of change, but should appreciate their context in the wider environment and their capacity to affect it in fundamental and significant ways. Some of the influences of peatlands on the environment occur over very long time periods, which makes it difficult to study, measure and experiment adequately, but this does not mean they are unimportant or any less real. In recent years, there has been much emphasis on carbon balance of peatlands and the relationship with climate. Given the move towards international agreements on limitations of greenhouse gases, this role will come under further scrutiny, especially at the national level where ultimately governments will have to act. Although the picture of current and predicted response to climate change is still blurred, it is abundantly clear that safeguarding the existing carbon pool and continued peat accumulation will be vital parts of peatland management strategies over the next century and beyond.

PART 4

RESOURCE MANAGEMENT

The earlier sections of this book have ranged from some of the basic concepts of peatland definitions and descriptions through to their functions and roles in landscape change at various levels. Most of this discussion has been concerned primarily with peatlands in their natural state, although I have used a number of examples from damaged peatlands to illustrate particular processes. For example, the response of peatland carbon cycles to future climate change where analogues can be found in artificially drained systems (Laine *et al*, 1996). Understanding of the functioning of peatlands under natural conditions is an essential prerequisite for predicting likely impacts of disturbance as well as for remedial management activity to mitigate damage. It is also fundamental to the management of the economic resource that peatlands represent. In this section we will move on to discussing the peatland resource in the broadest sense, in terms of economic, landscape, wildlife and functional values. I will draw on many of the principles and processes presented in previous sections of the book to do this and will assume some knowledge of these.

The first of the two chapters in Part 4 considers the different values of peatlands and examines the impacts of human activities on natural areas. Many human activities totally destroy peatlands as active ecosystems, or have such major impacts that other values are severely limited. However, other forms of exploitation have only a limited impact, or may simply alter the ecosystem so that it reaches a new equilibrium. In many cases, especially in western European mires, the modified mire system may be so long established in the historical period that it is regarded as being of equal or greater value to wildlife conservation than the original peatland from which it has been derived. In Chapter 11, we examine some of the ways in which such systems are maintained by active human intervention and management, as well as discussing the practice and potential of restoring severely damaged peatlands. The level of interest in the possibility of restoring peatlands after peat extraction has increased markedly over the last 20 years and there is now a substantial body of research on the subject. Clearly, where the extent of peatlands has been reduced to a minimum sustainable level for biodiversity and ecological functioning, these activities are essential for the future maintenance of the resource.

10

Values, Exploitation and Human Impacts

10.1 INTRODUCTION: PEATLAND VALUES

Many of the readers of this book will have the conservation value of peatlands in the forefront of their minds when they consider the value of peatland. Certainly peatlands are vitally important to the biodiversity of the planet and provide a haven for a wealth of unusual and specially adapted organisms. However, there are many other values of peatlands that need to be considered in managing the global peatland resource. Some of these values are competing and mutually exclusive, leading to confrontation between different interest groups. On other occasions, a number of values can be realised and maximised without conflict, or with only minor compromise between promoters of particular points of view. Here I will outline the different economic, wildlife and functional values of peatlands without any attempt to evaluate the relative importance of these in different locations. The second part of the chapter looks at the impacts of different economic uses on the other values of unexploited peatlands.

Many of the values of peatlands are similar to those that are associated with wetlands in general (Maltby, 1986; Mitsch and Gosselink, 2000), while others are more specific to peatlands. The obvious example of this is the value of peat as a fuel, a quality unique to peatlands. Other values are overlapping although differ in detail, such as educational values where different types of site provide particular educative experiences. The values of peatlands can be broadly divided into four main areas (Hughes and Heathwaite, 1995a):

(i) Biological and biochemical functions: provision of wildlife habitat and functions as both a source and sink for carbon, nutrients and other biochemical substances
(ii) Economic functions: for harvestable products such as peat for fuel and horticulture, forestry, reeds and other plants, livestock from grazing

(iii) Societal functions: aesthetic and landscape values, educational, research and cultural values, including the archaeological resource
(iv) Physical and hydrological functions: role in the hydrological cycle, altering patterns of runoff, recharge to aquifers, storage of flood flows in some peatlands.

10.2 ECONOMIC VALUES AND EXPLOITATION

10.2.1 Peat mining for fuel, horticulture and other uses

Peat itself is the most obvious and well-known economic product from peatlands and it has been put to a remarkable variety of uses over the years. It is not our intention here to explore the methods and technicalities of peat extraction and processing, and detailed information on these industries and on aspects of harvesting, storage and processing can be found in other texts (e.g. Göttlich et al, 1993; Lappalainen, 1996a) and throughout the soil science literature. As a fuel, peat has been used for at least the last two millennia, as recorded by Pliny the Elder in the middle of the first century AD (Göttlich et al, 1993). It is surprising that peat was used at all at a time when timber was still reasonably easy to obtain and suggests that it was probably used for a long time before this too. In mainland Europe, serious exploitation of peat as a fuel was well under way in the seventeenth century (Göttlich et al, 1993), and in Britain, rights of 'turbary' (i.e. the right to cut peat) were part of the many rules governing the use of common land where peatland and heathland were most abundant (Rackham, 1986). Although the beginnings of peat exploitation in North America were much later, the peat industry began there as early as 1864 (Warner and Buteau, 2000). The use of peat as a fuel and energy source is thus very long established in both the Old and the New Worlds. The use of peat as a domestic fuel has declined greatly over the last few decades of the twentieth century, but it is still maintained in many areas of northern Britain and Ireland, although commercially produced briquettes of peat have sometimes replaced turves cut by hand or machine. However, peat provides fuel for electricity generation in a number of countries and the overall use of peat as a fuel remains high, with a total annual production of 71 million m³, equivalent to approximately 7 Mt of oil in the energy generation sector (Table 10.1). For some countries energy generation from peat is a significant proportion of their total energy supply, approximately 10% in Ireland for example (Asplund, 1996).

The other major worldwide use of peat is in horticulture. Because of its physical and chemical properties, peat is an ideal growing medium and soil conditioner when used alone or blended with other materials. Schmilewsksi (1996) quotes six main properties of peat that make it especially well suited to economic large-scale production and horticultural use:

• The structure of *Sphagnum* leaves which provide good water-holding ability and good aeration simultaneously
• Low bulk density and the resulting ease and low cost of handling, use and transportation
• Low pH means that addition of lime can be used to adjust the acidity level to any desired value
• Low nutrient values allows adjustment to any value by addition of nutrients

Table 10.1 Production of energy peat in the year 1995, expressed as thousands of m³. Depending on the quality of the fuel peat, each m³ is the same as 0.08–0.15 tonnes of oil equivalent (toe) for power generation. Some significant users of fuel peat are not included in this table, such as China, which uses around 0.2 million toe (Asplund, 1996).

Country	Energy peat production (1000 m³)
Finland	24 000
Ireland	15 000
Bwlarus	12 000
Russia	9 000
Ukraine	4 000
Sweden	3 200
Estonia	1 800
Latvia	1 040
Germany	600
Lithuania	214
Great Britain	100
Total	71 000

• Free from any pests, pathogens and weed seeds
• Ease of processing, grading and blending

Although the horticultural industries, and especially the expansion in commercial market gardening in the post-war period, have provided the main use of these products, the amateur garden market has grown considerably in recent years. All the countries that currently produce significant amounts of fuel peat also produce horticultural peat since the two operations use many of the same techniques and often use complementary parts of the peat deposit. In general, the best horticultural peat comes from upper, less humified peat, whereas the more humified deeper peat often provides the most suitable source of fuel material. Other countries (e.g. Canada, USA, UK) concentrate more on horticultural peat where other sources of power generation are more economical or better established.

In addition to fuel and horticulture, there is a variety of other uses for peat. In central Europe, the therapeutic use of peat is well established through the use of peat baths (known as peat balneology). A slurry of peat and water is heated and the patient spends around 20 minutes immersed in the mixture. Variations on this treatment are claimed to be effective in treating conditions associated with rheumatism and gynaecological disorders, as well as a variety of other problems (Korhonen and Lüttig, 1996). A range of other minor uses of peat has been proposed and some put into commercial operation. There has been considerable research on the potential of peat to filter out chemicals and other toxic substances from various media because it provides a suitable medium for microbial growth and it has a high cation exchange capacity and a high moisture-holding capacity. Applications vary from the treatment of sewage sludge to biological air purification, oil absorption and heavy metal removal in waste water treatment (Mutka, 1996). A further industry that depends upon peatlands, although it is only interested in the uppermost parts of the peatland, is the *Sphagnum* moss industry. *Sphagnum* moss is mainly of use in horticulture, particularly for lining of growing containers where stronger structural properties than peat possesses are required.

10.2.2 Peatland for forestry and agriculture

Besides the extraction of peat for various uses, peatlands can provide excellent opportunities for forestry and agricultural land use. There are essentially two main approaches to both forestry and agricultural use of peatlands. First, the natural or semi-natural vegetation cover may be totally replaced with new trees or agricultural crops. The afforestation of treeless blanket mires with non-native tree species in western Britain and Ireland is a good example of this approach. The drainage and reclamation of peatlands for arable land or sown pasture also represent a complete alteration of the peatland ecosystem. Alternatively, the existing vegetation of a peatland may be modified without the introduction of new species, for example to provide better growth of native trees on forested peatlands or to improve the grazing quality of open sites for domesticated stock. In both forestry and agriculture, there are therefore different levels of disturbance that result from the use of the peat resource. Some of these practices and their impacts on the peatland system will be described below.

In addition to direct conversion from natural or semi-natural peatland vegetation to forestry or agriculture, mires that have been subject to extensive peat extraction may subsequently be used for both these purposes and others (Nyrönen, 1996). Considerable effort may be required to achieve the correct soil conditions for crop growth, including drainage and fertilisation to increase fertility and reduce acidity, while also considering the water quality of the outflow (e.g. Nilsson and Lundin, 1996).

10.2.3 Other economic values

Other economic values of peatlands beyond those of larger-scale commercial exploitation also exist, but are often restricted geographically. Use of peatlands for sporting activities is possibly one of the most widespread of these 'other' uses, with recreational shooting of wildfowl and other animals very common in both Europe and North America. The British country pursuits of deer stalking and grouse shooting take place over many areas that are at least partly composed of blanket mire. With regard to sport fishing, the presence of undamaged peatlands is sometimes an important control on river quality necessary for the reproduction and survival of fish such as trout and salmon. Throughout much of northern Europe, peatlands are an important source of wild berries, which are exploited to a great extent in the late summer and early autumn months and eaten fresh or kept for longer periods in preserves, liqueurs and other products.

10.3 WILDLIFE CONSERVATION VALUES

The combination of the very large areal extent of peatlands worldwide and their unique living conditions means that the wildlife value of peatlands is enormous. Despite the exploitation of the economic values of peatlands over many years, the very inhospitability of these habitats has meant that they are one of the few remaining near-natural ecosystems in many more populated areas of the world. The assessment of wildlife conservation values is a large subject in itself, but of the many attributes that are considered important in the assessment of individual sites (e.g. Usher, 1986), there are perhaps three key virtues of peatlands that deserve special mention: rarity, naturalness and fragility. These concepts, as with many other value assessments, are somewhat subjective and open to argument. Nevertheless, we can identify some generally applicable qualities of peatlands in these terms.

First, peatlands harbour many organisms that are unique and some that are endemic to mires in general or to specific mire regions. Good examples are plants such as some species of *Sphagnum* mosses and some of the carnivorous higher plants. Even where such organisms are not entirely unique to peatlands, their populations are largely dependent on them to maintain sustainable numbers. Beyond rarity of individual organisms, the peatland habitat itself is rare in a number of areas of the world, especially in Europe. Second, many of the world's peatlands are in a natural or near-natural state. Others have been subject to significant human intervention and would be better described as semi-natural (see Chapter 11). However, vast expanses of northern boreal peatlands and many temperate and tropical peatlands can be counted among the most natural ecosystems on earth. This aspect of value becomes more important in regions such as northwest Europe, where natural habitats are at a premium. Third, peatlands are fragile habitats and do not often recover quickly to their former state following major disturbance. Much effort is currently focused on attempting to restore severely damaged peatlands (see Chapter 11), but this is a slow and costly process and may not allow full recovery of every aspect of functional, wildlife and societal values. Beyond these three key criteria of rarity, naturalness and fragility, others are variously applicable to specific sites and regions and would need to be considered in wildlife conservation evaluation. These would include diversity, areal extent, representativeness and typicalness (Usher, 1986). The other societal values discussed below would normally be added to these in habitat conservation arguments.

10.4 FUNCTIONAL VALUES

Peatlands perform various functions in the wider environment, many of which we have discussed in some detail elsewhere in this book. Their role in the carbon cycle (Chapter 9) is probably the best example of one way in which peatlands perform a useful global function, not easy to appreciate in local terms or to evaluate in comparison with other values. The 'global peatland ecosystem' removes and stores carbon from the atmosphere as well as returning some carbon as methane and some as carbon dioxide. The hydrological functions of peatlands were discussed in Chapter 9 and include modification of water storage, runoff and river flow as well as alteration of water quality in the catchment. However, the hydrological functions of peatlands cannot be regarded as universally beneficial or deleterious. Undisturbed peatlands will in general have only a moderate influence on most aspects of hydrology, although they can serve both to increase and decrease water quality, for example. Unusual natural events or disturbance are the most likely factors to change this. For example, erosion or ploughing for forestry will significantly increase suspended sediment loads and fire (natural or anthropogenic) can produce increases in nutrient fluxes from peat-dominated catchments. Although the water storage capacity of most natural mires is limited, intact floodplain mires and other wetlands can be important in temporary flood mitigation.

10.5 VALUE TO SOCIETY

Societal values encompass many different aspects of the ways in which humans use peatlands. Increasing interest in wildlife and landscape as part of recreational activities means that peatlands and other natural areas have great value to a broader range of people than scientists and naturalists. Because of their 'naturalness' peatlands are often the only part of the landscape that could be considered 'wilderness' in lowland, agricultural landscapes in Europe, North America and some other regions. Peatlands also provide an important educational resource for many ages and different purposes. Again, it is in places where peatlands are most under threat in more heavily populated areas that these resources are most valuable as they are close to schools, colleges and universities. A third area of societal values covers the cultural, historical and archaeological features of peatlands. Much of the environmental evidence, such as that described in Chapter 6, is frequently used to inform the interpretation of past landscape change associated with human populations and activities and therefore provides a vital archaeological resource. In some cases, human artefacts and remains are found buried in peatlands, some of the most celebrated examples being those of bog bodies (Glob, 1969) and prehistoric trackways (Coles and Coles, 1986).

10.6 CONSERVATION AND 'WISE USE' OF PEATLANDS

In considering the overall value and management of the peatland resource at any scale, the potential values discussed briefly above clearly need to be weighed carefully. The calculation is not simple and depends on local and national considerations as much as the global perspective. There are many questions that need to be answered, such as:

(i)   How much natural peatland needs to be conserved and where should it be located?
(ii) How far should peat extraction be permitted and should it cease in some areas of the world where peatland has already suffered extensive damage?
(iii) Are the direct economic returns of forestry and other activities greater than the loss of other uses and values of the same areas?
(iv) Is low-input agricultural use of some peatlands sustainable?
(v)   To what extent do global values (such as carbon cycling) override those of local (often economic) needs?

Many of the values are interwoven, interdependent and some only become significant when combined with other elements of the landscape. For example, aesthetic values are usually partly dependent on other elements of the landscape. Even in peat-dominated areas, such as the Flow Country in northern Scotland, the vastness of the peat landscapes is emphasised by the isolated hills and the distant mountains of the west coast. Economically, income from tourism is unlikely to be purely dependent on peatlands, but also on other attractions and activities in the same area. The 'Peatlands Park' reserve, tourist and educational centre in Northern Ireland is apparently the second biggest attraction in the region (Stoneman and Brooks,

1997), but would be unlikely to attract non-local visitors without the other countryside and cultural attractions in the region.

Over much of the latter part of the twentieth century, many of the discussions over evaluation and use of the peatland resource were essentially 'environment-versus-economy' arguments with strongly entrenched views on both sides. These arguments have not disappeared but have perhaps been softened by some compromise on both sides and a realisation that to make progress there needs to be dialogue and understanding. One concept that has emerged in the last 10 years is that of 'wise use' of peatland resources. 'Wise use' encompasses the use, conservation and management of peatlands, taking into account all the values at both global and local scales. Thus, contrary to some suggestions that wise use is in opposition to conservation (Petterson, 1997), more recent moves are to develop inclusive planning policies based on this principle.

10.6.1 International conservation legislation and agreements

Most conservation legislation is enacted at the national level and particular countries will possess their own specific laws on the conservation and exploitation of peat resources. Clearly these will differ between nations in their details but they are guided by international legislation and agreements. Probably one of the most important pieces of legislation for peatlands is the Ramsar Convention on Wetlands of International Importance as it is the main international agreement on the conservation and use of wetlands specifically. Other worldwide agreements such as the United Nations Convention on Biodiversity are also relevant to the wildlife conservation of peatlands. Additionally, it is increasingly recognised that because peatlands have a direct bearing on the calculation of carbon dioxide sources and sinks for many countries, they will need to be considered in national inventories of greenhouse gases that are required by the Kyoto Protocol (Bergkamp and Orlando, 1999). The Kyoto Protocol arose out of the United Nations Framework Convention on Climate Change and outlines the main commitments of individual countries to the reduction of greenhouse gas emissions. Because many peatlands are net accumulators of carbon, and carbon trading between countries is currently permissible, for some countries peatlands will be important economically either in offsetting carbon emissions or in monetary terms by trade-offs with other nations.

Out of these various international agreements on habitats and issues other than peatlands is emerging a generally applicable policy on peatland use and conservation known as the Global Action Plan for Peatlands (GAPP) (IMCG, 2001). This is the culmination of a series of discussions at various scientific and planning meetings since 1994, including representatives of the peat industry, conservation bodies, scientists and various governments (e.g. North American Wetlands Conservation Council (Canada), 1996). The GAPP is still under discussion and is unlikely to be formally accepted and implemented until 2002 but will be the first global, legally binding agreement endorsed by all interested parties. The partners currently involved in the preparation of the plan include the International Peat Society (IPS), mainly representing the peat industry, and the International Mires Conservation Group (IMCG), concerned principally with the conservation of the peatland resource (Table 10.2), demonstrating the willingness of both 'sides' in the conservation arguments to come together for a common purpose. The vision statement associated with the GAPP includes reference to all the values mentioned in the preceding section:

> 'Recognition of the importance of peatlands to the maintenance of global diversity of ecosystems and species, the conservation of carbon vital to the world's climate system, and the wise use, conservation and management of natural resources for the benefit of people and the natural environment.'

A series of objectives and action themes arise from this general statement, to implement the GAPP at national and regional levels, but also concerning knowledge, research and infrastructural issues. Of course, the GAPP still has to be finalised and ratified, but it is already a major step forward in bringing together the

Table 10.2 Partners involved in the preparation and evaluation of the Global Peatland Action Plan (GAPP). Extracted from draft of the GAPP from IMCG (2001) (http:\\www.imcg.net)

Global Environmental Network (GEN)
Institute for Wetland Policy and Research (IWPR)
International Mire Conservation Group (IMCG)
International Peat Society (IPS)
IUCN Commission on Ecosystem Management (IUCN/CEM)
Ramsar Convention and its Contracting Parties
Society of Wetland Scientists (SWS)
Wetlands International (WI)

views of many individuals and organisations that until recently were in direct opposition to one another.

10.7 IMPACTS OF RECENT HUMAN DISTURBANCE: DRAINAGE AS A KEY PROCESS

If we want to manage the peatland resource with regard to all its values, we need to understand the ways in which exploitation of economic values affects the other values, especially those related to wildlife conservation and ecological/hydrological functioning. The rest of this chapter will review the impacts of human activities on peatlands and the ways in which we have destroyed and modified these ecosystems in the past; water is at the heart of many of these impacts.

Just as the amount and distribution of water are the key to the formation, functioning and ecology of intact mire systems, the exploitation of peatlands for peat extraction, agriculture or forestry depends on controlling the hydrology of the system. Usually, this means drainage in one form or another in order to reduce the water content of the surface peats so that they can be removed more easily or so that they provide better growing conditions for the economically valuable plants being grown. Some of the details of specific practices will be discussed in the following sections on peat extraction, forestry and agriculture; here we will simply review the effects of drainage in a general sense. The potential causes of drainage are shown schematically in Figure 10.1. Although the illustration is for a raised mire, the same principles apply to any mire type. Direct drainage is the most universal cause of peatland dehydration and may be implemented for peat extraction, forestry or agriculture. Removal of peat causes further potential losses either by increasing hydraulic gradients between cut and uncut areas or by reducing the resistance to downward seepage through to the mineral substrate. Further possibilities are reduced regional groundwater tables, perhaps from water abstraction, or increased drainage at the periphery of the mire for other purposes such as agriculture. Finally, invasion of other plant species may increase evapotranspiration and cause surface drying.

In terms of the physical effects of drainage, the main initial impact is to lower the water table by providing a route for increased surface runoff through the drainage ditches. This only affects the surface layers of peat at first, because of the very low hydraulic conductivity of the catotelm peat (Figure 10.2). In oceanic conditions where there is frequent precipitation, even the surface

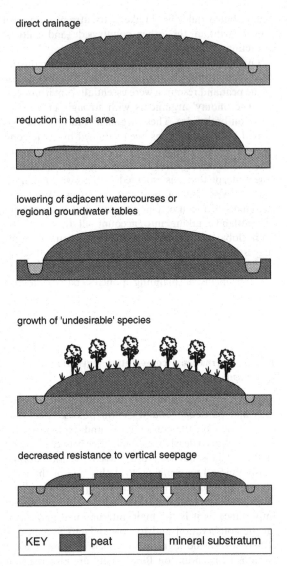

direct drainage

reduction in basal area

lowering of adjacent watercourses or regional groundwater tables

growth of 'undesirable' species

decreased resistance to vertical seepage

KEY   ▓ peat   ░ mineral substratum

Figure 10.1 Potential causes of dehydration in peatlands illustrated for a raised mire. Redrawn from Wheeler and Shaw (1995b) by permission of HMSO.

layers may remain quite wet, especially where drain spacing is wide. However, in dry conditions, lack of frequent precipitation recharge results in a decrease in water tables to much lower elevations than they would attain in natural conditions. In comparisons between drained and undrained areas of a raised mire in Finland, it can be seen that the variability of water table depth is much greater with drainage, even though on occasion the water table may be almost as high in the drained as in the undrained area (Figure 10.3).

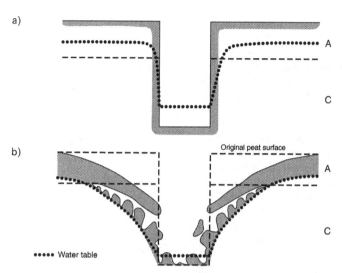

Figure 10.2 Changes in mire morphology following drainage as represented by the changing cross-section through a drainage ditch. The solid line marks the surface of the peat and the horizontal dashed line marks the change from the acrotelm (A) to the catotelm (C). The dotted line marks the water table. (a) Soon after construction of the drain showing the water table affected only close to the drain. (b) Following peat shrinkage and slumping, with the water table depressed further away from the drain and the surface of the peat lower than the original (marked by dashed lines). Redrawn from Ingram (1992).

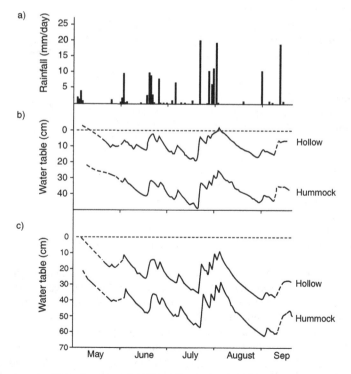

Figure 10.3 Changes in water table over the period May to September in 1976 on drained and undrained areas of a raised bog in Finland. (a) Daily rainfall, and water table changes in (b) undrained hummocks and hollows, (c) drained hummocks and hollows. Compiled from Lindholm and Markkula (1984).

When the catotelm peat begins to dry out, it shrinks and becomes irreversibly physically altered, producing cracking (Ingram, 1992). This can result in mass movement towards the drain, with slumping of surface peats and a change in the morphology of the drains from neat trenches to troughs with more inclined edges (Figure 10.2). With drier conditions within the peat, aerobic decomposition is accelerated, vertical subsidence and compaction can occur with the proportion of space occupied by solids increasing and ultimately a reduction in permeability and the amount of fast-draining pore space (Figure 10.4). The degree to which each of these processes occurs is a function of the depth and spacing of drains, peat composition and initial moisture status, as well as any vegetation changes imposed. For example, planting of trees on open mires may increase evapotranspiration and accelerate drying of the peat. Even in the absence of new planting, there may be positive feedbacks from changes in vegetation structure, such as increased growth of woody shrubs and trees already on the mire, that also increase evapotranspiration as they gain greater leaf area and total biomass.

Drainage also brings about changes in the chemistry of the surface peats, which can have an impact on nutrient availability to plants and affect the water quality of runoff. This is a more significant effect in fen peats, where a more substantial nutrient source is stored with the peat (Naucke et al, 1993). However, the responses of nutrient levels to drainage are not always straightforward and there are a number different effects and interactions that need to be considered. In general, nitrogen and phosphorus increase but potassium decreases after drainage, but this effect is variable depending on peat type and climatic conditions, with consequences for the most appropriate fertilisation regime for encouraging tree growth, for example (Sundström et al, 2000). Water quality changes in runoff following drainage are also strongly affected by local factors, but nutrients and pollutants can increase significantly in particular circumstances (see section 9.3, Figure 9.5, for example). Increased mercury levels in Finnish lakes have been attributed to drainage of forested peatlands using palaeoecological analyses of annually laminated sediments (Figure 10.5). There has been some suggestion that peatland drainage could be a significant source of pathogenic mycobacteria, but experimental work suggests this is unlikely (Iivanainen et al, 1999).

In the following sections we will refer back to some of these general responses to drainage and explore some of the results of specific drainage schemes associated

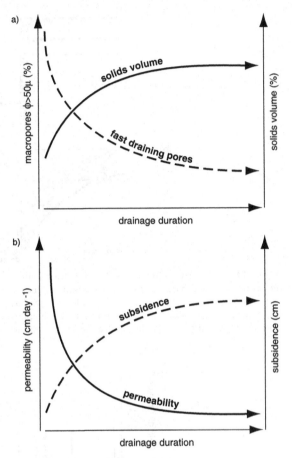

Figure 10.4 Schematic graphs of the changes in (a) peat texture and (b) subsidence and permeability over time following drainage. Redrawn from Heathwaite et al (1993a) with permission. © John Wiley & Sons Limited.

with various aspects of economic exploitation of peatlands.

10.8 PEAT EXTRACTION

The process of peat extraction, whether for horticulture or fuel, large or small scale, follows a basic series of operations. Drainage is a fundamental first step in order to begin the drying process to make the peat more manageable, and to remove excess surface water to make access to and from the peat much easier. There are then various methods of actually removing the peat, followed by management of the cutover area after extraction. The historical development of peat cutting and the details of various extraction techniques are dis-

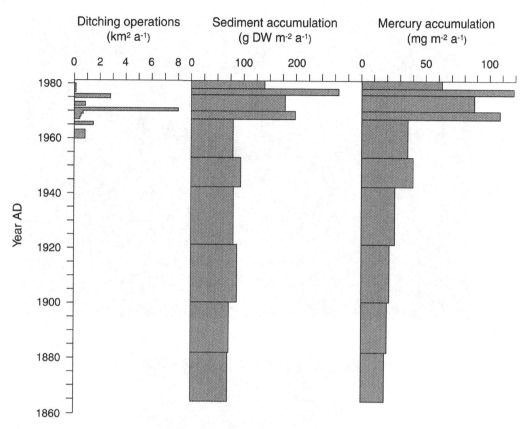

Figure 10.5 Drainage operations in the catchment of Lake Polvijärvi, Finland, plotted alongside mercury accumulation rates as measured by analyses of annually laminated lake sediments. The correspondence between the start of ditching operations and the major rise in mercury levels suggests that peatland drainage is the main cause of mercury pollution in the lake. Note also the rise in sedimentation rate at this time, suggesting increased suspended sediment load in the runoff from drained peatlands. Redrawn from Simola and Lodenius (1982) by permission of Springer-Verlag.

cussed by Göttlich *et al* (1993) but there are basically three main approaches: cutting of peat blocks (either by hand or machine), peat milling and 'sausage cutting'.

Block or turf cutting is the longest established of these techniques and is the approach most often associated with traditional domestic peat cutting in western Europe where families of small groups of people cut their own supply of fuel peat using hand tools. Domestic peat cutting typically involves working a peat face of perhaps 1–2 m height from above, gradually working back into the peatland (Figure 10.6). The steep hydraulic gradient resulting from the peat face provides enough drainage to begin to dewater the uncut peat and to provide a stable platform for working on. Cut peat turves are stacked to dry for several weeks before transportation back to the home. Several characteristics of this practice make it a much less damaging

activity than other approaches. First, it is generally only undertaken on a relatively small scale with the actual area of peatland cut in any one year being relatively small. Additionally, because surface peats are too fibrous to provide good fuel, the top turves are often laid or thrown into the base of the cutting which provides a source of vegetation regrowth in the cutover area. Kirkpatrick *et al* (1995) have suggested that good recovery of some plants occurs after around three years if turves are replaced carefully. Without this measure, the peat in the cutover areas can begin to crack, with potentially serious implications for peat erosion and the vegetation that could eventually colonise the area.

Although hand cutting is now declining in western Europe, it is being replaced in some areas by small-scale commercial fuel peat production using tractor-towed

(a)

(b)

Figure 10.6 (a) Hand-cut peat cuttings for domestic fuel. The peat bank is gradually cut back over the years, with sequential cuts into the peatland and downward. The cut turves are stacked behind the cutting before being removed and stacked close to the point of use. (b) Peat cut using a 'sausage' cutter. The cutting machine is towed behind a tractor to produce long round strips of peat that can be divided into lengths. These are left to dry on the surface or may be stacked prior to collection.

extrusion machines ('sausage cutting'). Here, the peat is extruded from below the surface and drawn up through slits in the surface vegetation to form 'sausages' of peat that can be left to dry and cut up into convenient sized lengths for domestic fires (Figure 10.6). There are rather few data on the extent and impact of this process but it is now widespread in Scotland and Ireland. The environmental impacts are potentially much greater than traditional hand cutting because of the much larger areas needed for the same annual production of peat. It is not clear how much of the peat is recoverable from an area using sausage cutting, as the gradual break-up of the surface vegetation from repeated cuts and machine travel may result in erosion or inoperable conditions over a period of time. Since the peat is extracted from underground, the effects on the surface vegetation and ecology may appear to be limited, but the total vegetation biomass and height are reduced and invertebrate populations are altered rapidly by the process (Todd *et al*, 1995).

Commercial extraction originally also used hand cutting, albeit on a much larger scale, but machines for cutting peat blocks (or 'sods') were already operational before the First World War (Göttlich *et al*, 1993). These are still in use in some places, although later machines are considerably more sophisticated than the early contraptions. The principle is the same as hand cutting in that blocks of peat are cut, stacked to dry and then transported off the peatland when the water content is sufficiently low. In mechanised block cutting, however, much larger areas are cut simultaneously, with surface vegetation removal over a large area to facilitate the operation of machinery. Blocks are cut and stacked as well as being turned periodically by machine on site to aid the drying process. The end result of peat extraction using block-cutting machines is a pattern of hollows separated by upstanding ridges (often called 'baulks'). Following abandonment, this pattern creates particular problems and opportunities for natural and assisted revegetation of cutover areas (see Chapter 11).

Sometimes commercial block cutting has reduced the extent of the intact surface, leaving fragments of the old peat surface surviving. The long-term effects on the remnant upstanding blocks of peatland can be significant, with reduced water table, peat shrinkage and noticeable changes to the vegetation. For example, Hammond *et al* (1990) found that the peat surface on Lodge Bog, County Kildare, Ireland, had declined by 0.20 m (in the centre) to 1.70 m (at the edge) on an area of intact bog since peripheral cutting began in the early 1950s. This was accompanied by reduced water tables

and the loss of plants such as various species of *Sphagnum*, *Drosera anglica*, *Eriophorum angustifolium* and *Rhynchospora alba*, typical of wetter bog conditions in the area. The hydrological impact of the removal of the edges of a raised bog results in a gradual alteration of the groundwater mound (see Chapter 3), such that an altered profile of the water table can be calculated from hydrological principles (Figure 10.7), although it seems likely that this would be a very long-term adjustment under constant climate conditions (Ingram, 1992). The general reduction in height of the groundwater mound and in the location of the highest point will lead to drying of the surface peat and the associated structural changes noted earlier.

In recent years, peat milling has become the dominant technique for large-scale peat extraction in terms of area covered and volume of material removed worldwide. Here, the surface is carefully prepared with drainage and levelling, as with commercial sod cutting. The drainage is crucial and takes around three years to accomplish as it involves progressively increasing drain depth to avoid the type of collapse illustrated in Figure 10.2. The drains are normally about 15 m apart (the width of most of the machinery) and joined laterally by outfalls to remove the water quickly. Drainage may also be further aided by cambering the inter-drain area (the 'field') so that water runs off more easily and surface vegetation and loose moss are removed completely. Once fully drained and levelled, the peat can take the heavier harvesting machinery. First the surface is milled using rotating drums fitted with pins. In this the surface 15 mm is cut and left to dry. The process is aided by turning with a harrow and, when the water content is sufficiently low, the peat is gathered into ridges in the centre of each field. The peat is harvested with a large vacuum harvester and then is transferred to stockpiles on or off site. Depending on the climatic conditions, there might be 10–12 harvest times during the summer months of a single year, representing a 150–200 mm layer of peat. As the peat depth decreases, supplementary drainage is required, mostly by deepening of the ditches between the fields at the end of the harvest season or at the start of the next. The storage and processing of peat is a complex and varied process, including for example the conversion to fuel briquettes or mixing with other materials for horticultural applications. The end results of large-scale commercial peat extraction are large areas of bare peat surfaces, normally around 0.5 m depth or sometimes much greater in the sub-peat depressions. The peatland is therefore almost completely destroyed at this point and there is little or no wildlife, functional or societal value remaining. In the past, such areas were regarded as being only fit for conversion to agricultural or forestry land and this may be the most appropriate post-harvest land use in some cases (Nyrönen, 1996). However, in recent years there have been many schemes attempting to restore wildlife and other values to cutover peatlands; these will be explored in Chapter 11.

Beyond the destruction of the peatland ecosystem (at least in the short term), there are a variety of other environmental impacts of peat extraction, including discharges of particulate carbon, DOC, nutrients and heavy metals, as well as carbon dioxide emissions from stockpiles and the ultimate breakdown of the extracted peat in use (Nyrönen, 1996). However, many of these effects are likely to be relatively minor and often measures can be taken to mitigate them. Despite this, objections have been made to proposals to develop large areas of the Florida peatlands for extraction largely on the basis of these potential

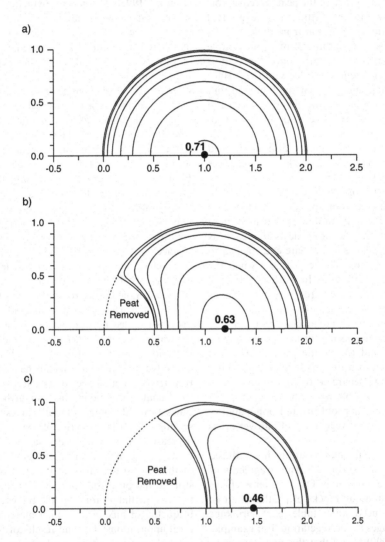

Figure 10.7   Predicted changes in the groundwater mound in a raised mire following peat cutting. (a) The groundwater mound of an intact raised mire on a flat surface with radius = 1. The highest point is at 0.71 units and is central. (b) Following removal of peat for half the original radius from a point on the left-hand edge. The highest point has declined to 0.63 and moved off centre. (c) Following removal of peat for a distance equal to the radius from the edge of the mire. The maximum height has declined further to 0.46 units and shifted further across to the right. See text for comment. Redrawn from Ingram (1992).

(d)

Figure 10.7   (d) Typical appearance of a raised bog that has been subject to peat cutting over part of its original area. Domestic peat cuttings have removed much of the peat at the edges of the site, leaving the central section uncut but subject to peripheral drainage and shrinkage. Photograph of Bank End Moss, Cumbria, England by Richard Lindsay.

external environmental effects rather than on grounds of wildlife preservation (Winkler and Dewitt, 1985). Carbon dioxide emissions from stockpiled peat are not insignificant and in Finland, for example, amount to about 19% of the emissions from burning of peat in power stations, or 2% of the total carbon dioxide released by fossil fuels (Ahlholm and Silvola, 1990).

Finally, in considering extractive industries and their impacts, it is worth mentioning the *Sphagnum* harvesting industry which is particularly active in Australasia, where *S. cristatum* is a significant peatland 'crop' for specialist horticultural applications. A recent study into the impacts and potential sustainability of the industry in Australia and New Zealand suggested that harvests were sustainable in some New Zealand sites where growth rates are sufficiently high and where the rate of picking is carefully controlled (Whinam and Buxton, 1997). However, the Australian sites were less productive and are thus unlikely to provide sustainable harvests in the long term. However, even where the growth of the *Sphagnum* is sufficient to tolerate controlled picking, there may be other ecological impacts

that have not been recorded. Despite this potential problem, *Sphagnum* harvesting certainly has less impact on these sites than peat extraction.

10.9 FORESTRY

There are two main approaches to forestry on peatlands. First, there is improvement of growing conditions for trees already present using drainage, fertilisation and occasionally pest control. Second, trees can be planted on previously unafforested areas, with site preparation including ploughing, as well as drainage. The former is practised mainly in northern boreal mires throughout Scandinavia and Canada, whereas the latter is more common in oceanic mires, especially those in western Europe and also as a post-harvest use of cutover peatlands. A third type of peatland forestry is logging activities in tropical peatlands, although this does not usually involve management activities to improve tree growth.

10.9.1 Naturally forested peatlands

The dominant species in forests on the northern boreal peatlands are mostly coniferous. Black spruce (*Picea mariana*) with some tamarack (*Larix laricina*) dominates in much of North America, and *Picea abies* (Norway spruce), *Pinus sylvestris* (Scots pine) and *Betula pubescens* (downy birch, a deciduous species) are the most common in Fenno-Scandinavia (Päivänen, 1996). All of these species are commercially useful, but tree growth rates are often limited by waterlogging and drainage has been used in many areas to provide greater economic returns. Early logging activities were selective cutting of larger timber on undisturbed peatlands in both North America and Fenno-Scandinavia and these would have had minimal environmental impact but are no longer commercially viable (Jeglum *et al*, 1982; Päivänen, 1996; Hökkä and Groot, 1999). In the post-war period, clear-cutting of larger areas has been the main harvesting practice, with natural regeneration, reseeding or restocking for the second and subsequent harvests. However, the most recent harvesting practices in some countries (known as 'shelterwood' systems) are designed to stagger removal of trees so that older trees provide shelter and seed sources for new growth (Hannerz and Hanell, 1997).

Drainage and fertilisation are the principal means of increasing the growth rate of trees on peatlands, but the most efficient combination of management depends on the trees present and the moisture and nutrient status of the peat on which they are growing. Drainage increases aeration and rooting depth of the trees, as well as promoting increased rates of mineralisation and nutrient availability. Nitrogen-mineralisation is often adequate for plant growth but fertilisation with phosphorus and potassium is used to optimise growth (Wells and Williams, 1996). All of these changes enhance growth and increased rooting depth helps reduce losses of trees due to windthrow. There have been many experiments to identify the best combination of ditch spacing and depth for specific areas, and practice has varied widely between geographic regions and at different times in the past. The total area of peatland drained for forestry worldwide is estimated at around 150 000 km^2 (Päivänen, 1996), although this includes afforested open mires as well as natural peatland forests. The growth response to drainage can be dramatic, although variable even within stands of trees. Macdonald and Yin (1999) found that trees that were smaller prior to drainage responded more vigorously to drainage than the larger, better growing trees (Figure 10.8). They suggested that the microsite variability within natural peatland tree stands (hummocks and hollows) controlled the initial growth rate of the trees but that this differential was lost with the drop in water tables following drainage. Growth rate increases are often delayed following drainage and in the first five years only small additional increments or sometimes slight decreases are recorded, as the trees adapt to the changed conditions, and probably put on new root growth rather than shoot growth initially (Sundström and Jeglum, 1992; Sundström, 1992b). In some sites, the cost of drainage may not be recouped by increased

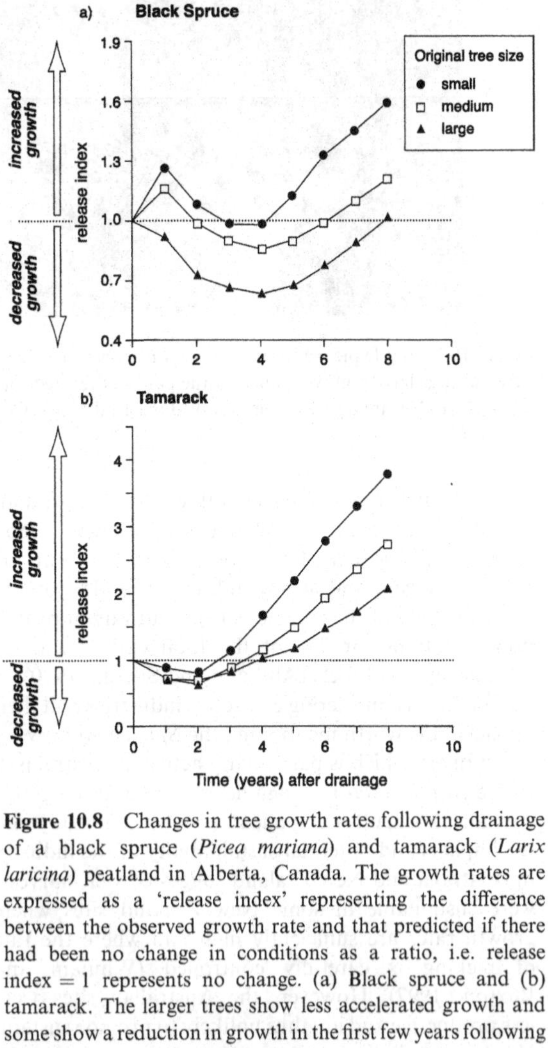

Figure 10.8 Changes in tree growth rates following drainage of a black spruce (*Picea mariana*) and tamarack (*Larix laricina*) peatland in Alberta, Canada. The growth rates are expressed as a 'release index' representing the difference between the observed growth rate and that predicted if there had been no change in conditions as a ratio, i.e. release index = 1 represents no change. (a) Black spruce and (b) tamarack. The larger trees show less accelerated growth and some show a reduction in growth in the first few years following drainage. Tamarack shows a much greater positive response than black spruce. Redrawn from Macdonald and Yin (1999) by permission of Blackwell Science Ltd.

yields. In Finland, where peatland forestry has been practised for many years, it is acknowledged that additional timber yields from draining very nutrient-poor sites do not justify the expense of drainage. Together with wildlife conservation concerns, this has led to little or no new drainage for peatland forestry in recent years.

While drainage can enhance tree growth, it may have significant impacts on the other functions of the peatland. Comparisons of drained and fertilised peatland forests with undisturbed mires show that plant community diversity is reduced by both drainage and fertilisation (Vasander, 1984) and certainly if the forest canopy becomes denser and conditions at the surface are drier, the ground flora and any shrub layer are almost bound to change composition. The rates of response of plants vary considerably. In Finnish mires drained over a 3–55-year period, tall sedges such as *Carex lasiocarpa* and *C. rostrata* disappeared rapidly after drainage and different *Sphagnum* species declined at variable rates, apparently dependent on their tolerance to shading (Laine *et al*, 1995). Structural changes in the vegetation have an impact on bird communities as well. Denser forest growth and the reduction in older trees with nesting holes and decayed wood can significantly alter bird populations in peatland forests in the long term (Vaisanen and Rauhala, 1983). The peat also shows significant structural changes, with increased bulk density and subsidence (Silins and Rothwell, 1998). Initially at least, ditching of peatlands for forestry can cause eutrophication of runoff and other pollution effects (see above and Figure 10.9). This is known to have an impact on lakes in the same catchments, with increases in sedimentation rates and lake productivity due to increased nutrient loads (Simola, 1983).

Other forestry practices that can have a wider impact on peatlands are those associated with thinning, harvesting and restocking. Up until the 1960s winter was the main harvest period in Canada when the frozen and snow-covered surface provides a solid base for machinery. Extending harvesting to the whole year to increase efficiency of use of workers and equipment causes much more disruption to soils and vegetation (Jeglum *et al*, 1982). Again effects are variable depending on vegetation type; Groot (1987) suggests that the level of damage is greatest in *Alnus rugosa* (speckled alder) herb-rich vegetation and least in *Ledum groenlandicum* (Labrador tea)-dominated vegetation in black spruce peatlands in Ontario, Canada. Although the hydrological and edaphic effects of disturbance during harvest are often assumed to be important, experimental studies suggest that most effects are minimal in comparison to the hydrological changes from removal of the forest cover. In the case of the Ontario peatlands, only the most intense disturbance produces significant reductions in the height of the peat surface, the depth of the aerobic layer and increases in water tables (Figure 10.9; Groot, 1998).

There are several possibilities for re-establishing tree growth on areas of harvested peatland. Natural regeneration from seed or remaining vegetative material is likely to be the cheapest option, but reseeding or restocking with seedlings are other possible methods (e.g. Sundström, 1992a; Groot and Adams, 1994). Black spruce is a major economic species in North America that is capable of regenerating vegetatively by layering of branches, especially in *Sphagnum*-dominated peatlands. This is known as 'advance growth' and careful logging can preserve it resulting in much faster regeneration following harvest (Groot, 1995; MacDonell and Groot, 1997). Harvest systems that preserve advance growth are also more likely to preserve a greater proportion of the ground flora, and to lead to less physical disturbance of the peat surface generally.

Following harvesting, many northern boreal peatlands become much wetter as a result of reduced evapotranspiration and physical impacts of harvesting (see above), a process known as 'watering up' (e.g. Roy *et al*, 1999). This causes considerable difficulties for seedling tree growth and further drainage may not necessarily ease the problem significantly. For example, Roy *et al* (1999) found that the provision of raised microsites for seedlings was more important than general drainage over a wide area in increasing seedling survival and growth rates. The other effect of opening up the forest canopy can be to promote growth of pioneer shrubs and other plants. In Canadian boreal peatlands this may mean rapid invasion of species such as *Populus tremuloides* (trembling aspen) and *Betula papyrifera* (white birch) in some sites, making it more difficult for the target coniferous species to re-establish (Roy *et al*, 2000). The ground flora will also be affected by the physical and hydrological changes following harvest. In a study of Norway spruce peatland forests in Sweden, the changes in ground flora were greater with clear-cutting than with shelterwood harvesting methods, and species with preferences for shady moist conditions survived much better where partial tree cover was left after cutting (Hannerz and Hanell, 1997). Partial harvest systems are also being examined for Canadian peatlands where potential benefits over clear-cutting are perceived as improved wildlife

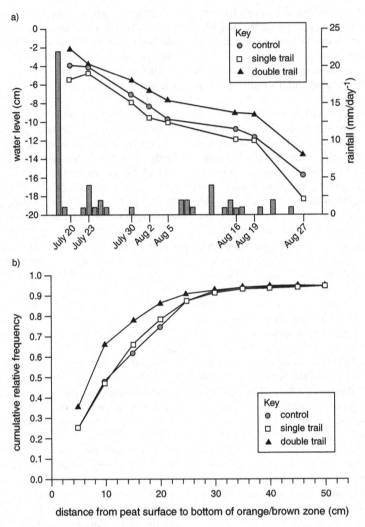

Figure 10.9 Hydrological and peat structural changes resulting from disturbance during tree harvesting on black spruce peatlands in Ontario, Canada, as indicated by (a) changes in water levels and (b) changes in the depth of the aerobic zone as measured by discoloration of steel rods over the year. The treatments are control (no summer traffic), single-trail skidder traffic (trails 15 m apart) and double-trail skidder traffic (trails 7.5 m apart). The single-skidder treatment disrupts approximately 16% of the surface while the double-skidder disrupts approximately 33%. Only the double-trail treatment significantly raised water tables and reduced the depth of the aerobic zone. Redrawn from Groot (1998) by permission of the Agricultural Institute of Canada.

habitat, maintenance of other ecological functions, and improved aesthetic and recreational values and opportunities (MacDonell and Groot, 1996; Groot, 1997).

10.9.2 Afforestation of open mires

Afforestation of open mires is especially common in the blanket and raised mires in oceanic western Europe. In the British Isles, forestry is one of the major land uses of lowland raised mires and also for many areas of upland blanket mires. In the 1980s, there was a rapid spread of forest plantations in the Flow Country of northern Scotland, and this area rapidly became the focus of a major conservation protest, attracting international condemnation as well as national opposition (Lindsay *et al*, 1988; Dierssen *et al*, 1990). However, this was perhaps the most extreme case in the 'trees versus bogs' debate and since that time forest planning and practice have been modified to take much greater account of wildlife and landscape values. In 1993 it was estimated that there were about 190 000 ha of forest on deep peat in Britain with a further 315 000 ha on shallower peat (< 45 cm deep) (Pyatt, 1993). Afforestation is a more fundamental alteration of a peatland system than management of naturally forested peatlands for several reasons. First, it involves a major change to the physical and hydrological conditions due to the combination of ploughing and drainage. Second, it is a major structural alteration to the vegetation, introducing a tree canopy where none existed before. Finally, it usually involves the introduction of non-native species, typically the North American *Picea sitchensis* (Sitka spruce) and *Pinus contorta* (lodgepole pine) in the British Isles.

Ploughing for forestry is a major disruption to the surface vegetation and hydrology. Most of the area is substantially modified, either becoming a trough or alternatively becoming the ridged-up surface where the ploughed material is turned over (Figure 10.10). Much of the surface vegetation is destroyed or buried and a large proportion of the ploughed area is exposed bare peat. One of the major concerns at this stage of forest development is the possibility of increased surface runoff and of changes to water quality of the runoff. In general, runoff can become more 'flashy' with afforestation as the water is transported to the streams more quickly, but this effect may be reduced after canopy closure (Pyatt, 1993), and varies considerably with site-specific factors of ditch density, peat type, vegetation cover and climate (David and Ledger, 1988). The plough furrows act as shallow drainage ditches to help lower the water table but additional drains are also installed to assist the flow of water away from the ploughed area. Most evidence suggests that the initial reduction in the water table due to ploughing and drainage is relatively small, but adequate to help seedling establishment, especially as the trees are generally planted on the ridge of peat turned over by the plough. Once the trees become more established, the water table sinks much further due to increased evapotranspiration. Over a period of years, the peat also shrinks and subsides, with large cracks developing in the surface and penetrating to some depth. Once a stand in a plantation forest has reached maturity, trees are normally removed by clear-cutting. Following harvest, runoff increases due to the reduced evapotranspiration. Results are variable depending on the local rainfall regime, but increases are generally of the order of 40% or more (Anderson *et al*, 1990).

Figure 10.10 Open blanket peatland ploughed for afforestation.

However, many of the plantations on deeper peat soils are yet to reach harvestable age and the impacts are therefore largely unknown.

The most important ecological impact of afforestation is the destruction of the open peatland habitat in the plantation areas. Within plantations there is little or no opportunity for the typical mire plant and animal species, and those that do remain are restricted to the open forest rides (narrow corridors) or other openings left within the forest (Wallace *et al*, 1992). This has been the main objection of wildlife conservationists to extensive plantations on blanket mire habitat in Britain, especially where the blanket mire is regarded as a prime international wildlife habitat (Figure 10.11). Bird life is perhaps even more severely affected than plant life, since many species depend on the blanket mire habitat to live and cannot adapt to forest plantations. After the first 10–15 years of tree growth with canopy closure, the moorland bird community is replaced by a woodland bird community (Stroud *et al*, 1987). Although bird numbers in plantation forests can be high, they are not the same as the rarer, more specialist populations that are found on open moorland. Areas such as the Flow Country peatlands of northern Scotland support large populations of key peatland species such as the dunlin, greenshank and golden plover, representing a significant proportion of the total European populations (Table 10.3). There are also considerable changes to invertebrate populations following afforestation. Reductions in total numbers of invertebrates of 57% were recorded in northern England, with shifts in the composition of ground beetle

faunas that were studied more intensively (Butterfield, 1992). This work also suggested that deeper peat sites were affected more severely than drier shallow peat sites, with characteristic species declining or being lost in afforestation.

Partly in response to wildlife conservation arguments but also in response to issues of landscape, recreational and aesthetic values, the forestry industry in many countries now has a broader outlook than simply maximising timber production. The Border Mires of Kielder Forest, England, is one area where large-scale commercial forestry is combined with wildlife and landscape values in a management system that aims to provide a truly multi-purpose forest. Kielder

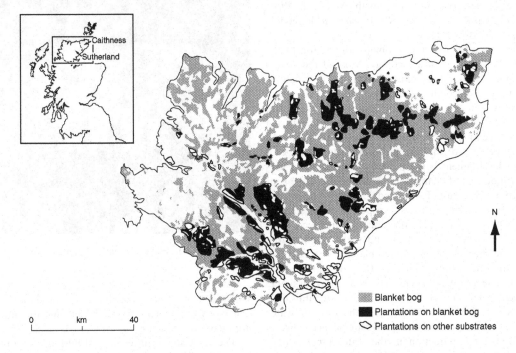

Figure 10.11 The extent of blanket bog in Caithness and Sutherland, Scotland, with areas of afforestation planted or approved for planting in 1988. Redrawn from Lindsay *et al* (1988) by permission of Scottish Natural Heritage.

Table 10.3 Estimated numbers of some key peatland bird species on the Flow Country peatlands of Caithness and Sutherland and the proportion of the total European populations they represent. Extracted from Stroud *et al* (1987).

Species	Estimated population (pairs)	Total British population (pairs)	Status elsewhere in Europe (EU countries)	Percentage of EU population in Caithness and Sutherland
Golden plover	3 980	22 600	<650 pairs	17
Dunlin	3 830	9 900	<100 pairs	35
Greenshank	630	960	Absent	66

Forest is one of the largest plantation forests in Europe at around 62 000 ha, with approximately 20% of this as unplanted land managed primarily for conservation, including just over 1000 ha of the Border Mires peatlands (Ogilvie, 1990). Up until the mid-1980s the management was mainly focused on timber production, and many areas of shallow peatlands were planted, leaving only remnant sites and parts of sites within the forest. Following concerns over the long-term changes in vegetation on the peatland sites (Smith and Charman, 1988; Chapman and Rose, 1991), a management group was formed composed of representatives of forestry and wildlife conservation agencies, the Northumberland National Park and university academics. In the last 10 years, changes to forest structure have been made to diversify the landscape and wildlife habitats. On the peatlands, trees have been removed from the edges of some affected sites and remedial work to raise water tables has been undertaken (Smith et al, 1995).

10.10 AGRICULTURAL RECLAMATION

Agricultural use of mires varies enormously in intensity and in the changes it brings about in the peatland system. At one extreme, peatlands can be 'reclaimed' purely for agricultural use, converting the peatland to nothing more than a substrate for crop growth. At the other extreme, peatlands may provide a proportion of the grazing land in an upland hill farm for part of the year. In this situation, most of the natural peatland flora and fauna coexist happily with domestic stock and the other functions and values remain largely intact. Many peatlands are therefore managed for the dual purposes of agriculture and wildlife conservation. In some cases the priority will be agriculture, with only incidental benefits for wildlife. However, for most of these sites, the priority is now wildlife conservation and the agricultural use is mainly a management tool to maintain and enhance the conservation interest. We will deal with these sites and management techniques in the first part of Chapter 11.

Reclamation of peatland for agricultural use probably represents the greatest proportion of peatland habitat loss over the whole historic period. Many of these areas would have been subject to peat extraction in the past and then reclaimed for agricultural use, while others have been drained and gradually converted to grassland or cultivated arable land use. Despite the total alteration of the natural peatland habitat, these sites retain some other values as they may maintain a palaeoecological archive beneath the surface and offer the only prospect of rehabilitation of peatland in some regions (Lindsay, 1995). Conversion of both fens and bogs to agricultural production has taken place over wide areas of Europe, especially in the Netherlands, Denmark, Germany and the British Isles. Fens were initially favoured, with large areas converted to pasture and drainage activities starting in the early medieval period, but some ombrotrophic bog reclamation also took place very early. For example, the old raised mire area of 'Waterland' in the Netherlands was largely settled and converted to agricultural production by the thirteenth century (Bos et al, 1988). Very large-scale conversion of mires to agricultural use took place in the seventeenth to nineteenth centuries throughout lowland Europe, usually combining peat extraction with agricultural exploitation (Göttlich et al, 1993). Consequently there are only limited areas of intact peatland remaining in many of these regions. Conversion of old cutaway peatland to grassland or other agricultural uses continues in some places in Europe (Curry et al, 1989).

In tropical areas, reclamation of peatlands continues and is almost certainly going to proceed for some time to come due to the need for increased food production. However, there are signs that the areas of exploitation and the methods used will be more carefully planned than they were in earlier periods in Europe. Existing recommendations suggest avoiding all the deeper peat areas, especially where these are ombrogenous, as pH and nutrient contents are very low and the depths of peat encountered are unmanageable (Hardjowigeno, 1997). Good planning with attention to the other values of the peatlands in tropical areas is also promoted (Widjaja-Ahdi, 1997; Notohadiprawiro, 1997). The problems with the use of reclaimed peat soils are similar to those encountered elsewhere, and various materials have to be used to increase pH and nutrient levels as well as using drainage to reduce the water content (bin Soewono, 1997). Saltwater intrusion in the coastal swamp areas may also be a problem against which there is no real defence and these areas are best left alone. Subsidence of the peat swamps reclaimed for agriculture is a major problem in some areas. In Malaysia, within the Western Johore Integrated Agricultural Development Project, total subsidence has reached almost 3 m since about 1960, with average rates of approximately 5 cm per year declining later to 2 cm per year (Wösten et al, 1997). Approximately 60% of the total subsidence was estimated to be due to oxidation and around 40% due to shrinkage. Subsidence is greatest at intermediate depths in the aerated zone, where peat is least resistant to decay and is away from the influence of the water table, which decreases decay rates for some of the time (Figure 10.12).

10.11 EFFECTS OF FRAGMENTATION

The combination of peat extraction, afforestation and agricultural reclamation in different combinations in various locations has resulted in large-scale fragmentation of the peatland habitat in some areas. The end result is 'islands' of intact peatland within anthropogenic landscapes dominated by farmland, urban areas or forest plantations. The worst affected areas are in more heavily populated parts in Europe and, to a lesser extent, North America, where these fragments are often preserved for their wildlife and aesthetic values. However, questions are being asked over how sustainable these reserves are likely to be given the fact that peatlands depend to some extent on the maintenance of the complete hydrological and ecological system.

Hydrology is often a key concern, particularly where the land surrounding a remnant peatland is lower than the peat surface either from peat extraction or subsidence following drainage. Peripheral drainage may dry out the remnant peatland, and so hydrological protec-

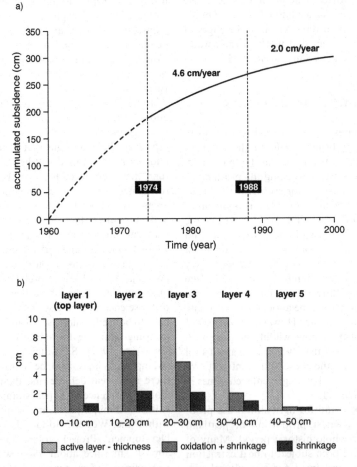

Figure 10.12   (a) Average subsidence rates over time for a tropical peat swamp in Malaysia. Data before 1974 are estimated. (b) The total volume reduction due to oxidation and shrinkage over a five-year period for one site in the area estimated from a model. The model divides the unsaturated peat into 10 cm thick layers, labelled 1 to 5. The left-hand bars represent the total depth of each layer – because the water table is at 47 cm, the lowest layer (5) is only 7 cm thick. The histograms indicate that total subsidence was greatest in the depths 10–20 and 20–30 cm, with approximately 64% of subsidence due to oxidation. Redrawn from Wösten *et al* (1997) by permission of Elsevier Science.

tion is crucial if is to survive in the long term. Raising and maintaining the water table is often the most critical task in peatland restoration and management (Chapter 11). For biological conservation, another issue is whether the viable populations of specialist plants and animals can survive. Fragmentation of peatlands in northern England has been associated with a decline in typical mire plant species (Smith and Charman, 1988). In this case, fragmentation by forest plantations has resulted in isolation of habitat, changes in the management regime of the sites and possible influence of peripheral drainage. The effects of fragmentation on bird populations are less predictable, with both negative and potentially positive impacts. In the case of peatland specialists such as the dunlin (*Calidris alpina*), a wading bird restricted to peatland nesting areas, fragmentation of peatlands will almost certainly lead to large numerical losses simply by removing the only suitable habitat for the birds' survival (Lavers and Haines-Young, 1996, 1997). For other birds, peatlands provide only partial support and they also use adjacent habitats (Calme and Desrochers, 2000). Some birds that are regionally rare, such as the ortolan bunting (*Emberiza hortulana*) in Norway, apparently depend on a combination of peatland and agricultural habitats to survive (Dale, 2000). Although the impacts of fragmentation on peatland systems can be severe, even small remnants may be worth some conservation effort since they provide refuges for smaller animals and other organisms that can survive on smaller areas (Borcard, 1997).

10.12 POLLUTION

Water-borne and airborne pollution may affect peatlands. Because many peatlands primarily depend on an atmospheric water supply, aerial pollution is potentially the most damaging, but runoff to some fens may also have a significant impact. Increased acidity in rainwater has occurred across a wide area of the world as a result of industrial pollution and is thought to have a significant effect on many ecosystems. However, given the natural acidity of many ombrotrophic bogs, the potential for further acidification as a result of acid rain may be limited (Clymo, 1984a). There are several causes of acidity in peatlands that make many of them naturally very acid (see Chapter 3), and the extent to which acid rain has decreased pH levels is not always clear. An extensive survey of peatland waters from ombrotrophic bogs in Britain suggested that the

decrease in pH due to acid rain could be of the order of 0.7–0.8 pH units. In addition, the greatest effect seems to be in areas subject to the most polluted rainfall rather than in those areas with the highest total acid loads (Figure 10.13). Although the effect on peatland water chemistry is measurable, other ecosystems in the same areas are likely to be much more sensitive and any assessment of critical loads should not be based on peatlands (Proctor and Maltby, 1998). Although much attention has been focused on oligotrophic peatlands, it may be rich fen systems that are suffering greatest impact from acid precipitation (Kooijman, 1992). Catchment liming has been carried out in an attempt to increase pH of watercourses, but although this can be successful in improving the quality of streams and rivers, it can have adverse effects on both plant and animal communities in naturally acid peatlands (Buckton and Ormerod, 1997).

A further possible effect of acid rain is to increase the supply of SO_4^{2-} and NO_3^-, which could increase plant growth rates, especially where N is a limiting nutrient. However, although growth increases in some taxa have been recorded following simulated acid rain, this only occurred for the first two to three years following acidification, and growth declined again thereafter (Rochefort et al, 1990). Taxa growing in positions isolated from the influence of any groundwater, where N is more limiting, may benefit to a greater extent than those in other locations (Rochefort and Vitt, 1988). In regions that have received high N loads in the past, the growth of *Sphagnum* is limited by P instead of N, so that additional inputs of N from pollution no longer stimulate plant growth (Aerts et al, 1992). Physiological changes in the plant communities can also cause unpredictable effects. Over time, *Sphagnum* appears to adapt to the additional supply of NO_3^- by losing its ability to retain N so efficiently, leading to greater NO_3^- availability in the peat and hence increased growth rates of higher plants (Woodin and Lee, 1987). Changes also occur within the peat itself, affecting the microbial communities and decay processes. Additional N in the system results in increased microbial biomasses (Gilbert et al, 1998), and as a result may cause increased decomposition and reduced rates of peat accumulation (Aerts et al, 1992).

Increased N supply to peatlands can be caused by factors other than aerial pollution, especially by runoff from agricultural and urban sources. Eutrophication of peatlands is usually also associated with increased P. Enrichment with nutrients from pollution is usually associated with changes in plant communities such as the switch from sawgrass (*Cladium jamaicense*) to

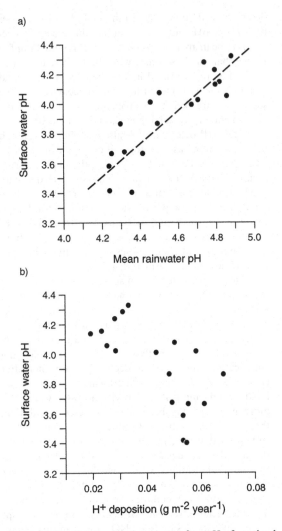

a)

b)

Figure 10.13 Relationship between surface pH of peatland waters and (a) mean rainwater pH, (b) total annual acid load, at 18 sites in Britain. Redrawn from Proctor and Maltby (1998) by permission of Blackwell Science Ltd.

cattail (*Typha domingensis*) communities in the Everglades of Florida (see Figure 3.18). Phosphorus levels appear to be more important in the Everglades than N, except where P loads are highest (Vaithiyanathan and Richardson, 1997). Plant community changes are closely related to the P gradient associated with proximity to the pollution source, with a switch from 'typical slough macrophytes' to other assemblages as P levels rise (Figure 10.14).

Industrial pollution since the mid-nineteenth century has not only caused acid rain, but has also resulted

in the deposition of significant quantities of heavy metals from fossil fuel burning for power generation, industrial machinery and transport. However, input to peatlands of some metals such as lead has a much longer history, with increases from early smelting activities from 3000 years ago or earlier if anthropogenic land disturbance as a source is also considered (e.g. Shotyk *et al*, 1998; see Chapter 6). Many peat profiles record the increasing input of heavy metals over the past century, especially from power stations. The specific effects of the heavy metals in historical records are difficult to separate from those of increased sulphur and nitrate pollution and increasing acidity. The loss of *Sphagnum* cover in heavily polluted areas

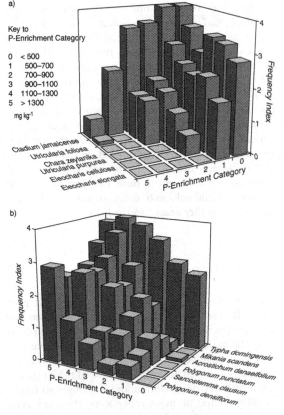

a)

b)

Figure 10.14 Changes in the frequency (expressed as an index) along the phosphorus pollution gradient in the Everglades peatlands, Florida. (a) The typical slough macrophytes found in less enriched areas, and (b) other macrophytes showing increasing frequency as P levels rise. Redrawn from Vaithiyanathan and Richardson (1999) by permission of the American Society of Agronomy, the Crop Science Society of America and the Soil Science Society of America.

is perhaps more likely to be a result of sulphur loading rather than heavy metal deposition (Lightowlers, 1988). Some studies on higher plants suggest that seed banks remain viable after suffering considerable pollution loads, although it is possible that seedling growth is reduced by heavy metal pollution (Huopalainen *et al*, 2001).

There are many other substances that fall on peat-lands in different areas of the world as a result of emis-sions from industrial sources. The impacts of industrial aerial pollution vary with the specific industrial process and the location of the mire. For example, alkaline fly ash was blamed for the disappearance of *Sphagnum* on mires in parts of Estonia, although it has started to recover following reduced emissions from the mid-1980s onwards (Karofeld, 1996). Other toxins such as PCBs are reported from peatlands, although the ecolo-gical impact is likely to be minimal (Himberg and Pakarinen, 1994).

10.13 RECREATION AND OTHER DISTURBANCE

Peatlands are not popular locations for regular recrea-tional activity when compared with other natural and semi-natural habitats such as mountains and uplands or coastal locations. Frequently the heaviest users of peatland habitats are scientists studying them! However, peatlands are likely to be much more sensi-tive to moderate disturbance than many other areas by virtue of their soft surface and low-growing plants, and there have been some studies examining the impacts of recreational access. Borcard and Matthey (1995) found that 10 minutes of experimental trampling repeated only three times a year for three years almost destroyed the cover of *Sphagnum recurvum* and *S. fuscum* on a Swiss raised bog. Furthermore, the related changes in soil fauna were also very dramatic. The prospects for recovery of the *S. recurvum* area were thought to be good, but the *S. fuscum* is unlikely to be able to respond to the relief from trampling pressure due to its isolation. In the Kosciusko National Park, an alpine area of Australia, *Sphagnum* growth was checked and other vegetation changes were noted in association with the development of a ski slope over the site (Clarke and Martin, 1999). Trampling effects on blanket bog in England are also thought to contribute to erosion, although they are not the only factor involved (Yalden and Yalden, 1988).

In particular settings, other forms of disturbance can be locally significant and any activity that disrupts the surface vegetation or peat can cause damage. In north-ern Manitoba, the installation and maintenance of hydroelectric transmission lines have resulted in impacts from tree clearance, vehicle access and use of herbicides (Magnusson and Stewart, 1987). In this case, the vegetation suffered considerable damage, with reduction in cover of many taxa and an increase in the area covered by bare peat surfaces. The change in vegetation was exacerbated by some non-target plant species disappearing in areas where the herbicide was used. Losses of peat arose from increased thawing and erosion in some of the most heavily used areas. A further unusual example is that of the impact of bomb-ing on Estonian mires (Karofeld, 1999). During the Soviet era, a large area of northern Estonia was used for military training including the use of live aerial bombing. The physical disruption damaged surface vegetation severely and the levels of heavy metals are increased considerably. Some recovery of vegetation has taken place following cessation of military activity in 1992.

10.14 LONG-TERM ANTHROPOGENIC DISTURBANCE

One of the impacts of human activity that we have not discussed in this chapter is the very long-term impact of landscape alteration by deforestation and land manage-ment that occurred throughout the developmental his-tory of many mire systems. The development of most existing mires throughout the world has taken place in the Holocene period alongside human settlement. There are only a few areas where we can be sure that peatlands developed in the complete absence of people. One such area is New Zealand (Newnham *et al*, 1999), although on a global basis this is the exception rather than the rule.

Palaeoecological studies have shown that many apparently pristine ecosystems have been subject to past disturbance (Huntley, 1991). Peatlands are no exception. Swamp forests in Sweden were thought to be long-established refugia from the influence of human activity until their forest history was reconstructed from pollen analysis (Segerstrom *et al*, 1994). This showed that the *Picea abies* (Norway spruce) cover has only developed in the last 300 years following earlier phases of deforestation and cultivation. Landscape disturbance by humans has also been impli-cated in the initiation of peatland growth in many areas. The best-known example is that of blanket peat development in the British Isles, but suggestions of a causal link between human presence and peat initiation have also been made for mires as far apart as Africa and southern Europe (see sections 4.6 and 4.8). In addition, major changes in mire systems occurred quite late in human history, particularly those associated with European settlement of the New World and especially the effect of forest clearance on some of the mire systems of North America. The evidence and detail of these impacts are covered in Chapter 4 and will not be repeated here. However, it is worth restating the point that human impact on peatlands is nothing new – it is the scale and rate of possible change that increased dramatically in the twentieth century. The destruction of large areas of peatland in Europe by enthusiastic drainage and reclamation from the sixteenth century onwards began this accelerated pace of change, but the development of mechanised techniques for peat extraction, drainage and forestry resulted in the rapid loss of total peat area over the past 100 years.

10.15 SUMMARY

We began this chapter by focusing on the diverse values of peatlands to the economy, to people and for intrinsic non-anthropocentric values. Value judgements on the correct management for peatlands are inevitably complex and subjective, and it would clearly be possible to make a case promoting only exploitation or conser-

vation for peatlands throughout the world. It would also be unrealistic and imbalanced to do so. Fortunately the dialogue among those concerned with the management of peatlands is now more active and co-operative than it has ever been, and the 'wise use' of peatlands is now a principle truly enshrined within such discussions. The second part of the chapter went on to consider the effects of human activities on peatlands, especially those resulting from exploitation of economic resource values. Many of the impacts occurred in the past and are unlikely to be repeated in quite the same way in the future. However, it is important to understand the nature and magnitude of these impacts if we are to develop effective strategies for remedial management of damaged systems and to predict responses to any future proposals for peatland development. Only a good understanding of these effects can provide the basis for a level-headed evaluation of planned development that can inform the careful balancing act that will inevitably take place between the various values involved. However, it is clear that for many reasons the maintenance of active peatland systems will be a priority for the vast majority of the global peatland resource. In the final chapter, we therefore move on to see how current management and restoration of sites for wildlife and habitat conservation are carried out, and what the future holds for peatland systems in the rest of the twenty-first century.

11

Conservation Management and Restoration

11.1 INTRODUCTION

This is the final chapter of this book, and in it we turn to the future of peatlands and the management of the global peatland resource. Despite possible future climate variability, the main processes we have described are unlikely to change markedly (although our understanding of them will) but the influences on them are already beginning to alter. The biggest shift over the past decade is one of attitudes and values. We have already noted the improved dialogue between 'exploiters' and 'preservers' of peatlands and the emergence of a better appreciation of the needs of both. To this can be added a broader acceptance and better understanding of the wider functions of peatlands, particularly of their role in the global carbon cycle. Together, these factors have forced reassessment of the relative importance of peatland values and provided much greater motivation for activities such as the restoration of damaged peatlands. Peatland restoration represents one of the most significant technical challenges in peatland management and a large amount of effort and finance is currently being directed towards this ultimate goal. However, there is a long tradition of more modest management of peatlands for conservation that concentrates much more on habitat maintenance than on the (re)creation of new peatlands. Many semi-natural peatlands are subject to grazing, removal of vegetation for hay or other uses, burning and even peat cutting, partly for economic gain but often purely as tools for conservation management. While management of wilderness peatlands usually means leaving them alone or at most restricting access, the emphasis in semi-natural sites is on retaining low-level disturbance to maintain particular successional conditions. While this is a well-established approach for many other habitats, it is not so widely appreciated as being necessary for peatlands because they are often perceived as wilderness areas. In the first part of this chapter we examine some fundamental concepts of the naturalness of peatlands and definitions of disturbance. This is followed by an exploration of the ways in which different management activities can be used to affect the ecology and functioning of peatland sites for nature conservation. Finally we take a look at some of the work that has been carried out on peatland restoration and examine the feasibility and progress of this ultimate challenge in wildlife conservation.

11.2 NATURALNESS, DISTURBANCE AND CONSERVATION

'There is a certain ambivalence in attitudes [to peatland management] amongst conservationists: when it is a fen, they often want to continue to damage it to maintain its interest, but when it is a bog, they usually tell us we should leave it alone.' (Wheeler, 1996, p. 30)

Managing disturbed peatlands must involve some deeper consideration of fundamental questions of what we wish to conserve and why. In the case of undamaged peatlands, the priority is clearly to maintain the system as it is, and prevention of disturbance or disruption to the system is the focus of management, if any is needed. However, many peatlands that are considered valuable for wildlife conservation and other functional values are partly the result of quite high levels of disturbance. The quote above highlights the apparently paradoxical opinion prevalent in conservation management of peatlands, in western Europe at least. Many of the remaining fens here have been created or at least substantially modified by past human activity and therefore need continued 'management' (or 'damage' to use Wheeler's blunt but accurate term) to maintain their current ecological functions. In contrast, many bogs do not depend in the same way on human activities, and if anything tend to be degraded by them.

Disturbance levels vary between very minor alteration of the natural functioning of the system to severe alteration of form and function. The gradient from completely natural to highly disturbed peatlands is described in Table 11.1. One important general point is that both past and present disturbance may need to be considered in any evaluation of the overall disturbance level. In previous chapters, we have seen how human disturbance hundreds or even thousands of years ago can alter the course of peatland development; this may have important implications for current management. Light grazing from low-density stocking of sheep or cattle on blanket mire within a mosaic of upland habitats could be considered a minor disturbance activity (see section 11.4). Moderate disturbance represents a more significant shift in one aspect of the peatland function or form. The change to *Sphagnum* bogs from fen systems in the kettle hole peatlands of Ontario following forest clearance would come into this category (see Chapter 4; Warner, 1993). Major disturbance involves changes in the form and functions of the peatland and changes to the ecology, particularly those involving reductions in diversity or changes in dominant species. One example of major disturbance is the afforestation of open peatlands where drainage and tree planting change both form and function in a significant way, yet some of the peatland flora and fauna can survive in unused areas of peatland, or even within planted areas in the early stages of growth (see Chapter 10). The most severe disturbance is that caused by peat extraction where the entire living ecosystem is removed and the peatland is reduced to the inert fossil peat mass. Following extraction, there remains only a barren thin peat that needs considerable management effort and expense to restore anything like a functioning peatland. Any peatland functions or values of such systems are only those that are produced as a result of human management. Of course, disturbance level is a continuous gradient and all peatlands will not fit neatly into

schemes such as that described in Table 11.1. Equally, over time, a site may move from one category to the next without any real change in disturbance levels. For example, overgrazing and pollution are known to have contributed to erosion of blanket mire in northern England, yet it is only over a long period of time that the level of disturbance is fully realised as more and more peat is removed, changing from moderate to major disturbance in the terms of Table 11.1.

It is also tempting to use the disturbance gradient as the *only* reflection of the value of the system and there are a number of examples of its use to evaluate the conservation importance of peatlands. For example, in an evaluation of Irish raised mires, site categories from A (highest) to D (lowest) were established primarily on the basis of the level of damage from drainage, peat cutting and burning (Cross, 1990; Table 11.2). This was supported by a scoring system for individual attributes of the sites such as the level of damage to hydrology, habitat diversity and presence of protected or rare species. The total amount of intact raised mire surfaces remaining has become an important statistic in discussions over the management of the raised mire resource in Ireland and elsewhere in Europe. Studies using disturbance ratings help in collecting these data. This particular study showed that only 23 000 ha out of an original 311 000 ha remained at the time (only 7%), showing the need to protect all remaining intact raised mires in Ireland.

However, while it may be appropriate to use the disturbance gradient as a way of grading sites in very specific circumstances over restricted areas, decisions over the priorities and aims of management are not always quite so straightforward. Even restoration of already severely damaged peatlands involves more complex value judgements. In the case of cutover peat-

Table 11.1 Classification of peatlands according to the level of disturbance. Expanded and adapted from Warner (1996).

Disturbance level	Typical character
Natural	No influence of human activity at any time in the past. Initiation and development processes have proceeded naturally, ecology and hydrology unaffected by current human activity
Minor disturbance	Some influence of humans in the distant past or very minor levels of recent or current disturbance. Peatlands retain same peatland type and form as they would have in the absence of disturbance
Moderate disturbance	Disturbance levels in the past or present sufficient to alter the type or form of peatland. A functioning peatland is retained but its structure may be changed and the functions are altered
Major disturbance	Human activity has altered the structure and form of the peatland. Changes have resulted in species impoverishment, major shift in hydrology, changes in the dominant species. Functions are significantly changed and compromised in one or more areas
Artificial	Damage has almost completely destroyed the original peatland and the existing peatland is entirely a result of human efforts to restore function and form to the site

Table 11.2 Site quality categories for Irish raised bogs based on a gradient of disturbance used for conservation management strategies (Cross, 1990).

Category	Generalised site description
A	High-quality or unique sites. Hydrology more or less intact. Extensive wet, soft and often quaking areas. *Sphagnum* cover extensive. Wide diversity of vegetation communities and plant species, particularly those associated with wet habitats. Mainly intact, or damage localised. If damage more extensive, a site of exceptional interest
Bi	Good-quality sites. Hydrology damaged but not seriously. Wet, soft and quaking areas and *Sphagnum* cover less well developed than A. Damage similar to or slightly more extensive than A but sometimes recently badly burned
Bi	Moderate-quality sites. Hydrology seriously damaged, but restoration possible. Extensive dry areas, but wet and soft areas still present. *Sphagnum* cover less extensive than A or Bi. Disturbance and damage more extensive, particularly from marginal peat cutting or from drainage or repeated burning
Biii	Poor-quality sites. Hydrology very seriously damaged and restoration difficult. Extensive dry areas, peat hard with little *Sphagnum* except for localised 'soft' spots. Extensive damage from cutting, drainage and repeated burning
C	Sites seriously damaged largely from hand cutting. Very little intact bog surface left
D	Sites drained and/or mechanically cut away or afforested over most of their area

lands, the assessment is whether the values of the restored peatland are adequate to outweigh the potential values of alternative uses such as agriculture or forestry. A peatland subject to restoration management is very highly disturbed, yet in terms of peatland conservation values, the position is clear: any peatland is better than no peatland, whatever the ultimate limitations of the restoration effort. Thus from a conservation viewpoint the decision on proceeding with restoration is relatively simple. For other sites, particularly those that have been subject to disturbance for many years and have become part of the ethnic character of the landscape and provide distinctive wildlife habitats in their own right, it is more difficult to see a clear rationale for management. For example, many fen habitats in western Europe are the result of management by grazing, reed cutting and very often peat cutting too (see section 11.4). Using disturbance level as the only criterion to assess the 'quality' of these sites would suggest that they are of relatively little value. They have suffered at least 'moderate disturbance' (Table 11.1) over many years with changes in the peatland type and functions as well as in vegetation composition, compared to their completely natural state. Yet such sites are rich in species that occur nowhere else in these regions mainly because the natural niches (such as river floodplains) of the species have been completely destroyed. They are also culturally important, as they reflect land use and a way of life that has persisted for many centuries. These are all good arguments for rating them among some of the most valuable habitats in the areas in which they occur and attempting to retain their key attributes through active management, even where this means further 'damage' to the sites.

11.3 MANAGEMENT OPTIONS FOR DISTURBED PEATLANDS

The possibilities for managing disturbed peatland sites fall into three main categories:

- Prevent further disturbance and aim to restore the peatland to its natural state using active management where necessary
- Maintain the disturbance as a means of preserving the status quo and retaining existing values of the site
- Allow disturbance to continue and then restore the site once the economic value of the site has been fully exploited

One could argue that the first of these options should be the priority in all cases. However, there are many philosophical and practical arguments against this. We have already used the example of some European fen systems to suggest that the current values of some sites may be greater than the values of the same site in its restored 'natural' state. Even if this philosophical position were altered there are often major practical obstacles to following through on the principle. In the case of some of the floodplain fens in East Anglia, England, the natural state may well be raised bog, but there would be insurmountable problems in attempting to re-establish true bog habitat on these sites. In other situations, the re-establishment of 'close-to-natural' conditions may be possible either by passive or active management. One example of such a possibility is that of treeless blanket mire in western Europe.

Oceanic blanket mire in western Europe is currently almost devoid of any trees apart from those that have been planted for commercial forestry. However, we know that in the past, there were trees growing on the peat surface in many locations, including in the far north of Scotland (Gear and Huntley, 1991). Equally, boreal mires in other regions less affected by grazing and burning retain some tree cover, especially at the drier margins. These facts could be used as an argument to allow afforestation of some blanket mires to restore more 'natural' vegetation. However, although the agricultural use of these peatlands probably does suppress tree and shrub growth to some extent, we know that oceanic blanket mires in other regions of the world are also largely treeless (e.g. in Newfoundland, Canada – Wells and Hirvonen, 1988). It is likely that some trees and shrubs would grow on the margins of some blanket mires in Britain and Ireland under completely natural conditions, but these would not be extensive and would certainly not be composed of the alien species used in commercial forestry in the region. Marginal growth of native species such as downy birch (*Betula pubescens*), rowan (*Sorbus aucuparia*) and Scots pine (*Pinus sylvestris*) would be a more realistic simulation of natural conditions. Over most areas of blanket mire, this is not currently being considered but there is no reason why it should not be, if control of grazing and burning can be reconciled with the agricultural needs of the areas concerned. Clearly in cases where arguments are being made for active management towards more natural conditions, it is important to have adequate evidence for the nature of the expected natural peatland system, as well as to have clear and well-reasoned management aims and philosophy. Neither is easy to achieve, and management of this kind should proceed cautiously, if at all.

The second option for management is to maintain the disturbance at an acceptable level to retain the existing character of the site. This option is widely practised in many semi-natural peatlands in Europe and to some extent elsewhere. Here the rationale is that the value of the disturbed site is greater than that of the state that it would be possible to achieve by management for greater naturalness. These management strategies will be discussed in section 11.4. The final option is to forget management in the short term and to wait until the economic value of the site is exploited before attempting to re-establish any kind of functioning peatland system. This is practised primarily for cut-over peatlands but may also be an option for post-harvest management of peatland forestry and potentially even for some areas reclaimed for agriculture. These options will be discussed towards the end of this chapter in section 11.5.

11.4 SEMI-NATURAL PEATLAND HABITAT MANAGEMENT

The above discussion has shown that there are peatlands where the maintenance of some disturbance is desirable, especially to achieve the most important aims of wildlife conservation in particular areas. Other management will be required in order to check the progress of damaging activities or processes. Most critical in this respect is the management of water in peatlands, but other factors such as the judicious use of grazing, cutting and burning may all be useful ways of preventing change taking place. In many cases management is directed at preventing natural successional change proceeding. For example, open fens may tend to become shrub dominated in the absence of grazing, mowing or burning as they become increasingly terrestrialised. The open fen habitat is often seen as more desirable for conservation because it harbours much more diverse and unusual plant, invertebrate and bird life. In the following section we will review the various practices used in the management of semi-natural peatlands and show how they are used to manipulate the conditions subtly.

11.4.1 Hydrological management

Throughout this book we have stressed the centrality of water and hydrological functioning to the peatland system. Without adequate water, the peatland will slowly but surely cease to exist at all. Initially, vegetation is affected and then the peat itself begins to shrink and decay (see Chapter 10). In damaged mires, enhanced drainage is often associated with human impact either on the mire itself or on the surrounding area (Figure 10.1). Consequently the management of water is central to maintaining proper conditions for the ecology of the system and peat growth. The fundamental problem is usually that there is an inadequate supply of water or that water loss is too rapid. The result in either case is the same: a low water table. Although it is possible to provide additional supplies of water by pumping water on to the site, most management concentrates on reducing the flow of water from the site.

Water loss can be prevented by damming or infilling of any drainage ditches, as well as by installing bunds (long embankments) to prevent runoff over a wider area. There are various types of dams and bunds as well as different approaches to installing them (Brooks, 1997). The simplest dams are sheets of plywood, metal or plastic pushed into the ground across

a)

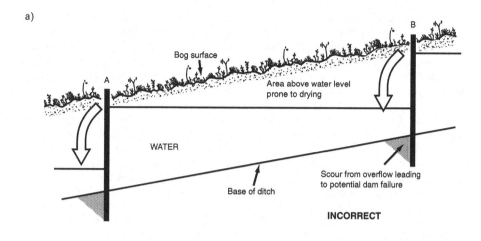

b)

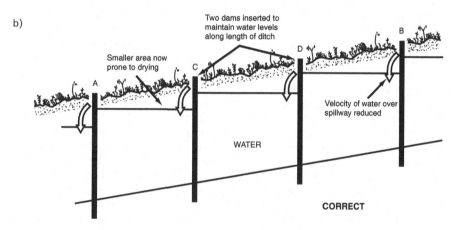

Figure 11.1 Principles of ditch blocking for mitigation of effects of drainage. (a) Dams need to be constructed so they provide adequate protection against seepage around and below the dam, and (b) arranged so that the change in water level between two adjacent dammed sections is not too high. This keeps the water pressure on individual dams at a reasonable level but also ensures that the water table is raised more evenly over a sloping surface. Redrawn from Brooks (1997). Crown Copyright. Reproduced by permission of HMSO.

the width of the drain and penetrating to some depth below the bottom of the drain. More complex dams of vertical planks or those sandwiching a peat filling between two separate boards can also be used. The two key objectives with damming are to raise the water level and to eventually block the drain by recolonisation of plants (Figure 11.1). The water level should also be raised evenly over the length of a drain, which entails making sure that drains are spaced sufficiently close together where there is any slope to the drain. Hydrological control will be discussed further in the section on peatland restoration.

11.4.2 Grazing

Grazing, mowing and burning are all ways of reducing the total standing biomass of the peatland. However, there are considerable differences in their effects, and each has to be used in particular ways to be sure of the final response. There is a fine line between management and damage with all these techniques. Using any of them too much or at the wrong time of year can be detrimental to the peatland.

Grazing is probably the most widespread agricultural use of peatlands, especially throughout

European mires, where wet grasslands, fens and upland peatlands are important marginal grazing land in many countries. Grazing and its impacts on any specific vegetation type is a complex process. Variations in the timing, frequency and intensity of grazing are fundamental. The timing is important especially in relation to the phenology (change in growth stage through time) of the plant species present. Plants subject to grazing just before flowering are unlikely to produce adequate seed for reproductive spread. Frequent and/or very intensive grazing may cause some plants to suffer sufficient damage that they are no longer able to compete with other plants in the community. The effects of grazing on plants are not only connected with removal of the growing shoots (and sometimes roots), but also with the physical effects of trampling and the nutrient loading from urine and faeces of the grazing animals. All of these variables can also be affected by the type of grazing animals, who vary in the degree to which they select different plants to eat and in the way they graze. Sheep nibble vegetation while cattle tend to rip the shoots from the plants, sometimes uprooting weakly rooted plants.

Grazing by sheep on blanket mire communities is one of the commonest forms of grazing on British peatlands and is probably essential to the existing character of many areas. Here there are clear preferences in plant species of the sheep, and the utilisation of different species varies throughout the year depending on what is available and what is most palatable. In one study, *Molinia caerulea* (purple moor grass) and *Trichophorum cespitosum* (deer sedge) were found to form the main part of the diet during the late spring and early summer, but these were replaced by *Calluna vulgaris* (heather) and *Eriophorum vaginatum* (hare's tail cotton grass) later in the year and especially through the winter months (Grant *et al*, 1976). This is partly due to availability (*Molinia* and *Trichophorum* are not available during the winter) and also palatibility – *Calluna* is eaten year-round but it forms a much greater proportion of the diet in winter since other more palatable species are not available. One of the problems in managing the grazing on sites such as these is how to avoid overgrazing. Overgrazing may not be immediately apparent as responses of many peatland plants are slow, but over time changes become apparent. In many blanket mires, *Calluna vulgaris* suffers long-term declines with high grazing pressure as shown by experimental monitoring (Grant *et al*, 1985) and from the response to cessation of grazing (Rawes, 1983). Figure 11.2 shows the changes in *Calluna* cover associated with different grazing levels over an 11-year

period. This also illustrates damage due to an increase in the extent of bare peat areas, which could eventually cause erosion and loss of peat. Palaeoecological studies reconstructing vegetation change over a period of several hundred years confirm suppositions that the rise to dominance of *Molinia caerulea* over *Calluna vulgaris* in Exmoor, England, is a recent phenomenon probably caused by increased grazing pressure (Chambers *et al*, 1999). Although the general problem of overgrazing on blanket mire in Britain is now clear, calculation of appropriate stocking levels to avoid undesirable vegetation change is not always straightforward due to variations in vegetation and growing conditions. In the Shetland Islands, problems have arisen because these local variations have not been adequately considered and because traditional calculation of stocking levels did not take account of an increase in the average size of the sheep kept (Birnie and Hulme, 1990). Equally, economic policies may cause conflict with environmental goals of blanket mire management. Often agricultural subsidies are paid per head of sheep, encouraging maximum use of upland grazing to the potential detriment of the vegetation in the long term (Pulteney, 1988; Birnie and Hulme, 1990). In contrast to most areas, some sites may be suffering from an absence of grazing. Isolation of the sites within the Kielder Forest has taken them out of the previously established management of light grazing and occasional burning. This may be the cause of loss of some of the plants typical of wetter conditions on ombrotrophic mires in the region (Chapman and Rose, 1991). Light grazing of such sites is therefore normally recommended management for nature conservation (Rowell, 1988).

Grazing management is not limited to European peatlands. Cattle have grazed some alpine peatlands in Australia for many years since European settlement. Despite the fact that the potential damage to species-rich sites was realised in the 1940s, grazing has continued and it is now apparent that the grazing pressure is sufficient to cause reduction of characteristic species with increases in bare peat and introduced species (Wahren *et al*, 1999). Here the issue is not really one of using grazing to maintain a peatland habitat but of controlling it in order to reduce damage, with the recommended option being exclusion of cattle. Grazing pressure from non-domesticated animals may also be a significant influence on peatland vegetation, although it is not normally one that is managed actively or thought to have reached damaging levels that could be considered 'non-natural'. Wild geese are important grazers of many peatlands in the Northern Hemisphere.

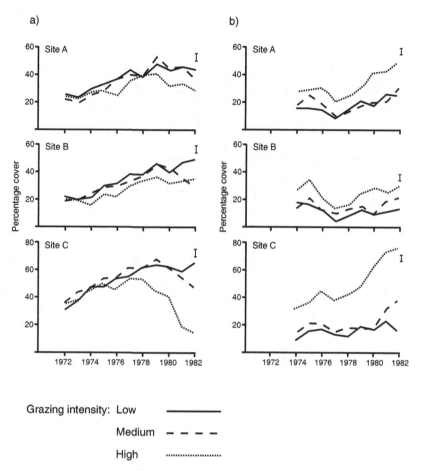

Figure 11.2   Changes in (a) the cover of *Calluna vulgaris* (heather) and (b) the extent of bare ground at three blanket mire sites in Scotland over an 11-year period with different sheep stocking rates. The vertical bars show standard error of the differences between individual plot means in 1982. Redrawn from Grant *et al* (1985) by permission of Blackwell Science Ltd.

In northern Canada, increased populations of the greater snow goose (*Chen caerulescens atlantica*) during the summer months have had a significant effect on vegetation of the tundra, including peatlands. The long-term effects of this are not yet known but the levels of grazing are currently limiting the total productivity of the graminoid species (Gauthier *et al*, 1996).

11.4.3 Mowing and cutting

Mowing or cutting of vegetation is an ancient method of harvesting a number of crops from semi-natural peatlands. Wet grasslands and fens produce good hay crops of nutritious material, plants such as reed (*Phragmites australis*) are harvested for roof thatching

and others are used as the raw material for basket weaving such as withies (the young shoots of willows, *Salix* spp.). These have all been much more widely practised in the past than they are now throughout Europe (e.g. Moen, 1995). The main effect of such practices is to remove biomass, retard the tendency of successional change in the vegetation and maintain plant community diversity by preventing complete dominance of a single competitive species. As a result these practices need to be continued if the stability and biodiversity of the ecosystem are to be maintained.

As with grazing, there are a number of variables that contribute to the overall effect of mowing on the vegetation. The timing of the cut is crucial with respect to the flowering and seed set times of annual or short-lived perennial species. For example, the milk parsley

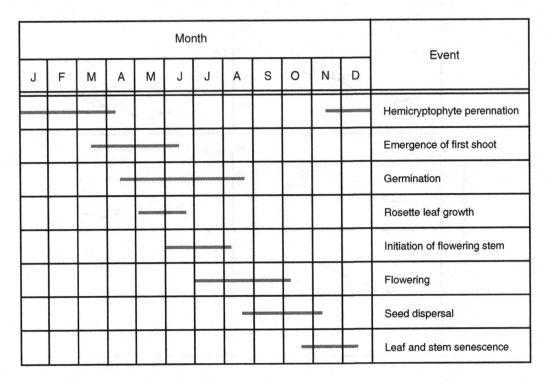

Month												Event
J	F	M	A	M	J	J	A	S	O	N	D	
												Hemicryptophyte perennation
												Emergence of first shoot
												Germination
												Rosette leaf growth
												Initiation of flowering stem
												Flowering
												Seed dispersal
												Leaf and stem senescence

Figure 11.3   The phenology (life history calendar) of *Peucedanum palustre* helps understand the effects of mowing at different times of year. Late summer mowing (after June) tends to prevent or reduce seed production as the flowering stem is removed. Mowing in the spring or early summer avoids this and the plant can reproduce more easily. Redrawn from Harvey and Meredith (1981) and Rowell (1988). Reproduced by permission of John Wiley & Sons, Ltd.

(*Peucedanum palustre*), an important and rare plant of the East Anglian fens in England, was badly reduced as a result of a late summer cutting regime, which damaged the flowering head of the plant as it emerged and matured (Figure 11.3). Changing to cutting during the spring avoided this problem and other alterations to the frequency of cutting produced further benefits for this plant (Harvey and Meredith, 1981; Rowell, 1988). As milk parsley is the food plant for a rare butterfly, the swallowtail (*Papilio machaon*), it is essential to successfully manage the plant effectively if the butterfly is also to flourish in the future.

Other variables that will alter the effect of mowing are the frequency and height of the cut. Plant phenology is also important here, but plant morphology can also be critical since low-growing species avoid the effect of a high cut height completely and may be placed at an advantage over higher-growing species. Graminoid species tend to tolerate mowing well, unless it is very frequent, but in local circumstances there are exceptions. Tyler (1984) suggests that *Calamagrostis*

varia is sensitive to mowing in Swedish fens, with other taxa such as *Carex hostiana* encouraged by regular mowing and others suffering with annually repeated mowing (e.g. *Cladium mariscus*, saw sedge). The major difference between mowing and grazing is that the physical effect of mowing is more predictable as there is no species selectivity and none of the microscale variability imposed by trampling and dunging patterns. Of course this evenness may be seen as a disadvantage as there are fewer opportunities for plant colonisation and growth.

Because of the importance of diversity to the conservation value of many fens, there has been particular interest in the effect of mowing regimes on the number of forb species represented in different vegetation types. At Wicken Fen, England, Rowell et al (1985) found that cutting at various times increased diversity in *Molinia* litter, especially where the cut was made in midsummer (Figure 11.4). The effect was similar in another vegetation type (*Calamagrostis* litter) but less marked with a cut taken at the end of the season.

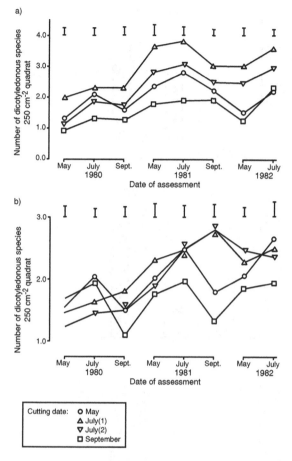

Figure 11.4 The effects of cutting on the diversity of dicoty-ledonous plant species in two fen vegetation types at Wicken Fen, England. The mean number of species present in a quadrat is given over three growing seasons after implementa-tion of cutting at the times of year shown. Cuts were made at the end of the months indicated. July (2) received an additional cut in May in year 2. Vertical bars show the standard errors of the mean. (a) Response in the *Molinia* litter vegetation type, (b) response in the *Calamagrostis* vegetation type. Redrawn from Rowell *et al* (1985) by permission of Blackwell Science Ltd.

Detailed field trials such as this one are the only real way of assessing the effect of mowing on particular vegetation types in specific sites, although some of the effects can be guessed at from a consideration of indi-vidual species ecology and the mowing regime used.

11.4.4 Burning

We have discussed fire in several other places in this book, particularly as a major natural influence on many boreal peatlands as well as a damaging impact of human activity where accidental fires are started (section 8.3). However, it is also used on a smaller scale as a management tool to maintain and modify peatland vegetation in both fen and bog systems, prin-cipally to help stimulate shoot growth for grazing by domestic animals and game.

Of all the peatland systems in the world, the best-known use of fire is probably that applied to blanket mires in Britain, and especially in Scotland. The principal use of fire in the British uplands is for the management of *Calluna vulgaris* (heather) which is encouraged by regular rotational burning (approxi-mately every 10–15 years). The main reason for the burning is to encourage a mosaic of different-aged heather stands to increase the carrying capacity for red grouse (*Lagopus lagopus scoticus*), the main game bird. Young shoots in recently burned areas provide food for the adults while taller, older stands provide nesting sites and cover for chicks. Although most of the management is carried out on shallow peat and podsolic soils, burning inevitably encroaches on to adjacent areas of deeper peat on blanket bogs. Perhaps partly because of the long cycles involved and the slow response of bog vegetation to change there is rather little unequivocal evidence for the bene-ficial or detrimental effects of light controlled burning on blanket bog. In general, advice is usually to avoid it because of potentially damaging impacts, especially if fire gets out of control (Rawes and Hobbs, 1979; Rowell, 1988; Brooks, 1997). The damaging effects of uncontrolled fires on blanket mires include loss of peat through combustion and erosion, as well as severe damage to plant and invertebrate communities and clearly should be avoided (Anderson, 1997a). However, the removal of dead plant litter by light burn-ing in the winter months plus the control of woody shrubs could help maintain a more open, wetter mire surface in some circumstances (Chapman and Rose, 1991).

Burning is used in fens to keep down shrub growth, although damage to other aspects of the ecology such as the invertebrate life have to be kept to a minimum with careful management of the frequency and extent of burn (Bennett and Friday, 1997). It is not only in European mires that fire is needed to maintain peat-lands in a particular state. In the prairies of North America, fire was a frequent event in the past, but fire protection schemes in more recent times have resulted in woody scrub invasion on fens in the midwestern United States (Bowles *et al*, 1996). However, it appears to be difficult to detect significant change in vegetation

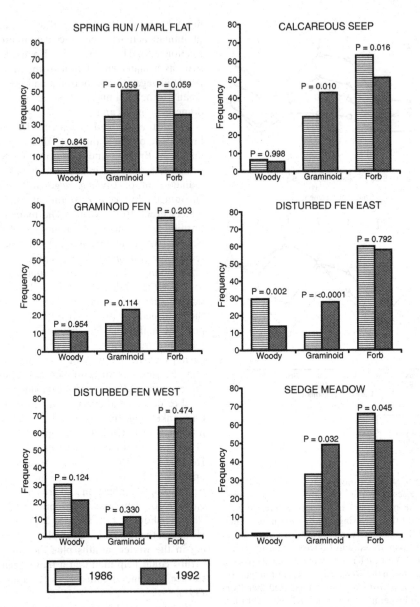

Figure 11.5 Changes in vegetation structure in six main plant communities in the Bluff Spring Fen in Illinois, USA, between 1986 and 1992 following annual burning. Only some of the recorded changes are statistically significant (shown by p values on each pair of bars), with notable increases in graminoid species frequency in several plant communities. The only significant reduction in woody plant taxa was in the disturbed fen area (East) where woody plant cover was one of the highest at the start of the monitoring period. Redrawn from Bowles *et al* (1996) by permission of the Natural Areas Association and Allen Press.

during monitoring of experimental burning over short (six-year) time periods, although a decline in woody plants and an increase in graminoid taxa can be shown (Figure 11.5). The control of woody plants such as *Cornus racemosa* will probably need additional management by cutting and herbicide use to increase

the open character of the vegetation of the worst affected sites (Bowles *et al*, 1996).

Another interesting example of the use of fire in New World peatlands is provided by a study of Crystal Fen in Maine, USA (Jacobson *et al*, 1991). Here, the construction of a railway line across the site in the late nineteenth century led to increased fire frequency because of sparks from steam engines. The railway line also impeded drainage and together these two effects led to a greater extent of open fen where woody plants could not establish themselves. This had a beneficial effect on some of the rarest plant species, which occupy wet, open fen in this region. Changes to the use of diesel engines after 1950 reduced fire frequency and the area of open fen began to contract. As a result it has been suggested that to preserve the rare fen plants burning management and other measures should be introduced to maintain open conditions within a mosaic of other peatland vegetation (Jacobson *et al*, 1991). This is a good example of the dilemma faced in management of semi-natural peatlands, made all the more apparent by the recent timing of the disturbance. Either the peatland can be allowed to revert to pre-late-nineteenth-century 'natural' conditions with the potential loss of some rare plants or it can be maintained in a 'disturbed' state to preserve biodiversity. In many European mires, the disturbance has continued for much longer, the likelihood of returning to near-natural conditions is much less, and the natural end point is not always clear. However, the philosophical arguments are in essence no different from those over the future of Crystal Fen, and sometimes result in the inconsistency in attitude suggested by the quote from Wheeler, above.

11.4.5 Other management activities

There are a number of other management activities on peatlands that are used for various reasons, although perhaps rather less commonly and sometimes for different reasons from those mentioned so far. Removal of woody plants by cutting and use of herbicide either before or afterwards is necessary where plants have become too large to be killed by fire or grazing. Scrub invasion is a particular problem on mires that have suffered drainage for any reason, and is sometimes one of the problems that have to be overcome with restoration of badly damaged peatlands (see below). Even in oceanic areas, an intact raised mire will have some woodland around the edges in the lagg zone. Drainage ditches allow colonisation of narrow strips of peatland alongside the drains. Various methods have been used to kill trees and scrub that are thought to reduce the conservation value of sites. Felling and/or treatment with herbicide is often recommended for larger trees, although ring-barking can also be effective, while seedlings can be hand pulled or cut by hand or with a brush cutter (Brooks, 1997). Disposal of the

timber and brushwood created by scrub removal can be an even greater problem than killing the unwanted plants in the first place. As with timber removal in forestry (Chapter 10), significant disturbance can be caused if cut trees and shrubs are removed by vehicle. Other methods range from burning or chipping on site, using remains to block ditches, to more elaborate operations removing larger timber by helicopter!

Perhaps one of the greatest apparent paradoxes in peatland management is the use of peat cutting to create and maintain particular conditions on some sites. Because many fen communities are transient phases in the autogenic succession, they inevitably have a limited lifetime. Cutting of peat may sometimes be the only effective method of reversing the successional processes to keep the plant communities most valuable for conservation. Many fens were created by peat cutting in the first place and so it is not surprising that they require peat cutting to be maintained (Giller and Wheeler, 1986a). The scale of past peat usage in some areas is surprising: Wheeler and Shaw (1995a) suggest that in some areas of the East Anglian fens rates may have been as high as the equivalent of 1 m of peat removed from 1 ha of fen every 50 years. Of course, though collective rates may have been high, this was not modern large-scale commercial peat extraction, but would have occurred in a much more patchy pattern over time. Abandoned cuttings would have been recolonised by plants from adjacent fens and other wetlands in the area, leading to the emergence of a diverse pattern of fens at different stages of succession. Figure 11.6 shows the kind of changes that could occur within abandoned fen peat cuttings. The type of plant community that establishes is mainly dependent upon the water level within the peat pit. In turn, this is determined by the location of the cutting in relation to the groundwater table, and the nature of drainage management on adjacent sites. In deeper water, colonisation would be limited to tall emergent plants (*Phragmites australis*, reed and *Typha angustifolia*, reedmace), while those with a wet surface and little standing water would be colonised by a greater range of fen and wet grassland taxa. Hydrological conditions may not have been stable but could have become flooded after an initial phase of moist conditions, following a local change in management. This could lead to the formation of a floating raft of vegetation, producing more rapid terrestrialisation of the pit. In terms of management of fens using peat cutting, it could suggest that a period of drainage of the cutting would help promote initial establishment of plant cover (Wheeler and Shaw, 1995a).

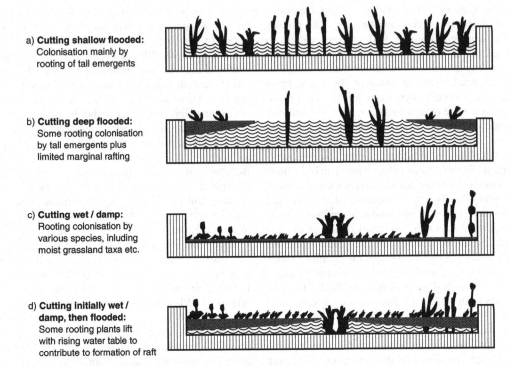

a) **Cutting shallow flooded:**
Colonisation mainly by
rooting of tall emergents

b) **Cutting deep flooded:**
Some rooting colonisation
by tall emergents plus
limited marginal rafting

c) **Cutting wet / damp:**
Rooting colonisation by
various species, inluding
moist grassland taxa etc.

d) **Cutting initially wet /
damp, then flooded:**
Some rooting plants lift
with rising water table to
contribute to formation of raft

Figure 11.6   Possible mechanisms for recolonisation of abandoned peat cuttings in the East Anglian fens. The two flooded cuttings allow only the establishment of taller emergent plants, with some marginal extension of a vegetation raft from the edges of the intact surface (a and b). Drained cuttings with moist conditions at the surface may have been colonised more rapidly with a diversity of plant species (c). With a rise in the water table and subsequent flooding, this vegetation would form a floating mat and produce a complete plant cover more rapidly (d). Redrawn from Wheeler and Shaw (1995a) with permission. © John Wiley & Sons Limited.

11.5 RESTORATION AND REHABILITATION

There is a long history of the creation of wetlands in various parts of the world, both for economic reasons and for wildlife habitat creation (Mitsch and Gosselink, 2000). However, until recently rather few of the 'new' wetlands have been peatlands, perhaps partly because of the difficulties and long periods required to achieve possible success (Mitsch and Wilson, 1996). Some of the first restoration work was carried out in the Netherlands in the Engbertsdijksvenen area where peat mining licences began to expire in 1953, from which point on it was managed by the state with the aim of re-establishing ombrotrophic bog vegetation (Schouwenaars, 1992). Further work was carried out in the Low Countries and Germany after this but peatland restoration remained a relatively minor part of peatland conservation and management activities until the 1980s (Akkerman, 1982; Heathwaite *et al*,

1993a). In the last 20 years there has been a major increase in interest in peatland restoration, partly associated with a more general rise in consciousness of the need to go beyond *preservation* to *recreation* in wildlife conservation. However, restoration of cutover raised mires in particular has also been promoted and funded by the peat industry as a response to environmental critics and the need to build a more sustainable peat industry. While restoration of peat extraction sites has been the focus of the greatest interest, other restoration research has been carried out on sites damaged by drainage, erosion, forestry, agricultural reclamation or even simple abandonment (in the case of fens). In some of these cases, there is a fine line between more radical management and true restoration. For the purpose of this chapter, we have included any form of management that is attempting to bring about significant *change* rather than simply *maintaining* a peatland ecosystem.

11.5.1 Definitions and objectives

'Restoration' is the term most commonly used for attempting to bring peatlands back to functioning ecosystems. Wheeler (1995) suggests that 'restoration' may well encompass a variety of eventual objectives including either the restoration of the site to its original state or to a former state. However, it is not always clear what the 'original' state is, because peatlands change naturally over time through succession (Chapter 7). It would perhaps be valid to attempt to restore a mire to one of its many earlier seral stages. On the other hand, with raised mires, the conservation concern is usually with the scarcity of the final seral stage (the ombrotrophic bog habitat), and the objective would normally be to achieve a similar vegetation cover at least. Perhaps the ideal in restoration ought to be to return the peatland to the condition it would have been in now without disturbance. Clearly this will never be achieved. Even if the exact conditions of surface vegetation, ecology and hydrological functioning could be reproduced (a very tough task as we shall see), at the very least the stratigraphic record would certainly be disrupted. Given the difficulties of complete restoration, it is better to have more modest but achievable aims in a restoration programme, including much wider wildlife conservation goals. Other terms that have been used for creation of peatland habitats are 'rehabilitation' or 'renaturation'. These terms are variably used and most of the time appear to imply a wider range of acceptable objectives than restoration, including the creation of peatland habitat that may never have existed on the site at any time in the past. Such activities can still be important to peatland conservation as they still produce 'new' peatland habitat from previously seriously damaged areas. The term 'regeneration' is usually used to imply complete restoration of the natural contemporary vegetation of the site (Pfadenhauer and Klötzli, 1996), but should not be confused with the natural autogenic successional processes used to describe renewed peat growth following dry periods (Chapter 7).

The overall rationale for restoration is similar to that for wildlife conservation as a whole. The difference is merely one of scale. We are used to thinking of financing and promoting the conservation of existing peatlands. Restoration simply takes it one step further in using much greater resources to reverse earlier destruction and recreate habitat that has some conservation value, because of the species, ecosystems or functions it possesses. It could be argued that such resources would be better employed in conservation of existing

sites, where these are still in need of further protection. From a wildlife conservation point of view, it would clearly be madness to use large sums of money to restore one site while other undamaged sites in the same area were in need of protection. However, we should not forget political and economic drivers for peatland restoration. For example, they allow the peat industry to demonstrate their commitment to the environment by showing they are willing and able to replace what they have taken away (e.g. Keys, 1992).

The detailed objectives of restoration are extremely variable and are affected by our (subjective) views on the relative values of the systems that we are restoring and of those that it would be possible to achieve. In cases where the damage has not totally destroyed the original system such as afforested blanket mire or an abandoned and overgrown fen, identifying the conservation objectives may be quite straightforward, especially if there are remnants of the original habitat close by. Clearly most of the time we would wish to recreate a peatland as close as possible to that which we would expect in the absence of damage. This is often the easiest to achieve, and satisfies several objectives for wildlife, landscape and functional values. Decisions are more complex when the 'blank sheet' of an excavated raised bog is presented. The ultimate objective could be to restore the site to a fully functional raised mire but, given the rarity of fen habitat in the same regions and the relative ease of restoration, a poor fen system may become the target. Wheeler (1995) suggests four considerations when deciding on the objectives of restoration:

(i)   Feasibility: what type of peatland and associated vegetation is it scientifically and economically possible to produce? Our 'ideal' peatland may be impossible to achieve due to changes in climate, water supply or the surrounding landscape
(ii) Former character: how well do we know what the peatland was like in the past and which stage of development should we aim for?
(iii) Rarity: should we prioritise rare habitats or rare species? It may not be possible to achieve both in the same site
(iv) Scarcity of opportunity: some end points may be relatively easy to achieve in a variety of locations; others are possible in only a few circumstances

There are not always easy answers when all these considerations are summed up, but they do at least promote a thorough exploration of the rationale for taking radical and expensive measures to restoration.

As mentioned earlier, the greatest efforts have been made on raised mire restoration and a discussion of these forms the largest part of the next section. However, many other damaged mire sites are currently undergoing restoration work of various kinds and we will give a short account of some of the key issues in these areas.

11.6 RESTORATION OF CUTOVER OMBROTROPHIC MIRES

The restoration of a mire that has, to all intents and purposes, been largely removed is clearly an enormous undertaking yet it is one that is exercising more peat-land scientists than ever before. There are conflicting opinions on the extent of success, often due to inexact objectives being specified and different site types being considered. In the following discussion we will examine the evidence for past responses following abandonment of peat cutting and the techniques and extent of success in restoring large-scale peat extraction sites.

11.6.1 Past natural regeneration

We know that peat extraction from raised mires has occurred for centuries and possibly millennia. Some of these sites have remained undisturbed for a considerable time since peat was removed and we can learn a great deal from the pathways and rates of change that are recorded in the palaeoecological record and sometimes also in the documentary record. When considering past natural recovery from cutting, it seems appropriate to talk about 'regeneration' rather than 'restoration' since there is no human intention behind the process. There are important differences between techniques of peat cutting that have a bearing on the potential and problems of restoration in cutover raised mires: hand cuttings, commercial block cutting and commercial peat milling all produce very different post-harvest conditions. Almost all of the evidence for past natural regeneration comes from small-scale hand cuttings. While this limits the application of these ideas to very large-scale restoration programmes, it does at least show how fast plant successions can take place given the right conditions.

There is now good evidence from stratigraphic surveys in old cuttings that vegetation recovery in excavated peatlands is quite possible over long periods of time (>100 years). The initial colonisation by *Sphagnum* can be quite rapid – certainly less than 25 years (Robert *et al*, 1999). Joosten (1995) reviews the palaeoecological evidence for past recovery in small-scale peat cuttings and shows that there are many examples of regrowth of plants typical of ombrotrophic conditions that have occurred in old hand-cut peat workings. One area of peat cuttings in the Netherlands abandoned in 1857 shows recolonisation by *Sphagnum papillosum* followed by a more mixed mire community in the last 50 years. After 130 years, the vegetation is almost entirely composed of ombrotrophic peatland species (Joosten 1985). Figure 11.7 shows an example of the detailed sequence of change at this site, which also shows the relatively rapid rates of accumulation that can be achieved. Rates of peat accumulation are highly variable and appear to be quite unpredictable with ranges of $100-490$ g m^{-2} yr^{-1} quoted from a variety of studies (Joosten, 1995). Similar rates of vegetation recovery are recorded for hand cuttings on three Swiss peatlands with development of ombrotrophic vegetation by between 60 and 100 years (Buttler *et al*, 1996). Interestingly, these profiles also showed that the microfauna also recolonised rapidly, with a variety of testate amoebae communities indicative of acid bog conditions at a very early stage in recovery.

The almost complete recovery of some sites to ombrotrophic vegetation does not mean that natural regeneration has always taken place or indeed that the changes have all been towards appropriate vegetation types. Where no regeneration has taken place, there will be no palaeoecological evidence and peat wastage and total disappearance of a recognisable peatland may have occurred. Old block-cut peat workings on Thorne Waste, England, show more limited recovery after approximately 65 years, with poor fen species more common than true raised bog taxa (Smart *et al*, 1989). In this case, this is not necessarily seen as a 'failure' to regenerate, as many of these plants would have been present within the lagg fen around the original natural site, a habitat that has disappeared through subsequent drainage and peat cutting (Smart *et al*, 1986).

Comparisons of existing vegetation on previously cut and uncut areas also suggest that natural regeneration is often more limited on some sites, especially those that have been milled and vacuum harvested rather than being block cut. At Cacouna Station, a raised bog in Quebec, Canada, the recolonisation of some taxa has been slow on areas abandoned from the 1970s onwards (Lavoie and Rochefort, 1996). On block-cut areas, vegetation cover was approximately 50% but although the plants were typical peatland species, *Sphagnum* cover was still much less than on

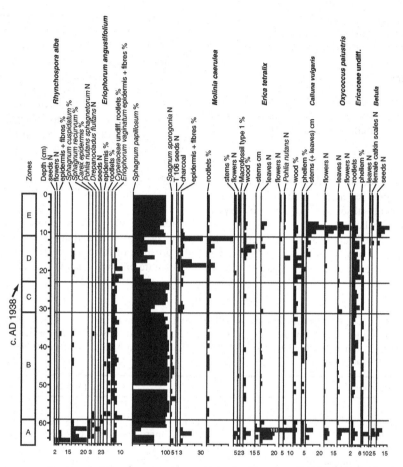

Figure 11.7 Plant community changes following natural regeneration in old peat cuttings in the Netherlands as reflected in plant macrofossils. The dominant plant is *Sphagnum papillosum* but a more diverse plant community has developed since the horizon dated at AD 1938. Redrawn from Joosten (1985) by permission of Elsevier Science.

the undamaged areas even after at least 25 years. On the vacuum harvested areas that were not abandoned until much later, the most abundant plant was a species of birch tree, *Betula populifolia*, and other peatland taxa were largely absent after the five-year recovery period. At Stormossen, Sweden, peat pits abandoned 50 years ago show much better recovery with good vegetation cover, although plant community diversity was much lower than in an undamaged site (Soro *et al*, 1999). Diversity in some of the other old cuttings in the same region was much higher but only as a result of poor fen and wooded peatland taxa being present. Animal life also changes following abandonment of peat cuttings. Again in block-cut sites, natural recovery appears to be better than for vacuum harvested sites. In peatlands in southern Quebec, bird diversity and abundance were similar in undamaged and abandoned

block-cut sites after about 20 years, but both were lower on abandoned vacuum harvested sites (Desrochers *et al*, 1998).

Although it seems clear that peat cuttings can regenerate successfully from natural processes, this probably requires a number of important preconditions. First, that the excavations are relatively small, hand-cut and probably a small proportion of the total mire expanse. Second, that there are local sources of plant propagules from which new vegetation can develop. Finally, the hydrological conditions are almost certainly crucial. For the old cuttings where regeneration has been successful, water tables at around the surface may have been partly serendipitous, but are more likely within shallow, fragmented cuttings. Although raised mire vegetation recovery is possible to some extent, it will therefore not occur in large flat expanses of areas excavated by peat milling. Conditions on these sites need to be manipulated to encourage plant growth. This is the main focus of the rest of this section.

11.6.2 Constraints

The peatland remaining after peat extraction is not a favourable place for the regrowth of many peatland species unless it is carefully managed. The chief constraints are:

(i) Hydrological – the water table is often too low and fluctuates too much. Also the water storage capacity is much lower than in an intact bog and losses to mineral subsoil may be greater if the peat depth is very shallow

(ii) Biological – there is no seed or spore bank in the sterile peat. New plants must come in from other areas of the same site or further afield

(iii) Chemical – the water supply may be too nutrient rich for bog plants and the bare peat surface may provide additional nutrient input (where for example it has been cut down to fen peat horizon). Release of sulphates may also make the surface water more acid than is ideal

(iv) Structural – the peat is compact and may be a poor seed bed as well as affecting the hydrological properties adversely (see above)

(v) Microclimatic – the peat surface has no protection from plant cover and temperatures and moisture levels fluctuate a lot with the diurnal cycle. Wind speeds may also be high

In addition, there may be other external constraints in terms of legal, practical and financial issues that make the prospect of restoration more or less favourable or necessitate different aims (Table 11.3).

Table 11.3 Constraints and considerations affecting possible restoration options for raised mires. Some of these constraints will mean it is impractical to attempt full 'restoration' of raised mire vegetation, and other options including rehabilitation to other peatland or wetland habitats may be considered more achievable. Modified from Wheeler and Shaw (1995b). See text for further detail on practical constraints.

Consideration or constraint	Impact
Legal	
Planning legislation and policies	Affects most appropriate use via planning conditions and costs
Conservation legislation	Affects most appropriate use and acceptable methods to achieve this
Water resource legislation (impoundment, abstraction, discharge)	Affects methods
Neighbouring landowners	May constrain possible after-use and methods
Other (rights of way, grazing rights, public utilities)	May constrain possible after-use and methods
Practical	
Topography (site surface and relationship with surrounding landscape)	Feasibility for retaining water and methods
Hydrology	Requires information on causes and processes to design appropriate methods
Peat depths and types	Affects hydrological, chemical and physical conditions
Existing flora and fauna	Availability of recolonist species, current wildlife interest may need to be preserved (even if it is not typical of raised bogs)
Peat archive	Preservation of intact profiles/archaeological interest constrains methods
Chemistry	Base status of substrate and water supply. Air pollution may limit possible recolonist taxa
Access	Machine access usually required
Practical assistance	Requires expertise and availability of machinery, survey, monitoring, management
Financial	Costs may include: compensation, survey and feasibility studies, management and machinery operations and maintenance of conditions after initial work. Grants may offset costs

Assuming there are no serious legal or financial constraints, the conditions needed for restoration of raised mire vegetation to be a serious possibility are (Rowell, 1988):

(i) There should be a large area of peat where the drainage does not cut into the mineral substrate.
(ii) There should be at least 50–100 cm of compressed, humified peat
(iii) It should be possible to exclude all sources of nutrient enrichment (air- and water-borne)
(iv) There should be a buffer zone between the site and agricultural land
(v) A source for plant colonisation should exist locally

If these conditions can be met, a variety of methods is available for beginning the process of restoration. Here we will consider techniques and impacts of hydrological management and vegetation recolonisation.

11.6.3 Hydrological management

Establishing the correct hydrological conditions is critical for re-establishment of peatland vegetation on a cutover peat surface. The chief requirements are to achieve a high and stable water table that keeps the surface saturated throughout the year and for the water supply to be of suitable quality (unpolluted rainwater). The specific measures that are needed to achieve these objectives in any one site are highly variable, but they can include recontouring and shaping the site, ditch blocking and/or filling, sealing the edges of the site, and pumping of additional water from a reservoir. Various publications deal with the detail of these processes, including Rowell (1988), Wheeler and Shaw (1995b) and Stoneman and Brooks (1997), as well as numerous research papers that deal with very specific aspects and experiments, some of which are mentioned below.

In older abandoned cuttings, the shape of the remnant mire may not be suitable for even rewetting of the site; there are often upstanding blocks and other rapid changes in elevation that make for some very dry locations and other naturally very wet locations. Recontouring of the site smoothes out these uneven changes, as well as being used to construct impoundments for water where the surface has continuous large flat areas. In mires which have been very severely reduced in lateral extent, some recontouring has been aimed at achieving a water table similar to that predicted from the groundwater mound theory (see

Chapter 3). This can be by design or may also occur by default from allowing natural wastage and slumping of the peat mass to occur, but it is likely to allow more thorough recovery of a pseudo-natural hydrological system more rapidly (Bragg, 1995). More recently, it has become common policy, and sometimes a condition of planning permission for peat extraction, to leave the site in an appropriate physical shape for restoration. This may involve changes either to where and how much peat is extracted or to the peat company being responsible for the larger-scale recontouring work.

A key activity is usually blocking or preferably filling of any ditches to reduce water loss. The principles are similar to those for blocking of ditches in any peatland site as regards the spacing and design of dams (see above, Figure 11.1). The effect of damming alone, however, may not be adequate to achieve the correct hydrological conditions for restoration of the vegetation. One study monitoring changes following damming at two sites in Cumbria, north-west England, shows that although water tables are raised and fluctuations are moderated a little, they continue to be much more variable than natural bog surfaces. Flooding is more frequent and dry periods more extreme in the dammed cutover peatland than in undamaged bog (Figure 11.8). In the same study, there is some suggestion that even over the first year, the dammed mire surface is rising, presumably as a result of swelling due to some degree of more permanent rewetting of the peat (Mawby, 1995). Similar results have been found elsewhere although the rise in water table is not always so pronounced (Price, 1996).

Ditch blocking alone is often unlikely to be sufficient, partly because of the more limited storage capacity of the surface peat layers and also because vertical seepage may also be occurring (Schouwenaars, 1995). To increase the storage of water at the surface to maintain wet conditions during summer months, bunds enclosing shallowly inundated areas of peat are often used or adapted from existing upstanding baulks on block-cut sites. These have to be carefully constructed to allow only shallow water to cover the surface in winter while preventing the water table falling too low in the summer. The enclosed areas also help keep water tables in the separating ridges reasonably high throughout the year (Schouwenaars, 1995). Various possibilities exist, including excavation of shallow ponds with sloping sides to encourage recolonisation (Figure 11.9; Rowell, 1988). In situations where seepage rates are high or rainfall is marginal for keeping water tables close to the surface, pumping of additional water from elsewhere may be considered. At Thorne

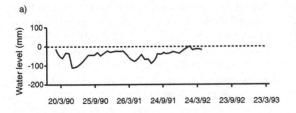

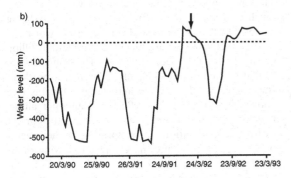

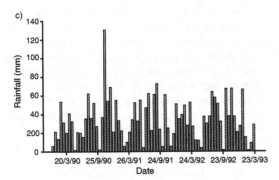

Figure 11.8 Changes in water level in (a) an area of undisturbed *Sphagnum*-dominated bog on Glasson Moss, Cumbria, (b) cutover bog on Wedholme Flow, Cumbria, where ditches were dammed in February 1992 (shown by vertical arrow). Rainfall totals are shown in (c). Changes in the elevation of the peat surface were also recorded for both sites. At Wedholme Flow, the level of the peat surface appears to be rising slightly following damming. Redrawn from Mawby (1995) with permission. © John Wiley & Sons Limited.

Moors, in Yorkshire, England, this was carried out as rainfall is very low (583 mm yr^{-1}), yet it was still inadequate to overcome losses from vertical and lateral seepage (Heathwaite, 1995). Pumping of additional water is often also a last resort as it may be difficult to obtain a source of water sufficiently low in nutrients and of low pH. Downward seepage can be a particular problem in areas where there is only a shallow cover of peat remaining or especially where the peat is on a well-

drained substrate rather than clay. In some of the sites in the Netherlands that are on a sandy substrate, the downward seepage can be as great as the annual moisture surplus (Schouwenaars, 1993).

11.6.4 Re-establishing vegetation

The main aim of hydrological management is of course to provide the correct conditions for the establishment of appropriate plant communities. Another issue is whether there are sources of seeds and spores from which plants can colonise newly exposed surfaces. In smaller-scale cuttings, remnants of surface vegetation may be present that could represent a source of plant material. However, these remnants often have different vegetation from that on undisturbed bogs due to the effects of peripheral drainage (e.g. Poulin *et al*, 1999), so that they may not supply all the species required. On larger expanses of bare peat this is even less likely. Any seed bank is likely to be dead, and the influx of seeds and their successful germination is variable (Salonen, 1987), so reintroduction of plants is almost always an important way of assisting restoration. This is discussed in relation to *Sphagnum* moss below.

Assuming adequate hydrological control can be achieved, one of the other key factors determining which species are likely to be able to establish on a cutover surface is the chemical environment. Cutover peat surfaces are highly variable in terms of their botanical composition and in whether they are cut to fen or bog peat. It is largely a product of the peat extraction process, dependent upon the main product of the peat industry at the site and the stage in extraction at which the site was abandoned. The dramatic differences in growing conditions that can occur within bog sites are shown by Figure 11.10. In this study, the conditions on four cutover sites were compared with those on adjacent undisturbed bogs. The undisturbed bogs had very low pH with low available nutrients in both the peat and surface water. Due to extraction proceeding to different stratigraphic levels in each of the cutover bogs, the water and soil chemistry of the cutover surfaces was highly variable. Only one site (from central Quebec) showed no difference with the undisturbed site for most variables, probably due to the more limited depth of peat extraction, leaving bog peat still exposed at the surface. Values recorded at the other sites were more similar to conditions expected in poor fen, or even moderate-rich fen in the Alberta site. It probably would not be advisable to attempt to revegetate all these sites

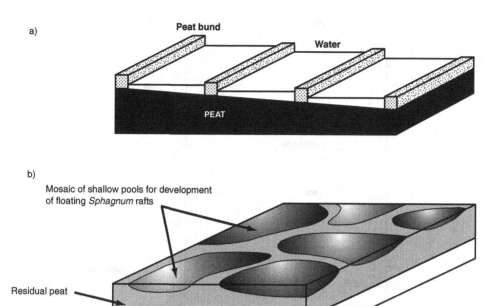

Figure 11.9 Small, shallow lagoons are often used as a first step in the establishment of peatland vegetation on cutover mire surfaces. (a) On a sloping site, parallel strip lagoons can be constructed to retain the water. (b) On a flat surface a mosaic of shallow pools can be excavated providing suitable conditions for *Sphagnum* establishment. The size of any lagoons is normally kept to less than 20 m in width to prevent excessive wave action and erosion, and the depth is usually only around 50 cm maximum, although it is variable depending on the site. Redrawn from Money (1995) with permission. © John Wiley & Sons Limited.

with the same ombrotrophic bog species, as least at first. In the long term, growth of some fen *Sphagna* on the more nutrient-rich sites could help promote acidification and a succession to bog conditions (Wind-Mulder *et al*, 1996).

Sphagnum is often regarded as the key taxon to establish, partly because it is a characteristic component of ombrotrophic mire vegetation but also because it helps to re-establish a functioning acrotelm, to stabilise the hydrochemical conditions (see Chapter 3), and also to moderate the microclimatic fluctuations. *Sphagnum* recolonisation is therefore often one of the key goals in raised mire restoration. However, it should be remembered that although this is an important goal, it is really only the beginnings of a new raised mire plant community. Nevertheless, understanding and controlling *Sphagnum* re-establishment remains important and there has been much experimental work on the subject, summarised by Rochefort (2000; see also Figure 11.11). The key activities are those of rewetting the surface (as discussed above), reintroducing *Sphagnum*, and mulching to moderate the microclimate. Once *Sphagnum* is established, there is a series

of feedbacks to the surface that aid further *Sphagnum* spread as well as the colonisation by other plants and the development of more complete vegetation cover.

The most important factor is whether there is a viable source of *Sphagnum* already present. In most recently abandoned peat fields that are fully worked out, no living *Sphagnum* is likely to be present in any appreciable quantities. It must therefore be reintroduced from another site. This could be done by using spores, but more often small fragments of vegetative material are used, known as 'diaspores'. There have been a number of experiments carried out on techniques of *Sphagnum* re-establishment from diaspores, especially for Canadian peatlands through the work of the Peatland Ecology Research Group at Université Laval, Quebec, Canada. The choice of species is important as different species have different capacities for regeneration from diaspores and of survival on the bare peat surface. Ideally one would want to establish species similar to those likely to occur in natural conditions but it may be better to aim for initial establishment of any species that will survive and spread effectively. Greenhouse experiments suggest that all

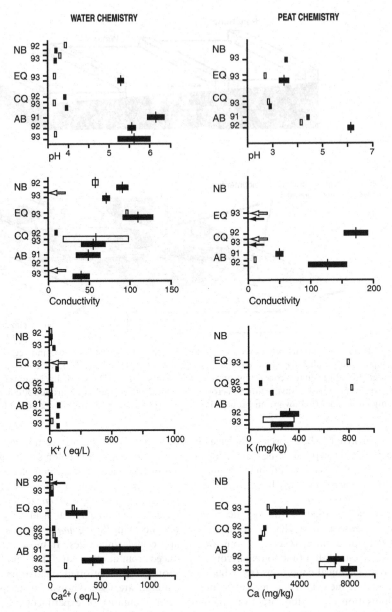

Figure 11.10   Differences in peat and water chemistry of cutover bog sites compared with their undisturbed counterparts for four sites in Canada, shown as annual means ± standard errors. The sites were in New Brunswick (NB), Eastern Quebec (EQ), Central Quebec (CQ) and Alberta (AB). The white bars represent the undisturbed bogs and the filled bars are the cutover bogs. Narrow bars reflect low standard error values. Arrows also represent low values. Redrawn from Wind-Mulder *et al* (1996) by permission of Elsevier Science.

species perform better with increased surface wetness and humidity, with the exception of *S. fuscum*, but that there were no major differences between species in overall growth rates (Campeau and Rochefort, 1996). Others have suggested that particular species are more

suitable for growth on bare peat surfaces. Grosvernier *et al* (1997) suggest that *S. fallax* is the most suitable colonist for oceanic sites in Europe, especially where hydrological control is poor and diaspores are exposed to greater fluctuations in moisture conditions. *S. fallax*

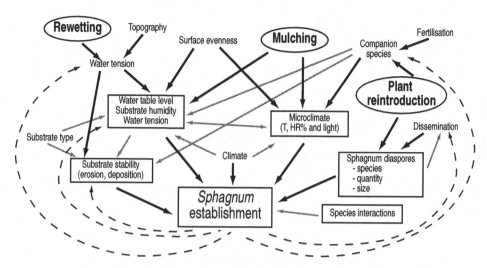

Figure 11.11   Flow chart summarising the main factors affecting *Sphagnum* establishment on bare peat surfaces. Thick arrows indicate relationships substantiated by experimental data and thin arrows are more speculative relationships. Dashed arrows are possible feedbacks. Redrawn from Rochefort (2000) by permission of *The Bryologist*.

was also found to be particularly tolerant of drying and heating compared to *S. fuscum* and *S. magellanicum* in another experiment (Sagot and Rochefort, 1996). The lesson seems to be that the choice of *Sphagnum* species is limited unless hydrological control is good. Most diaspore fragments appear to be able to survive for a period of two weeks without water but beyond this time viability is poor (Figure 11.12). Given this susceptibility of diaspores to drying, it seems a little surprising that more work on the use of spores has not been carried out, as they have a much longer viable life, extending to years or even decades in some conditions (Sundberg and Rydin, 2000). They may be of use where hydrological management is particularly difficult and clearly play a role in the dispersion of *Sphagnum* to sites where seeding with diaspores is not used.

The hydrological management measures already mentioned are the key techniques in rewetting, but there are still some questions over what constitute the optimal conditions for establishing *Sphagnum* cover. Conventionally, most monitoring of hydrological conditions is done using water table measurements, but for the plants, water table may be less important than soil moisture, which determines how much water is available for use. Price (1997) found that soil moisture was not much different between drained and ditch-blocked sites. This was apparently due to problems of high bulk density at the blocked site associated with the long period since abandonment of the site. Price (1996) has also pointed out that poor *Sphagnum* regeneration may

also be caused by its inability to extract water from the peat, even when this moisture is present, due to high matric suction in the peat. Moisture contents in bare peat surfaces can be higher than in *Sphagnum* hummocks on undisturbed peatlands, but structural differences in the peat compared with established *Sphagnum* moss mean that the water cannot be accessed by colonising *Sphagnum*. Thus we should be careful in assuming that high water tables and/or soil moisture will necessarily mean that *Sphagnum* can be successfully reintroduced to a bare peat surface. Other measures may be necessary to provide adequate water that is available for use by the plants.

The use of open water reservoirs is apparently successful in promoting the correct hydrological conditions for *Sphagnum* growth on adjacent baulks, also enhanced by closer spacing of open water areas compared to normal drain spacing (LaRose *et al*, 1997). Mulches have also been shown to keep the growing surface of the peat cooler and wetter during the dry summer period, leading to enhanced *Sphagnum* recolonisation (Figure 11.13). Other physical measures that appear to assist regeneration are protection using shading techniques (Buttler *et al*, 1998; Rochefort and Bastien, 1998) and surface reprofiling and surface flow management (Bugnon *et al,* 1997). Other plants can also play an important role in assisting *Sphagnum* recolonisation, either by using existing plants or by planting new specimens. Following rewetting of Danes Moss in England, *Sphagnum cuspidatum* devel-

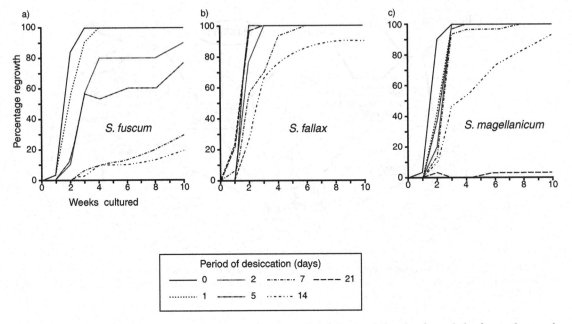

Figure 11.12   Regrowth of *Sphagnum* diaspores over a 10-week period, following desiccation for periods of up to three weeks. (a) *S. fuscum*, (b) *S. fallax*, (c) *S. magellanicum*. The regeneration success is expressed as the number of stems showing signs of regrowth after the specified time. Three replicates of 10 stems each were tested for each period. Redrawn from Sagot and Rochefort (1996) by permission of A.D.A.C. and *Cryptogamie*.

opment was most successful in areas of waterlogged *Molinia caerulea* tussocks and also appeared to be helped in areas where *Juncus effusus* litter was present to help support it. *S. cuspidatum* disappeared from other areas that were inundated with standing water, perhaps because of wave action in addition to the lack of a structural support for their growth (Meade, 1992). Ferland and Rochefort (1997) report increased success of *Sphagnum* when *Eriophorum angustifolium* was used as a companion planted species. Two species of *Eriophorum* (*E. spissum* and *E. angustifolium*) already growing on abandoned peat workings also promoted *Sphagnum* recolonisation (Boudreau and Rochefort, 1998).

Much of the focus in restoration of raised mires is on the restoration of hydrological conditions and vegetation cover. Rather few projects have been so concerned with the various animal groups that form part of a functioning peatland system. Perhaps this is justifiable for an activity that is still in its infancy; until successful vegetation recovery can be achieved surely we should not expect the animal life to return. However, it is interesting to see what has been achieved and what might be possible from looking at animal populations on damaged and undamaged sites. Inevitably responses

to post-harvest abandonment and restoration activities are complex due to the interaction with other habitats and the dependence on plant structure for many birds and invertebrates. Given the need for transplantation, one of the issues is whether invertebrate populations are also transferred in this process or whether they can recolonise by themselves once suitable habitats exist. Certainly recolonisation seems unlikely in some cases, such as the fly species associated with pitcher plants (*Sarracenia purpurea*). It is also possible that some invertebrates are adversely affected by restoration management, especially during the early rewetting phases when significant disturbance and change occur on the site. The large heath butterfly (*Coenonympha tullia*) is an important species for wildlife conservation on Fenn's and Whixall Moss on the English/Welsh border, yet it is likely to suffer severely with excessive rewetting of its host plant *Eriophorum vaginatum* (Joy and Pullin, 1997). Ideally such considerations should form part of the pre-restoration management plan but this is usually hampered by a lack of knowledge. Some insects will be more mobile and adaptable, and clearly it will be advantageous to have a local population on a natural bog close by to increase the chances of recolonisation. However, so little work has been done

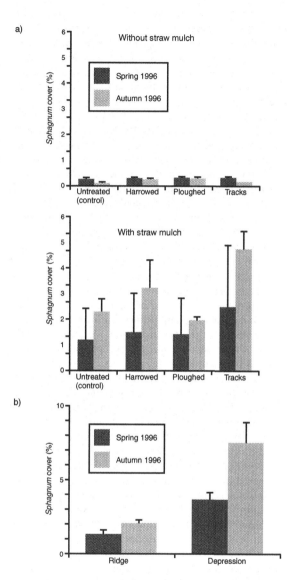

Figure 11.13 Enhanced *Sphagnum* recolonisation using a straw mulch. The graphs show the extent of *Sphagnum* cover (mean ± standard error) at approximately 12 months (spring 1996) and 18 months (autumn 1996) after spreading of *Sphagnum* plant fragments on the surface. (a) Comparison between various surface preparations with and without a straw mulch, showing that the mulch increased *Sphagnum* recolonisation in all areas. (b) Comparisons between different microtopographic areas within a zone constructed into linear ridges and depressions (approximately 1 m wide, 1 m apart and 0.2 m deep). *Sphagnum* cover increased more in the depressions than on the ridges and this was further enhanced by the application of the mulch. Redrawn from Price *et al* (1998) by permission of Elsevier Science.

on this aspect of peatland restoration that the impacts and possibilities of management are poorly known at present.

11.7 RESTORING OTHER DAMAGED SYSTEMS

11.7.1 Fen restoration

A huge effort has been directed at restoring worked-out ombrotrophic peat deposits, but there are numerous other peatland sites that are in need of restorative management to improve their wildlife conservation value. Fen restoration forms the largest of the other restoration activities on peatlands. Given that the destruction of fen habitats is at least as severe as that of ombrotrophic mires, especially in western Europe, it is perhaps surprising that there has not been more attention given to fen restoration in the past. One of the reasons why fen peatlands have not appeared to be so prominent is that many of the plant communities found on moderate-rich fens can also be found on mineral-based wetlands. As a result, fen restoration is also considered within much of the non-peat-forming wetland literature (Grootjans and van Diggelen, 1995). For our purposes we are mainly concerned with the restoration of fully functioning peatlands, which of necessity must include peat accumulation. Of course some of the mineral wetlands that have been recreated may eventually form depths of peat, but this is highly variable between sites and the future prospects for developing a peat-forming system are mostly unknown. In terms of restoring some aspects of the ecology of fen peatlands, mineral wetlands are still valuable for the plant communities they conserve as well as for the biodiversity of plants and animals they contain. Fen restoration is a much broader subject than raised bog restoration, because of the enormous variability in hydrological conditions, plant communities and management strategies (see section 11.4 above). The literature is therefore scattered, but a number of excellent case studies and reviews are contained in Wheeler *et al* (1995).

Fens that are in need of restoration have been affected by a variety of factors, from simple abandonment allowing successional processes to lead to a changed plant community, through to complete destruction from agricultural reclamation or drainage. The *creation* of new fens where there are not even remnants of former systems may even be considered possible in some instances. The considerations and constraints for fen restoration mirror those for raised bog restoration to some extent, especially in terms of the legal and financial considerations. The detail of the practical constraints may be different but there are still three main factors that need to be considered: water quantity, water quality and sources of plant propagules. The desired end point is often more variable and the broader concept of fen often allows greater latitude in restoration objectives than in raised mire restoration, although it may also require more careful planning and manipulation of some factors such as water quality (see below). The existing conservation interest (including the peat archive) and topographic constraints may also be important in some sites (Table 11.3) and may constrain restoration plans based on vegetation. For example, in a survey of invertebrates, unmanaged fen vegetation was found to harbour a significant number of rare species not present on actively managed areas of much higher botanical interest (Foster and Procter, 1995).

As with ombrotrophic mire restoration, fen restoration requires good hydrological control, especially where the surrounding landscape has been drained and in extreme cases may have subsided significantly already. Interaction between declining water tables and changes in other management practices such as mowing often complicate the assessment of cause and effect of vegetation change (e.g. Rowell, 1986). In decisions on restoration management, a clear understanding of changes in all controlling factors as well as the associated vegetation development is required to establish suitable management options. Decreased water tables are undoubtedly one of the major causes of damage to fen systems, either through intentional drainage of the fen or from a general decrease in the groundwater table (Figure 10.1). Although the most obvious long-term impacts are on vegetation, other groups of organisms are also often dependent on particular water table conditions. Bird populations may be especially sensitive to hydrological changes that affect winter feeding and nesting conditions (Green and Robins, 1993).

Similar techniques to those discussed previously can be used to maintain high water tables, but greater use of pumped water may be necessary as many systems will require greater input of water than can be provided from direct precipitation alone. Use of pumped groundwater can maintain higher and more stable water tables as well as keeping pH at a more appropriate level than can be achieved by conservation of precipitation input alone (Best et al, 1995). Water quality of the additional water supply is clearly critical as excess nutrients could result in a more eutrophic system than is desired. Equally, without additional nutrients

and especially without adequate base elements such as Ca^{2+}, some systems may be in danger of undergoing acidification with the development of Sphagnum-dominated vegetation (see below). Obtaining the correct nutrient and water table levels is therefore a balancing act that needs careful manipulation and monitoring to achieve particular conditions for vegetation development. In ombrotrophic bog restoration, the decision is simple, as only water directly from precipitation is suitable for use. In fen restoration, it is more complex and data on the nutrient and base status of inputs are essential. However, it is possible to achieve similar hydrochemical conditions in managed restored systems to those found in natural fen landscapes (Wassen, 1996).

Eutrophication can be as harmful as unmanaged succession and acidification and, after actual destruction of fens, is probably the greatest cause of severe disturbance to fens. Eutrophication is a result of increased atmospheric nitrogen input and pollution of groundwater river water and surface water, but is also promoted by release of internal nutrient sources which may be more important than the external causes in some cases (Koerselman and Verhoeven, 1995). With eutrophication, nitrogen or phosphorus levels, or more commonly both, are increased. Restoration management aims to reduce the availability of the nutrient that controls plant growth and then to reduce availability of both N and P so that all major plant nutrients have low availability (Figure 11.14). In the first phase, plant growth is reduced, providing opportunities for less competitive species to colonise. With co-limitation of N and P, a more stable condition is achieved that is not controlled by only a single nutrient, and which is more suitable for the development of sustainable low-nutrient fen vegetation. Reducing inputs of N and/or P may be adequate in some cases to restore a fen damaged by eutrophication but it is often necessary to also reduce availability or remove nutrients already contained within the peat or vegetation. Table 11.4 suggests the possible strategies for achieving this and reflects the differential response of N and P availability to different management options, mainly due to different biogeochemical processes. There may also be other considerations, such as the loss of the seed bank with removal of surface peats (Koerselman and Verhoeven, 1995). Bakker and Olff (1995) describe the restoration of a fen meadow by ceasing fertiliser application and then continued cutting of vegetation for haymaking. Here, nitrogen content of the soil and the rate of nitrification were both significantly reduced over a period of around 50 years, and were associated with desirable changes in vegetation.

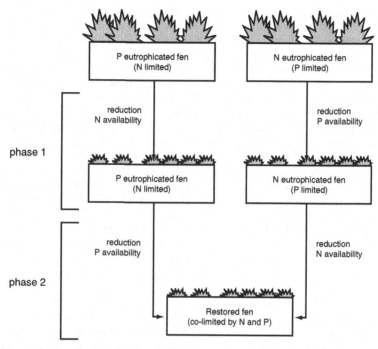

Figure 11.14 Pathways of vegetation development for restoration management of fens affected by eutrophication. Phase 1 management reduces plant growth by reducing the availability of the limiting plant nutrient, and phase 2 aims to achieve co-limitation of nutrients by reducing availability of the excess nutrient. Redrawn from Koerselman and Verhoeven (1995) with permission. © John Wiley & Sons Limited. See text for details.

Table 11.4 Possible management strategies to reduce nutrient availability in fens affected by eutrophication. Source: Koerselman and Verhoeven (1995)

Management	N	P
Reduce nutrient inputs	Yes	Yes
Restore high groundwater level	Yes	No?[1]
Sod removal	Yes	Yes
Mowing	No?[2]	Yes
Restore discharge of clean Ca/Fe-rich groundwater	No	Yes

[1] May increase P availability because of interaction between P and Fe hydroxides.
[2] Only effective if standing stocks of N exceed inputs, especially in areas affected by high N-inputs in precipitation.

Acidification can also present a serious threat to the maintenance of some fen systems – often this is a natural successional trend towards a bog system. If the aim of conservation management is to maintain the fen in a state of plagioclimax, the successional trend needs to be retarded. The balance between fen and bog appears to be finely balanced, with embryonic bog formation taking place within only a few decades where the surface vegetation becomes disconnected from the base-rich groundwater, especially by the formation of floating mats (van Diggelen *et al*, 1996). Acidification can also be accelerated by drainage, especially where the supply of enriched water from runoff or groundwater is reduced, increasing the influence of precipitation chemistry on the base status of the peatland. Drainage of the Lieper Posse, a calcareous groundwater-fed fen in north-eastern Germany, during the early nineteenth century, is thought to have led to the development of an area of *Pinus sylvestris–Ledum palustre* bog over part of the site (van Diggelen *et al*, 1991). In floodplain mires, such as those around the river Jegrznia in north-eastern Poland, drainage and lack of regular flooding reduce the nutrient input to the system and may ultimately lead to reduced base status and changes in nutrient limitations (de Mars *et al*, 1996). The restoration of systems affected by acidification will require very varied approaches. The primary aim of any restoration would be to restore the correct hydrochemical conditions. In the case of systems suffering from removal of enriched water sources, measures may need to be designed to prevent the loss of groundwater and surface runoff and raise water levels, decreasing the relative influence of rainwater chemistry (van Diggelen et al, 1991). Conversely, where floating mats of acidic vegetation have developed, excess nutrient-poor water may need to be removed by improved drainage (Beltman et al, 1996b). Supplemented by removal of the surface peat and vegetation (approximately the top 25 cm), this has proved successful in restoring pH levels in Dutch rich fens from values of pH 4 back to around 6.5 (Beltman et al, 1995). In such situations a clear understanding of water flows and hydrochemical variability, together with site history and vegetation change, are essential if successful restoration management is to be planned and implemented.

There are generally fewer problems with plant propagules in fen systems than there are in cutover bogs, probably because in most cases there are existing seed banks or remnant patches of vegetation on-site, whereas on cutover bogs Sphagnum in particular may have been completely eliminated. However, in some situations this can be a concern and active management is needed to reintroduce appropriate plant species. Research suggests that fen seed banks are only persistent for around five years after removal of vegetation cover for agricultural use, with seed survival especially low where conversion to arable production has occurred (Maas and Schopp-Guth, 1995). Studies on other fen wetlands suggest that seed survival is likely to differ between species, and it may be necessary to consider transplantation of some taxa in restoration management (Galatowitsch and van der Walk, 1995; Brock and Britton, 1995). One of the few documented examples of plant reintroductions in fen restoration is that for the peatlands in the Rocky Mountains of Colorado (Cooper and MacDonald, 2000). Abandoned mined peatlands in this region have much lower species diversity than equivalent undisturbed systems and in particular lack some of the dominant species of sedges and willows that would be expected to occur. Reintroduction is seen as the only viable way of re-establishing the main species to provide the basis for future peat growth and restoration of the system in the long term.

11.7.2 Drained and afforested mires

While cutover ombrotrophic mires and degraded fens are the main sites where restoration has been focused, there are other ombrotrophic peatlands where major management has been carried out to assist vegetation recovery. There has been particular interest in restora-

tion of mires affected by afforestation in Europe, but much of the management is still at an early stage and responses are difficult to assess. Most nations still rightly prioritise the protection of undamaged systems but in the UK since the early 1990s, there has also been a move towards active restoration management of blanket and raised mire sites planted with alien tree species. Management generally includes removal of trees as well as hydrological management by ditch blocking and projects are under way in northern England and Scotland to convert drained and forested peatlands back to open blanket mire. Burlton (1997) describes the range of management options for the Border Mires in Northumberland, England. Tree removal is prioritised on areas where trees are growing poorly or are still at a very early stage of growth, but may not be immediately undertaken where trees are larger, especially if they are approaching harvestable size. In other cases, quite large trees have been removed in order to begin restoration (Brooks and Stoneman, 1997). There is rather little evidence of the rate and extent of success following these operations as yet, but anecdotal evidence suggests that Sphagnum recolonisation is taking place. Complete removal of trees may not be necessary in some cases, and experimental work is being carried out to establish whether at least some of the material can be used to help reduce the efficiency of drains or simply be left on the surface, without limiting vegetation recovery (Wilkie et al, 1997). Where naturally forested mires are affected by drainage, a more moderate approach to restoration may be possible, using ditch blockage as the primary means of restoring the system (Vasander et al, 1992).

11.8 THE FUTURE FOR PEATLANDS IN THE TWENTY-FIRST CENTURY

It seems appropriate in the final few pages of this book to look to the future and attempt some predictions of the direction in which peatland science and conservation will go in the twenty-first century. To do this we first need to summarise some of the key past trends. Looking back over the past 40 years or so, we have seen very substantial improvements in our scientific understanding of the functioning of peatlands, right from basic peat accumulation processes through to the ecology and hydrology of various systems around the world. We know more about the long-term development of peatlands and the various autogenic and allogenic influences that control the nature and rate of change through time. The human impacts on mire systems are also better understood and can be placed in the context of much improved information on past changes and existing resources. However, progress in peatland science has not been constant in coverage or rate of change over the years. We have seen interests and advances in the various subdisciplines wax and wane, and changes in the amount and nature of work being carried out in different geographic regions. One of the best examples of a rapid development in peatland science is the work on carbon cycling which has seen an explosion of activity and results since the spectre of 'global warming' was taken seriously in the early 1990s. Perhaps due to the variability in national funding priorities, it has been the work on Finnish and Canadian peatlands that has most notably taken this forward (see Chapter 9).

In some ways, the shifts in awareness of peatlands and attitudes to peatland values are even more remarkable than the progress in the scientific realm. Until relatively recently, those involved with managing the peatland resource split into two diametrically opposed camps of 'conservationists' and 'exploiters'. At the same time, public awareness of peatland management issues was minimal. Early campaigns in the late 1980s and early 1990s by conservation organisations helped raise this awareness to unprecedented levels, especially in western Europe. One particularly vigorous campaign was mounted against the use of horticultural peat from lowland raised mires in Britain (e.g. Royal Society of Nature Conservation, 1990). While in some ways this simply fanned the flames of the existing conservation–exploitation arguments, it also brought the debate into a wider political and scientific arena and forced greater discussion on the issue of peatland conservation. Because the peat industry is an international business, the debate also spread well beyond the initial geographical foci to areas where the pressures on the peat resource were less intense. Some compromises on both sides have subsequently seen some convergence of opinion or at least increased communication between the parties involved. Peatland restoration research and implementation funded by peat producers is one area that has grown largely out of this improvement in relations. Likewise, forestry companies such as Forest Enterprise in the UK are involved with rehabilitation of peatlands as part of their remit. Undoubtedly these changes in emphasis in peatland resource management are also assisted by a general trend towards greater environmental concern.

So what of the future? It is not easy to predict future trends in any area of science and peatland research is no exception. We can, however, identify some key gaps in understanding that need to be filled to maintain the forward momentum. A key area is still the description and quantification of the global peatland resource. Much work has been done in many areas of the world (e.g. Lappalainen, 1996a) but the detail of this information is poor. The answer to how much peat there is on earth still needs refining, especially for tropical and subtropical areas. Beyond this basic question, the nature of the peat and peatlands is poorly described and details of the plant and animal ecology have not been studied for many areas. The evaluation of the total peatland resource is not helped by differences between countries in peat classification terminology, which is inconsistent and confusing in some respects. In terms of thematic issues within peatland research, the possibilities are various. Given the potential role of peatlands in national carbon budgets and the need to provide inventories of carbon sources and sinks for international agreements on climate change, carbon cycling will continue to be an important area of work. Perhaps research on carbon dynamics needs to move towards greater precision in predictions of responses to future changes and also begin to develop practical approaches to long-term monitoring of responses to climate changes. Many fundamental questions on peatland chemistry and hydrology also exist, not least with respect to nutrient dynamics and nutrient limitation in different systems. Many issues also remain to be addressed in the palaeoecological record; although there has been some exploitation of this source of data on past environmental change, increasingly sophisticated techniques for analysis of the archive are certain to yield new and exciting discoveries in the future. In terms of management, there is likely to be a continued interest in peatland restoration and perhaps increasingly in *creation*, too. While much progress has been made, it is still not possible to claim that we can manipulate and restore most peatlands to anything like their original state. The question of whether the beginnings of dialogue on 'wise use' of peatlands can be sustained and developed is perhaps the most critical one for the future of the global peatland resource. While conservationists need to accept that in the short term, at least, the properties of peat are uniquely suitable for some applications, the peat industry should also recognise the increased potential of peat alternatives. Hyperbole on both sides will need to be moderated to continue realistic and meaningful discussion. Governments and international regulatory bodies will ultimately have to weigh up the economic, environmental and societal values, to implement the ideals of 'wise use'.

References

Aaby, B. 1976. Cyclic climatic variations over the past 5500 years reflected in raised bogs. *Nature* **263**, 281–284.

Aaby, B. 1986. Palaeoecological studies of mires. In: *Handbook of Holocene Palaeoecology and Palaeohydrology* (ed. by B.E. Berglund), pp. 145–164. Wiley, Chichester.

Aaby, B. and Berglund, B.E. 1986a. Characterisation of peat and lake deposits. In: *Handbook of Holocene Palaeoecology and Palaeohydrology* (ed. by B.E. Berglund), pp. 231–246. Wiley, Chichester.

Aaby, B. and Berglund, B.E. 1986b. Sampling techniques for lakes and bogs. In: *Handbook of Holocene Palaeoecology and Palaeohydrology* (ed. by B.E. Berglund), pp. 181–194. Wiley, Chichester.

Aaby, B. and Tauber, H. 1975. Rates of peat formation in relation to local environment as shown by studies of a raised bog in Denmark. *Boreas* **4**, 1–17.

Aario, L. 1932. Pflanzentopographische und palaogeographische Mooruntersuchungen in N. Sakakunta. *Fennia* **55**, 1–179.

Aaviskoo, K. 1993. Changes of plant cover and land-use types (1950s to 1980s) in 3 mire reserves and their neighbourhood in Estonia. *Landscape Ecology* **8**, 287–301.

Aaviskoo, K., Kadarik, H. and Masing, V. 1997. *Aerial views and close-up pictures of 30 Estonian mires*. Ministry of the Environment, Tallinn.

Adams, J.M., Faure, H., Faure-Denard, L., McGlade, J.M. and Woodward, F.I. 1990. Increases in terrestrial carbon storage from the Last Glacial Maximum to the present. *Nature* **348**, 711–714.

Adger, W.N., Brown, K., Shiel, R.S. and Whitby, M.C. 1992. Carbon dynamics of land use in Great Britain. *Journal of Environmental Management* **36**, 117–133.

Aerts, R. and de Caluwe, H. 1997. Nutritional and plant-mediated controls on leaf litter decomposition of *Carex* species. *Ecology* **78**, 244–260.

Aerts, R. and de Caluwe, H. 1999. Nitrogen deposition effects on carbon dioxide and methane emissions from temperate peatland soils. *Oikos* **84**, 44–54.

Aerts, R., Verhoeven, J.T.A. and Whigham, D.F. 1999. Plant-mediated controls on nutrient cycling in temperate fens and bogs. *Ecology* **80**, 2170–2181.

Aerts, R., Wallen, B. and Malmer, N. 1992. Growth-limiting nutrients in *Sphagnum*-dominated bogs subject to low and high atmospheric nitrogen supply. *Journal of Ecology* **80**, 131–140.

Agnew, A.D.Q., Rapson, G.L., Sykes, M.T. and Wilson, J.B. 1993a. The functional ecology of *Empodisma minus* (Hook f) Johnson and cutler in New Zealand ombrotrophic mires. *New Phytologist* **124**, 703–710.

Agnew, A.D.Q., Wilson, J.B. and Sykes, M.T. 1993b. A vegetation switch as the cause of a forest mire ecotone in New Zealand. *Journal of Vegetation Science* **4**, 273–278.

Ahlholm, U. and Silvola, J. 1990. CO_2 release from peat-harvested peatlands and stockpiles. 2: Posters, 1–12. 1990. Association of Finnish Peat Industries and University of Jyväskylä. Peat 90 – versatile peat: International Conference on Peat Production and Use. 11 June 1990.

Akkerman, R. (ed.) 1982. *Regeneration von Hochmooren. Berichte des Moor-Symposions Vechta, June 1980. Inform. Naturschutz Landschaftspflege No. 3.* BSH, Wardenburg.

Albinsson, C. 1997. Niche relations and association analysis of southern Swedish mire hepatics. *Journal of Bryology* **19**, 409–424.

Almendinger, J.C., Almendinger, J.E. and Glaser, P.H. 1986. Topographic fluctuations across a spring fen and raised bog in the Lost River Peatland, Northern Minnesota. *Journal of Ecology*, **74**, 393–401.

Almendinger, J.E. and Leete, J.H. 1998a. Peat characteristics and groundwater geochemistry of calcareous fens in the Minnesota River Basin, USA. *Biogeochemistry* **43**, 17–41.

Almendinger, J.E. and Leete, J.H. 1998b. Regional and local hydrogeology of calcareous fens in the Minnesota River Basin, USA. *Wetlands* **18**, 184–202.

Almquist-Jacobson, H. and Foster, D.R. 1995. Toward an integrated model for raised-bog development: theory and field evidence. *Ecology* **76**, 2503–2516.

Anderson, A.R., Pyatt, D.G. and Stannard, J.P. 1990. The effects of clearfelling a sitka spruce stand on the water balance of a peaty gley soil at Kershope Forest, Cumbria. *Forestry* **63**, 51–71.

Anderson, D.E. 1998. A reconstruction of Holocene climatic changes from peat bogs in north-west Scotland. *Boreas* **27**, 208–224.

Anderson, D.S., Davis, R.B. and Janssens, J.A. 1995. Relationships of bryophytes and lichens to environmental gradients in Maine peatlands. *Vegetatio* **120**, 147–159.

Anderson, J.A.R. 1983. The tropical peat swamps of western Malesia. In: *Ecosystems of the World 4B, Mire: Swamp, Bog, Fen and Moor* (ed. by A.J.P. Gore), pp. 181–199. Elsevier, Amsterdam.

Anderson, P. 1997a. Fire damage on blanket mires. In: *Blanket Mire Degradation: Causes, Consequences and Challenges* (ed. by J.H. Tallis, R. Meade and P.D. Hulme), pp. 16–28. Macaulay Land Use Research Institute, Aberdeen.

Anderson, R. 1997. Hydrology and rainfall. In: *Conserving Bogs: the Management Handbook* (ed. by R. Stoneman and S. Brooks), pp. 65–72. The Stationery Office, Edinburgh.

Anon 1986. For peats sake, clean up our brown water. *New Scientist*, **109**, 18.

Appleby, P.G. and Oldfield, F. 1978. The calculation of lead-210 dates assuming a constant rate of supply of unsupported ^{210}Pb to the sediment. *Catena* **5**, 1–8.

Armentano, T.V. and Menges, E.S. 1986. Patterns of change in the carbon balance of organic soil – wetlands of the temperate zone. *Journal of Ecology* **74**, 755–774.

Armstrong, A.C. 1995. Hydrological model of peat-mound form with vertically varying hydraulic conductivity. *Earth Surface Processes and Landforms* **20**, 473–477.

Armstrong, A.C. 1996. Reply: Peat mounds with non-uniform properties. *Earth Surface Processes and Landforms* **21**, 769–771.

Arseneault, D. and Payette, S. 1997. Reconstruction of millennial forest dynamics from tree remains in a subarctic tree line peatland. *Ecology* **78**, 1873–1883.

Asplund, D. 1996. Energy use of peat. In: *Global Peat Resources* (ed. by E. Lappalainen), pp. 319–325. International Peat Society, Finland.

Aucour, A.M., Hillairemarcel, C. and Bonnefille, R. 1994. Late Quaternary biomass changes from c-13 measurements in a highland peatbog from equatorial Africa (Burundi). *Quaternary Research* **41**, 225–233.

Averdieck, F.R., Hayen, H., Heathwaite, A.L. and Willkomm, H. 1993. The chronology of mire development. In: *Mires: Process, Exploitation and Conservation* (ed. by A.L. Heathwaite and Kh. Göttlich), pp. 123–170. Wiley, Chichester.

Backéus, I. 1985. Aboveground production and growth dynamics of vascular bog plants in central Sweden. *Acta Phytogeographica Suecica* **74**, 1–98.

Backéus, I. 1988. Weather variables as predictors of *Sphagnum* growth on a bog. *Holarctic Ecology* **11**, 146–150.

Backéus, I. 1989. Flarks in the Maloti, Lesotho. *Geografiska Annaler* **71**, 105–111.

Backéus, I. 1991. The cyclic regeneration on bogs – a hypothesis that became an established truth. *Striae* **31**, 33–35.

Baden, W. and Eggelsman, R. 1963. Zur Durchlässigkeit von Moorböden. *Zeitschrift für Kulturtechnik und Flurbereinigung* **4**, 226–254.

Baird, A.J. 1995. Hydrological investigations of soil water and groundwater processes in wetlands. In: *Hydrology and Hydrochemistry of British Wetlands* (ed. by J. Hughes and A.L. Heathwaite), pp. 111–129. Wiley, Chichester.

Baird, A.J. and Gaffney, S.W. 1995. A partial explanation of the dependency of hydraulic conductivity on positive pore water pressure in peat soils. *Earth Surface Processes and Landforms* **20**, 561–566.

Baird, A.J. and Gaffney, S.W. 1996. Discussion: 'Hydrological model of peat-mound form with vertically varying hydraulic conductivity' by Adrian C. Armstrong. *Earth Surface Processes and Landforms* **21**, 765–767.

Baird, A.J. and Heathwaite, L. 1997. Water movement in undamaged blanket peats. In: *Blanket Mire Degradation: Causes, Consequences and Challenges* (ed. by J.H. Tallis, R. Meade and P.D. Hulme), pp. 128–139. Macaulay Land Use Research Institute, Aberdeen.

Baird, A.J. and Wilby, R.L. (eds.) 1999. *Eco-Hydrology: Plants and Water in Terrestrial and Aquatic Environments*. London, Routledge.

Baker, A., Caseldine, C.J., Gilmour, M.A., Charman, D., Proctor, C.J., Hawkesworth, C.J. and Phillips, N. 1999. Stalagmite luminescence and peat humification records of palaeomoisture for the last 2500 years. *Earth and Planetary Science Letters* **165**, 157–162.

Bakker, J.P. and Olff, H. 1995. Nutrient dynamics during restoration of fen meadows by haymaking without fertiliser application. In: *Restoration of Temperate Wetlands* (ed. by B.D. Wheeler, S.C. Shaw, W.J. Fojt and R.A. Robertson), pp. 143–166. Wiley, Chichester.

Bakker, S.A., Jasperse, C. and Verhoeven, J.T.A. 1997. Accumulation rates of organic matter associated with different successional stages from open water to carr forest in former turbaries. *Plant Ecology* **129**, 113–120.

Bannister, P. 1964a. The water relations of certain heath plants with reference to their ecological amplitude. II Field studies. *Journal of Ecology* **52**, 481–497.

Bannister, P. 1964b. The water relations of certain heathland plants with reference to their ecological amplitude. III General conclusions. *Journal of Ecology* **52**, 499–509.

Barber, K., Dumayne-Peaty, L., Hughes, P., Mauquoy, D. and Scaife, R. 1998. Replicability and variability of the recent macrofossil and proxy-climate record from raised bogs: field stratigraphy and macrofossil data from Bolton Fell Moss and Walton Moss, Cumbria, England. *Journal of Quaternary Science* **13**, 515–528.

Barber, K.E. 1976. History of vegetation. In: *Methods in Plant Ecology* (ed. by S.B. Chapman), pp. 5–83, Blackwell, Oxford.

Barber, K.E. 1981. *Peat Stratigraphy and Climatic Change*. A.A. Balkema, Rotterdam.

Barber, K.E. 1984. A large capacity Russian pattern sediment sampler. *Quaternary Newsletter* **44**, 28–31.

Barber, K.E., Battarbee, R.W., Brooks, S.J., Eglinton, G., Haworth, E.Y., Oldfield, F., Stevenson, A.C., Thompson, R., Appleby, P.G., Austin, W.E.N., Cameron, N.G., Ficken, K.J., Golding, P., Harkness, D.D., Holmes, J.A., Hutchinson, R., Lishman, J.P., Maddy, D., Pinder, L.C.V., Rose, N.L. and Stoneman, R.E. 1999. Proxy records of climate change in the UK over the last two millennia: documented change and sedimentary records from lakes and bogs. *Journal of the Geological Society* **156**, 369–380.

Barber, K.E., Chambers, F.M., Maddy, D., Stoneman, R. and Brew, J.S. 1994. A sensitive high resolution record of late

Holocene climatic change from a raised bog in northern England. *Holocene* **4**, 198–205.

Barbier, D. and Visset, L. 1997. Logne, a peat bog of European ecological interest in the Massif Armorican, western France: bog development, vegetation and land-use history. *Vegetation History and Archaeobotany* **6**, 69–77.

Batzer, D.P. and Wissinger, S.A. 1996. Ecology of insect communities in nontidal wetlands. *Annual Review of Entomology* **41**, 75–100.

Bauer, M.R. and Yavitt, J.B. 1996. Processes and mechanisms controlling consumption of CFC-11 and CFC-12 by peat from a conifer-swamp and black spruce–tamarack bog in New York State. *Chemosphere* **32**, 759–768.

Bayley, S.E., Schindler, D.W., Beaty, K.G., Parker, B.R. and Stainton, M.P. 1992a. Effects of multiple fires on nutrient yields from streams draining boreal forest and fen watersheds – nitrogen and phosphorus. *Canadian Journal of Fisheries and Aquatic Sciences* **49**, 584–596.

Bayley, S.E., Schindler, D.W., Parker, B.R., Stainton, M.P. and Beaty, K.G. 1992b. Effects of forest-fire and drought on acidity of a base-poor boreal forest stream – similarities between climatic warming and acidic precipitation. *Biogeochemistry* **17**, 191–204.

Bell, C.J.E., Finlayson, B.L. and Kershaw, A.P. 1989. Pollen analysis and dynamics of a peat deposit in Carnarvon national park, central Queensland. *Australian Journal of Ecology* **14**, 449–456.

Bell, J.N.B. and Tallis, J.H. 1974. The response of *Empetrum nigrum* to different mire water regimes, with special reference to Wybunbury Moss, Cheshire and Featherbed Moss, Derbyshire. *Journal of Ecology* **62**, 75–95.

Bell, M. and Walker, M.J.C. 1992. *Late Quaternary Environmental Change: Physical and Human Perspectives.* Longman, London.

Bellamy, D.J. 1968. An ecological approach to the classification of European mires. In: *Proceedings of the Third International Peat Congress*, Québec (ed. by C. Lafleur and J. Butler), pp. 74–79.

Belland, R.J. and Vitt, D.H. 1995. Bryophyte vegetation patterns along environmental gradients in continental bogs. *Écoscience* **2**, 395–407.

Beltman, B., Kooijman, A.M., Rouwenhorst, G. and Vankerkhoven, M. 1996a. Nutrient availability and plant-growth limitation in blanket mires in Ireland. *Biology and Environment – Proceedings of The Royal Irish Academy* **96B**, 77–87.

Beltman, B., van den Broek, T. and Bloemen, S. 1995. Restoration of acidified rich-fen ecosystems in the Vechtplassen area: successes and failures. In: *Restoration of Temperate Wetlands* (ed. by B.D. Wheeler, S.C. Shaw, W.J. Fojt and R.A. Robertson), pp. 273–286. Wiley, Chichester.

Beltman, B., van den Broek, T., vanMaanen, K. and Vaneveld, K. 1996b. Measures to develop a rich-fen wetland landscape with a full range of successional stages. *Ecological Engineering* **7**, 299–313.

Belyea, L.R. 1996. Separating the effects of litter quality and microenvironment on decomposition rates in a patterned peatland. *Oikos* **77**, 529–539.

Belyea, L.R. 1999. A novel indicator of reducing conditions and water-table depth in mires. *Functional Ecology* **13**, 431–434.

Belyea, L.R. and Clymo, R.S. 1998. Do hollows control the rate of peat bog growth? In: *Patterned Mires and Mire Pools* (ed. by V. Standen, J.H. Tallis and R. Meade), pp. 55–65. British Ecological Society, London.

Belyea, L.R. and Warner, B.G. 1994. Dating of the near-surface layer of a peatland in northwestern Ontario, Canada. *Boreas* **23**, 259–269.

Belyea, L.R. and Warner, B.G. 1996. Temporal scale and the accumulation of peat in a *Sphagnum* bog. *Canadian Journal of Botany* **74**, 366–377.

Bennett, K.D., Boreham, S., Sharp, M.J. and Switsur, V.R. 1992. Holocene history of environment, vegetation and human settlement on Catta Ness, Lunnasting, Shetland. *Journal of Ecology* **80**, 241–273.

Bennett, P.C., Siegel, D.I., Hill, B.M. and Glaser, P.H. 1991. Fate of silicate minerals in a peat bog. *Geology* **19**, 328–331.

Bennett, T.J. and Friday, L.E. 1997. Reed-beds. In: *Wicken Fen: the Making of a Wetland Nature Reserve* (ed. by L.E. Friday). Harley Books, Colchester.

Bergkamp, G. and Orlando, B. 1999. *Wetlands and Climate Change: Exploring Collaboration Between the Convention on Wetlands (Ramsar, Iran, 1971) and the UN Framework Convention on Climate Change.* IUCN.

Berglund, B.E. 1986. *Handbook of Holocene Palaeoecology and Palaeohydrology.* Wiley, Chichester.

Berglund, B.E., Birks, H.J.B., Ralska-Jasiewiczowa, M. and Wright, H.E. (eds.) 1996. *Palaeoecological events during the last 15000 years: Regional syntheses of palaeoecological studies of lakes and mires in Europe.* John Wiley, Chichester.

Best, E.P.H. and Jacobs, F.H.H. 1997. The influence of raised water table levels on carbon dioxide and methane production in ditch-dissected peat grasslands in the Netherlands. *Ecological Engineering* **8**, 129–144.

Best, E.P.H., van der Schaaf, S. and Oomes, M.J.M. 1995. Responses of restored grassland ditch vegetation to hydrological changes, 1989–1992. *Vegetatio* **116**, 107–122.

Beyens, L. 1985. On the Subboreal climate of the Belgian Campine as deduced from diatom and testate amoebae analysis. *Review of Paleobotany and Palynology* **46**, 9–31.

Billings, W.D. 1987. Carbon balance of Alaskan tundra and taiga ecosystems: past, present and future. *Quaternary Science Reviews* **6**, 165–177.

bin Soewono, S. 1997. Fertility management for sustainable agriculture on tropical ombrogenous peat. In: *Biodiversity and Sustainability of Tropical Peatlands* (ed. by J.O. Reiley and S.E. Page), pp. 315–320. Samara Publishing, Cardigan.

Birks, H.H. 1970. Studies in the vegetational history of Scotland. I. A pollen diagram from Abernethy forest, Inverness-shire. *Journal of Ecology* **58**, 827–846.

Birks, H.J.B. 1995. Quantitative palaeoenvironmental reconstructions. In: *Statistical Modelling of Quaternary Science*

Data (ed. by D. Maddy and J.S. Brew), pp. 161–254. Technical Guide No. 5, Quaternary Research Association, London.

Birks, H.J.B. and Birks, H.H. 1980. *Quaternary Palaeoecology*. Edward Arnold, London.

Birnie, R.V. and Hulme, P.D. 1990. Overgrazing of peatland vegetation in Shetland. *Scottish Geographical Magazine* **106**, 28–36.

Bjork, S. and Digerfeldt, G. 1991. Development and degradation, redevelopment and preservation of Jamaican wetlands. *Ambio* **20**, 276–284.

Blackford, J. 1993. Peat bogs as sources of proxy climatic data: past approaches and future research. In: *Climate Change and Human Impact on the Landscape* (ed. by F.M. Chambers), pp. 47–56. Chapman and Hall, London.

Blackford, J. 2000. Palaeoclimatic records from peat bogs. *Trends in Ecology and Evolution* **15**, 193–198.

Blackford, J.J. and Chambers, F.M. 1993. Determining the degree of peat decomposition in peat-based palaeoclimatic studies. *International Peat Journal* **5**, 7–24.

Blackford, J.J. and Chambers, F.M. 1995. Proxy climate record for the last 1000 years from Irish blanket peat and a possible link to solar variability. *Earth and Planetary Science Letters* **133**, 145–150.

Blackford, J.J., Edwards, K.J., Dugmore, A.J.C. and Buckland, P.C. 1992. Icelandic volcanic ash and the mid-Holocene Scots pine (*Pinus sylvestris*) pollen decline in northern Scotland. *Holocene* **2**, 260–265.

Blyakharchuk, T.A. and Sulerzhitsky, L.D. 1999. Holocene vegetational and climatic changes in the forest zone of Western Siberia according to pollen records from the extrazonal palsa bog Bugristoye. *Holocene* **9**, 621–628.

Blytt, A. 1876. *Essay on the Immigration of the Norwegian Flora during Alternating Rainy and Dry Periods*. Cammermeyer, Kristiana.

Boatman, D.J. 1983. The Silver Flowe Nature Reserve, Galloway, Scotland. *Journal of Biogeography* **10**, 163–274.

Bobrov, A.A., Charman, D.J. and Warner, B.G. 1999. Ecology of testate amoebae (Protozoa: Rhizopoda) on peatlands in western Russia with special attention to niche separation in closely related taxa. *Protist* **150**, 125–136.

Borcard, D. 1997. Oribatid mites in peat bog remnants of the Swiss Jura mountains. *Abh. Ber. Naturkundemus. Görlitz* **69**, 19–23.

Borcard, D. and Matthey, W. 1995. Effect of a controlled trampling of *Sphagnum* mosses on their Oribatid mite assemblages (Acari, Oribatei). *Pedobiologia* **39**, 219–230.

Bord na Móna 1984. *Fuel Peat in Developing Countries*. Study report prepared for the World Bank, Dublin.

Bos, J.M., van Geel, B. and Pals, J.P. 1988. Waterland – environmental and economic changes in a Dutch bog area, 1000 A.D. to 2000 A.D. In: *The Cultural Landscape: Past Present and Future* (ed. by H.H. Birks, H.J.B. Birks, P.E. Kaland and D. Moe), pp. 321–331. Cambridge University Press, Cambridge.

Botch, M.S., Kobak, K.I., Vinson, T.S. and Kolchugina, T.P. 1995. Carbon pools and accumulation in peatlands of the Former Soviet Union. *Global Biogeochemical Cycles* **9**, 37–46.

Botch, M.S. and Masing, V.V. 1983. Mire systems in the USSR. In: *Ecosystems of the World 4B, Mire: Swamp, Bog, Fen and Moor* (ed. by A.J.P. Gore), pp. 95–152. Elsevier, Amsterdam.

Boudreau, S. and Rochefort, L. 1998. Restoration of post-mined peatlands: effect of vascular pioneer species on *Sphagnum* establishment. *Peatland Restoration and Reclamation* 1–18 July, pp. 39–43.

Bouwmann, A.F. 1990. Global distribution of the major soils and land cover types. In: *Soils and the Greenhouse Effect* (ed. by A.F. Bouwmann), pp. 33–59. John Wiley, Chichester.

Bowles, M., Mcbride, J., Stoynoff, N. and Johnson, K. 1996. Temporal changes in vegetation composition and structure in a fire-managed prairie fen. *Natural Areas Journal* **16**, 275–288.

Boyd, W.E. 1982. Sub surface formation of charcoal: an unexplained event in peat. *Quaternary Newsletter* **38**, 15–16.

Boyle, J.F., Mackay, A.W., Rose, N.L. and Appleby, P.G. 1998. Sediment heavy metal record in Lake Baikal: natural and anthropogenic sources. *Journal of Paleolimnology* **20**, 135–150.

Boyle, J.F., Rose, N.L., Bennion, H., Yang, H. and Appleby, P.G. 1999. Environmental impacts in the Jianghan plain: evidence from lake sediments. *Water Air and Soil Pollution* **112**, 21–40.

Bradbury, I.K. and Grace, J. 1983. Primary production in wetlands. In: *Mires: Swamp, Bog, Fen and Moor. Ecosystems of the World 4A* (ed. by A.J.P. Gore), pp. 285–310. Elsevier, Amsterdam.

Bradshaw, R. and McGee, E. 1988. The extent and time course of mountain blanket peat erosion in Ireland. *New Phytologist* **108**, 219–224.

Bragazza, L., Alber, R. and Gerdol, R. 1998. Seasonal chemistry of pore water in hummocks and hollows in a poor-mire in the southern Alps (Italy). *Wetlands* **18**, 320–328.

Bragg, O.M. 1995. Towards an ecohydrological basis for raised mire restoration. In: *Restoration of Temperate Wetlands* (ed. by B.D. Wheeler, S.C. Shaw, W.J. Fojt and R.A. Robertson), pp. 305–314. Wiley, Chichester.

Bragg, O.M. 1997. Understanding ombrogenous mires of the temperate zone and the tropics: an ecohydrologist's viewpoint. In: *Biodiversity and Sustainability of Tropical Peatlands* (ed. by J.O. Reiley and S.E. Page), pp. 135–146. Samara Publishing, Cardigan.

Bragg, O.M., Hulme, P.D., Ingram, H.A.P., Johnston, J.P. and Wilson, A.I.A. 1994. A maximum–minimum recorder for shallow water tables, developed for ecohydrological studies on mires. *Journal of Applied Ecology* **31**, 589–592.

Branfireun, B.A., Heyes, A. and Roulet, N.T. 1996. The hydrology and methylmercury dynamics of a Precambrian shield headwater peatland. *Water Resources Research* **32**, 1785–1794.

Branfireun, B.A. and Roulet, N.T. 1998. The baseflow and

storm flow hydrology of a Precambrian shield headwater peatland. *Hydrological Processes* **12**, 57–72.

Breen, C.M., Heeg, J. and Seaman, M. 1993. Wetlands of Africa: South Africa. In: *Wetlands of the World: Inventory, Ecology and Management* (ed. by D. Whigham, D. Dykyjova and S. Hejny), pp. 79–110. Kluwer, Dordrecht.

Brenninkmeijer, C.A.M., Vangeel, B. and Mook, W.G. 1982. Variations in the D/H and O-18/O-16 ratios in cellulose extracted from a peat bog core. *Earth and Planetary Science Letters* **61**, 283–290.

Bridge, M.C., Haggart, B.A. and Lowe, J.J. 1990. The history and palaeoclimatic significance of subfossil remains of *Pinus sylvestris* in blanket peats from Scotland. *Journal of Ecology* **78**, 77–99.

Bridgham, S.D., Pastor, J., Janssens, J.A., Chapin, C. and Malterer, T.J. 1996. Multiple limiting gradients in peatlands – a call for a new paradigm. *Wetlands* **16**, 45–65.

Bridgham, S.D. and Richardson, C.J. 1993. Hydrology and nutrient gradients in North-Carolina peatlands. *Wetlands* **13**, 207–218.

Bridgham, S.D., Richardson, C.J., Maltby, E. and Faulkner, S.P. 1991. Cellulose decay in natural and disturbed peatlands in North Carolina. *Journal of Environmental Quality* **20**, 695–701.

Brock, M.A. and Britton, D.L. 1995. The role of seed banks in the revegetation of Australian temporary wetlands. In: *Restoration of Temperate Wetlands* (ed. by B.D. Wheeler, S.C. Shaw, W.J. Fojt and R.A. Robertson), pp. 183–188. Wiley, Chichester.

Bromley, J. and Robinson, M. 1995. Groundwater in raised mire systems: models, mounds and myths. In: *Hydrology and Hydrochemistry of British Wetlands* (ed. by J. Hughes and L. Heathwaite), pp. 45–109. Wiley, Chichester.

Bronk-Ramsay, C. 2000. *OxCal Program v3.5.* University of Oxford Radiocarbon Accelerator Unit.

Brooks, S. 1997. Methods and techniques for management. In: *Conserving Bogs: the Management Handbook* (ed. by R. Stoneman and S. Brooks), pp. 89–149. The Stationery Office, Edinburgh.

Brooks, S. and Stoneman, R. 1997. Tree removal at Langlands Moss. In: *Conserving Peatlands* (ed. by L. Parkyn, R.E. Stoneman and H.A.P. Ingram), pp. 315–322. CAB International, Wallingford.

Brown, T.A., Farwell, G.W., Grootes, P.M. and Schmidt, F.H. 1992. Radiocarbon AMS dating of pollen extracted from peat samples. *Radiocarbon* **34**, 550–556.

Brugam, R.B. and Swain, P. 2000. Diatom indicators of peatland development at Pogonia Bog Pond, Minnesota, USA. *Holocene* **10**, 453–464.

Bubier, J., Costello, A., Moore, T.R., Roulet, N.T. and Savage, K. 1993. Microtopography and methane flux in boreal peatlands, Northern Ontario, Canada. *Canadian Journal of Botany* **71**, 1056–1063.

Bubier, J.L. and Moore, T.R. 1994. An ecological perspective on methane emissions from northern wetlands. *Trends in Ecology and Evolution* **9**, 460–464.

Buckton, S.T. and Ormerod, S.J. 1997. Effects of liming on the Coleoptera, Hemiptera, Araneae and Opiliones of catchment wetlands in Wales. *Biological Conservation* **79**, 43–57.

Bugnon, J.L., Rochefort, L. and Price, J.S. 1997. Field experiment of *Sphagnum* reintroduction on a dry abandoned peatland in Eastern Canada. *Wetlands* **17**, 513–517.

Bunting, M.J. 1996. The development of heathland in Orkney, Scotland – pollen records from Loch-of-Knithcen (Rousay) and Loch-of-Torness (Hoy). *Holocene* **6**, 193–212.

Bunting, M.J., Morgan, C.R., Van Bakel, M. and Warner, B.G. 1998. Pre-European settlement conditions and human disturbance of a coniferous swamp in southern Ontario. *Canadian Journal of Botany* **76**, 1770–1779.

Bunting, M.J., Warner, B.G. and Aravena, R. 1996. Late Quaternary vegetation dynamics and hydroseral development in a *Thuja occidentalis* swamp in southern Ontario. *Canadian Journal of Earth Science* **33**, 1439–1456.

Bunting, M.J., Warner, B.G. and Morgan, C.R. 1997. Interpreting pollen diagrams from wetlands: pollen representation in surface samples from Oil Well bog, southern Ontario. *Canadian Journal of Botany* **76**, 1780–1797.

Burlton, B. 1997. The Border Mires approach. In: *Conserving Peatlands* (ed. by L. Parkyn, R.E. Stoneman and H.A.P. Ingram), pp. 271–279. CAB International, Wallingford.

Burt, T.P. 1995. The role of wetlands in runoff generation from headwater catchments. In: *Hydrology and Hydrochemistry of British Wetlands* (ed. by J. Hughes and A.L. Heathwaite), pp. 21–38. Wiley, Chichester.

Bussières, B., Payette, S. and Filion, L. 1996. Late Holocene deforestation and peat formation in Charlevoix highlands: onset of the subalpine and alpine belts. *Géographie Physique et Quaternaire* **50**, 257–269.

Butterfield, J. 1992. The effect of conifer plantations on the invertebrate communities of peat moorland. In: *Peatland Ecosystems and Man: an Impact Assessment* (ed. by O.M. Bragg, P.D. Hulme, H.A.P. Ingram and R.A. Robertson), pp. 309–315. International Peat Society, Finland.

Buttler, A., Dinel, H. and Levesque, P.E.M. 1994. Effects of physical, chemical and botanical characteristics of peat on carbon gas fluxes. *Soil Science* **158**, 365–374.

Buttler, A., Grosvernier, P. and Matthey, Y. 1998. Development of *Sphagnum fallax* diaspores on bare peat with implications for the restoration of cut-over bogs. *Journal of Applied Ecology* **35**, 800–810.

Buttler, A., Warner, B.G., Grosvernier, P. and Matthey, Y. 1996. Vertical patterns of testate amoebae (Protozoa: Rhizopoda) and peat forming vegetation on cutover bogs in the Jura, Switzerland. *New Phytologist* **134**, 371–382.

Butzer, K.W. 1982. *Archaeology as Human Ecology.* Cambridge University Press, New York.

Calme, S. and Desrochers, A. 2000. Biogeographic aspects of the distribution of bird species breeding in Quebec's peatlands. *Journal of Biogeography* **27**, 725–732.

Camill, P. 1999a. Patterns of boreal permafrost peatland vegetation across environmental gradients sensitive to climate warming. *Canadian Journal of Botany* **77**, 721–733.

Camill, P. 1999b. Peat accumulation and succession following permafrost thaw in the boreal peatlands of Manitoba, Canada. *Écoscience* **6**, 592–602.

Camill, P. 2000. How much do local factors matter for predicting transient ecosystem dynamics? Suggestions from permafrost formation in boreal peatlands. *Global Change Biology* **6**, 169–182.

Camill, P. and Clark, J.S. 1998. Climate change disequilibrium of boreal permafrost peatlands caused by local processes. *American Naturalist* **151**, 207–222.

Campbell, D.I. and Williamson, J.L. 1997. Evaporation from a raised peat bog. *Journal of Hydrology* **193**, 142–160.

Campbell, E.O. 1964. The restiad peat bogs at Motumaoho and Moanatuatua. *Transactions of the Royal Society of New Zealand* **2**, 219–227.

Campbell, E.O. 1983. Mires of Australasia. In: *Ecosystems of the World 4B. Mires: Swamp, Bog, Fen and Moor* (ed. by A.J.P. Gore), pp. 153–180. Amsterdam, Elsevier.

Campbell, I.D., Campbell,C., Yu, Z.C., Vitt, D.H. and Apps, M.J. 2000. Millennial-scale rhythms in peatlands in the western interior of Canada and in the global carbon cycle. *Quaternary Research* **54**, 155–158.

Campeau, S. and Rochefort, L. 1996. *Sphagnum* regeneration on bare peat surfaces: field greenhouse experiments. *Journal of Applied Ecology* **33**, 599–608.

Cannell, M.G.R., Milne, R., Hargreaves, K.J., Brown, T.A.W., Cruickshank, M.M., Bradley, R.I., Spencer, T., Hope, D., Billett, M.F., Adger, W.N. and Subak, S. 1999. National inventories of terrestrial carbon sources and sinks: the UK experience. *Climatic Change* **42**, 505–530.

Caseldine, C.J. 1981. Surface pollen studies across Bankhead Moss, Fife, Scotland. *Journal of Biogeography* **8**, 7–26.

Caseldine, C.J., Baker, A., Charman, D.J. and Hendon, D. 2000. A comparative study of optical properties of NaOH peat extracts: implications for humification studies. *Holocene* **10**, 649–658.

Caseldine, C.J. and Hatton, J. 1993. The development of high moorland on Dartmoor: fire and the influence of Mesolithic activity on vegetation change. In: *Climate Change and Human Impact on the Landscape* (ed. by F.M. Chambers), pp. 119–132. Chapman and Hall, London.

Casparie, W.A. and Streefkerk, J.G. 1992. Climatological, stratigraphic and palaeo-ecological aspects of mire development. In: *Fens and Bogs in the Netherlands: Vegetation, History, Nutrient Dynamics and Conservation* (ed. by J.T.A. Verhoeven), pp. 81–129. Kluwer, Dordrecht.

ChagueGoff, C. and Fyfe, W.S. 1996. Geochemical and petrographic characteristics of a domed bog, Nova Scotia – a modern analog for temperate coal deposits. *Organic Geochemistry* **24**, 141–158.

Chai, X. 1980. Peat in China. *Proceedings of the 6th International Peat Congress*, Duluth, USA. 16–20.

Chambers, F.M. 1981. Date of blanket peat initiation in South Wales. *Quaternary Newsletter* **35**, 24–29.

Chambers, F.M. 1982. 2 Radiocarbon-dated pollen diagrams from high-altitude blanket peats in South Wales. *Journal of Ecology* **70**, 445–459.

Chambers, F.M. 1983a. 3 Radiocarbon-dated pollen diagrams from upland peats northwest of Merthyr Tydfil, South Wales. *Journal of Ecology* **71**, 475–487.

Chambers, F. 1983b. New applications of palaeoecological techniques: integrating evidence of arable activity in pollen peat and soil stratigraphies, Cefn Graenog, north Wales. In: *Integrating the Subsistence Economy* (ed. by M. Jones), BAR International Series 181, Chapter 5, 16 pp.

Chambers, F.M., Barber, K.E., Maddy, D. and Brew, J. 1997. A 5500-year proxy-climate and vegetation record from blanket mire at Talla Moss, Borders, Scotland. *Holocene* **7**, 391–399.

Chambers, F.M. and Elliott, L. 1989. Spread and expansion of Alnus Mill, in the British Isles: timing, agencies and possible vectors. *Journal of Biogeography* **16**, 541–550.

Chambers, F.M., Mauquoy, D. and Todd, P.A. 1999. Recent rise to dominance of *Molinia caerulea* in environmentally sensitive areas: new perspectives from palaeoecological data. *Journal of Applied Ecology* **36**, 719–733.

Chanton, J.P., Bauer, J.E., Glaser, P.A., Siegel, D.I., Kelley, C.A., Tyler, S.C., Romanowicz, E.H. and Lazrus, A. 1995. Radiocarbon evidence for the substrates supporting methane formation within northern Minnesota peatlands. *Geochimica et Cosmochimica Acta* **59**, 3663–3668.

Chapin, F.S., Moilanen, L. and Kielland, K. 1993. Preferential use of organic nitrogen for growth by a nonmycorrhizal arctic sedge. *Nature* **361**, 150–153.

Chapman, S.B. 1964. The ecology of Coom Rigg Moss, Northumberland. I Stratigraphy and present vegetation. *Journal of Ecology* **52**, 299–313.

Chapman, S.B. and Rose, R.J. 1991. Changes in the vegetation at Coom Rigg Moss National Nature Reserve within the period 1958–86. *Journal of Applied Ecology* **28**, 140–153.

Charman, D.J. 1992a. Blanket mire formation at the Cross Lochs, Sutherland, northern Scotland. *Boreas* **21**, 53–72.

Charman, D.J. 1992b. Relationship between testate amoebae (Protozoa: rhizopoda) and microenvironmental parameters on a forested peatland in northeastern Ontario. *Canadian Journal of Zoology* **70**, 2474–2482.

Charman, D.J. 1993. Patterned fens in Scotland? Evidence from vegetation and water chemistry. *Journal of Vegetation Science* **4**, 543–552.

Charman, D.J. 1994. Patterned fen development in northern Scotland: developing a hypothesis from palaeoecological data. *Journal of Quaternary Science* **9**, 285–297.

Charman, D.J. 1995. Patterned fen development in northern Scotland – hypothesis-testing and comparison with ombrotrophic blanket peats. *Journal of Quaternary Science* **10**, 327–342.

Charman, D.J. 1998. Pool development on patterned fens in Scotland. In: *Patterned Mires and Mire Pools* (ed. by V. Standen, J.H. Tallis and R. Meade), pp. 39–54. British Ecological Society, London.

Charman, D.J., Aravena, R. and Warner, B.G. 1994. Carbon dynamics in a forested peatland in north-eastern Ontario, Canada. *Journal of Ecology* **82**, 55–62.

Charman, D.J., Caseldine, C., Baker, A., Gearey, B., Hatton, J. and Proctor, C.J. 2001. Palaeohydrological records from peat profiles and speleothems in Sutherland, northwest Scotland. *Quaternary Research* **55**, 223–234.

Charman, D.J. and Hendon, D. 2000. Long-term changes in soil water tables over the past 4500 years: relationships with climate and North Atlantic atmospheric circulation and sea surface temperature. *Climatic Change* **47**, 45–59.

Charman, D.J., Hendon, D. and Packman, S. 1999. Multiproxy surface wetness records from replicate cores on an ombrotrophic mire: implications for Holocene palaeoclimate records. *Journal of Quaternary Science* **14**, 451–463.

Charman, D.J., Hendon, D. and Woodland, W. 2000. *The Identification of Peatland Testate Amoebae*. Quaternary Research Association Technical guide no. 9, London. 147 pp.

Charman, D.J. and Smith, R.S. 1992. Forestry and blanket mires of Kielder Forest, northern England: long term effects of vegetation. In: *Peatland Ecosystems and Man: an Impact Assessment* (ed. by O.M. Bragg, P.D. Hulme, H.A.P. Ingram and R.A. Robertson), pp. 226–230. International Peat Society, Finland.

Charman, D.J. and Warner, B.G. 1992. Species–environment relationships of testate amoebae on a forested peatland in northeastern Ontario, Canada. *Canadian Journal of Zoology* **70**, 2474–2482.

Charman, D.J. and Warner, B.G. 1997. The ecology of testate amoebae (Protozoa: Rhizopoda) in oceanic peatlands in Newfoundland, Canada: modelling hydrological relationships for palaeoenvironmental reconstruction. *Écoscience* **4**, 555–562.

Chason, D.B. and Siegel, D.I. 1986. Hydraulic conductivity and related physical properties of peat, Lost River peatland, northern Minnesota. *Soil Science* **142**, 91–99.

Christensen, T.R. and Cox, P. 1995. Response of methane emission from arctic tundra to climatic-change – results from a model simulation. *Tellus Series B – Chemical and Physical Meteorology* **47**, 301–309.

Clark, J.S. 1988. Particle motion and the theory of charcoal analysis: source area, transport, deposition and sampling. *Quaternary Research* **30**, 67–80.

Clark, R.L. 1982. Point count estimation of charcoal in pollen preparations and thin sections of sediments. *Pollen et Spores* **24**, 523–535.

Clarke, P.J. and Martin, A.R.H. 1999. *Sphagnum* peatlands of Kosciusko National Park in relation to altitude, time and disturbance. *Australian Journal of Botany* **47**, 519–536.

Clarkson, B.R. 1997. Vegetation recovery following fire in two Waikato peatlands at Whangamarino and Moanatuatua, New Zealand. *New Zealand Journal of Botany* **35**, 167–179.

Clements, F.E. 1916. *Plant Succession: an Analysis of the Development of Vegetation*. Publication 242, Carnegie Institute of Washington, Washington, DC.

Clymo, R.S. 1965. Experiments on breakdown of *Sphagnum* in two bogs. *Journal of Ecology* **53**, 747–757.

Clymo, R.S. 1973. The growth of *Sphagnum*: some effects of environment. *Journal of Ecology* **61**, 849–869.

Clymo, R.S. 1983. Peat. In: *Mires: Swamp, Bog, Fen and Moor. Ecosystems of the World 4A* (ed. by A.J.P. Gore), pp. 159–224. Elsevier, Amsterdam.

Clymo, R.S. 1984a. *Sphagnum*-dominated peat bog – a naturally acid ecosystem. *Philosophical Transactions of the Royal Society of London Series* **B305**, 487–499.

Clymo, R.S. 1984b. The limits to peat bog growth. *Philosophical Transactions of the Royal Society of London* **B303**, 605–654.

Clymo, R.S. 1988. A high resolution sampler of surface peat. *Functional Ecology* **2**, 425–431.

Clymo. R.S. 1991. Peat growth. In: *Quaternary Landscapes* (ed. by L.K.C. Shane and E.J. Cushing), pp. 76–112. Belhaven, London.

Clymo, R.S. 1992a. Models of peat growth. *Suo* **43**, 127–136.

Clymo, R.S. 1992b. Productivity and decomposition of peatland ecosystems. In: *Peatland Ecosystems and Man: an Impact Assessment* (ed. by O.M. Bragg, P.D. Hulme, H.A.P. Ingram and R.A. Robertson), pp. 3–16. International Peat Society, Finland.

Clymo, R.S. and Hayward, P.M. 1982. The ecology of *Sphagnum*. In: *Bryophyte Ecology* (ed. by A.J.E. Smith), pp. 229–289. Chapman and Hall, London.

Clymo, R.S., Oldfield, F., Appleby, P.G., Pearson, G.W., Ratnesar, P. and Richardson, N. 1990. The record of atmospheric deposition on a rainwater-dependent peatland. *Philosophical Transactions of the Royal Society of London Series* **B327**, 331–338.

Clymo, R.S. and Reddaway, E.J.F. 1971. Productivity of *Sphagnum* (bog moss) and peat accumulation. *Hydrobiologia* **12**, 181–192.

Clymo, R.S., Turunen, J. and Tolonen, K. 1998. Carbon accumulation in peatland. *Oikos* **81**, 368–388.

Cohen, A.D., Casagrande, D.J., Andrejko, M.J., Best, G.R. (eds.) 1984. *The Okefenokee Swamp: Its Natural History, Geology and Geochemistry*. Wetland Surveys, Los Alamos, New Mexico.

Cohen, A.D., Raymond, R., Ramirez, A., Morales, Z. and Ponce, F. 1989. The Changuinola peat deposit of northwestern Panama – a tropical, back-barrier, peat (coal)-forming environment. *International Journal of Coal Geology* **12**, 157–192.

Coles, B. 1992. Further thoughts on the impact of beaver on temperate landscapes. In: *Alluvial Archaeology in Britain* (ed. by S. Needham and M.G. Macklin), pp. 93–99. Oxbow Monographs 27, Oxford.

Coles, B. and Coles, J. 1986. *Sweet Track to Glastonbury*. Thames and Hudson, London.

Colhoun, E.A., Vandegeer, G. and Fitzsimons, S.J. 1991. Late glacial and Holocene vegetation history at Governor Bog, King Valley, western Tasmania, Australia. *Journal of Quaternary Science* **6**, 55–66.

Comeau, P.L. and Bellamy, D.J. 1986. An ecological interpretation of the chemistry of mire waters from selected sites in eastern Canada. *Canadian Journal of Botany* **64**, 2576–2581.

Conrad, R. 1989. In: *Exchange of Trace Gases between Terrestrial Ecosystems and the Atmosphere* (ed. by M.O. Andreae, and D.S. Schimel), pp. 39–58. Wiley, Chichester.

Conway, V. M. 1954. Stratigraphy and pollen analysis of southern Pennine blanket peats. *Journal of Ecology* **42**, 117–147.

Cooper, D.J. 1996. Water and soil chemistry, floristics, and phytosociology of the extreme rich High Creek Fen, in South Park, Colorado, USA. *Canadian Journal of Botany* **74**, 1801–1811.

Cooper, D.J. and MacDonald, L.H. 2000. Restoring the vegetation of mined peatlands in the southern Rocky Mountains of Colorado, USA. *Restoration Ecology* **8**, 103–111.

Cooper, S.R., Huvane, J., Vaithiyanathan, P. and Richardson, C.J. 1999. Calibration of diatoms along a nutrient gradient in Florida Everglades Water Conservation Area-2A, USA. *Journal of Paleolimnology* **22**, 413–437.

Coulson, J.C., Bauer, L., Butterfield, J., Downie, I., Cranna, L. and Smith, C. 1995. The invertebrates of the northern Scottish Flows, and a comparison with other peatland habitats. In: *Heaths and Moorland: Cultural Landscapes* (ed. by D.B.A. Thompson, A.J. Hester and M.B. Usher), pp. 74–94. Scottish Natural Heritage, Edinburgh.

Coulson, J.C. and Butterfield, J. 1978. An investigation of the biotic factors determining the rates of plant decomposition of blanket bog. *Journal of Ecology* **66**, 631–650.

Coulson, J.C. and Butterfield, J.E.C. 1985. The invertebrate communities of peat and upland grassland in the north of England and some conservation implications. *Biological Conservation* **34**, 197–220.

Coulson, J.C., Downie, I.S. and Butterfield, J.E.L. 1998. The invertebrate fauna of lowland mires in Britain: comparison with high altitude mires and evidence of an east–west gradient. In: *Patterned Mires and Mire Pools* (ed. by V. Standen, J.H. Tallis and R. Meade), pp. 170–181. British Ecological Society, London.

Cowardin, L.M., Carter, V., Golet, F.C. and Laroe, E.T. 1979. *Classification of Wetlands and Deepwater Habitats of the United States*. US Department of the Interior, Washington, DC.

Craft, C.B. and Richardson, C.J. 1997. Relationships between soil nutrients and plant species composition in Everglades peatlands. *Journal of Environmental Quality* **26**, 224–232.

Craft, C.B., Vymazal, J. and Richardson, C.J. 1995. Response of Everglades plant-communities to nitrogen and phosphorus additions. *Wetlands* **15**, 258–271.

Cresser, M.S., Dawod, A.M. and Rees, R.M. 1997. Influence of precipitation composition on the chemistry of streams draining from peat examined using Na:Ca:Mg ratio. *Water Research* **31**, 2253–2260.

Cronberg, G. 1986. Blue-green algae, green algae and chrysophyceae in sediments. In: *Handbook of Holocene Palaeoecology and Palaeohydrology* (ed. by B.E. Berglund), pp. 507–526. Wiley, Chichester.

Cross, J.R. 1990. Survey and selection of peatland sites for conservation in the Republic of Ireland. In: *Ecology and Conservation of Irish Peatlands* (ed. by G.J. Doyle), pp. 175–188. Royal Irish Academy, Dublin.

Cruickshank, M.M. and Tomlinson, R.W. 1996. Application of CORINE land cover methodology to the UK – some issues raised from Northern Ireland. *Global Ecology and Biogeography Letters*, **5**, 235–248.

Cruikshank, J.G. and Cruikshank, N.M. 1981. The development of humus-iron podsol profiles linked by radiocarbon dating and pollen analysis to vegetation history. *Oikos* **36**, 238–253.

Cruise, G.M. 1990. Holocene peat initiation in the Ligurian Apennines, Northern Italy. *Review of Palaeobotany and Palynology* **63**, 173–182.

Curry, J.P., Boyle, K.E. and Farrell, E.P. 1989. The invertebrate fauna of reclaimed cutaway peat in central Ireland and its influence on soil fertility and plant-growth. *Agriculture Ecosystems and Environment* **27**, 217–225.

Dahl, T.E. 1990. *Wetland losses in the United States 1780s to 1980s*. US Fish and Wildlife Service, Washington, DC.

Dale, S. 2000. The importance of farmland for Ortolan Buntings nesting on raised peat bogs. *Ornis Fennica* **77**, 17–25.

Damman, A.W.H. 1986. Hydrology, development, and biogeochemistry of ombrogenous peat bogs with special reference to nutrient relocation in a western Newfoundland bog. *Canadian Journal of Botany* **64**, 384–394.

Damman, A.W.H. and French, T.W. 1987. *The Ecology of Peat Bogs of the Glaciated Northeastern United States: a Community Profile*. US Fish and Wildlife Service Biological Report 85(7.16), pp. 1–100. Washington.

Daniels, R.E. and Eddy, A. 1985. *Handbook of European Sphagna*. Institute of Terrestrial Ecology, Monks Wood.

David, J.S. and Ledger, D.C. 1988. Runoff generation in a plough-drained peat bog in southern Scotland. *Journal of Hydrology* **99**, 187–199.

Davis, A.M. 1984. Ombrotrophic peatlands in Newfoundland, Canada – their origins, development and trans-Atlantic affinities. *Chemical Geology* **44**, 287–309.

Davis, A.M. 1988. Toward a perspective on paludification. *The Canadian Geographer* **32**, 76–85.

Dehmer, J. 1993. Petrology and organic geochemistry of peat samples from a raised bog in Kalimantan (Borneo). *Organic Geochemistry* **20**, 349–362.

De Mars, H. and Wassen, M.J. 1999. Redox potentials in relation to water levels in different mire types in the Netherlands and Poland. *Plant Ecology* **140**, 41–51.

De Mars, H., Wassen, M.J. and Peeters, W.H.M. 1996. The effect of drainage and management on peat chemistry and nutrient deficiency in the former Jegrznia Floodplain (NE Poland). *Vegetatio* **126**, 59–72.

Demello, W.Z. and Hines, M.E. 1994. Application of static and dynamic enclosures for determining dimethyl sulfide and carbonyl sulfide exchange in *Sphagnum* peatlands – implications for the magnitude and direction of flux.

Journal of Geophysical Research – Atmospheres **99**, 14601–14607.

Denny, P. 1993a. Wetlands of Africa: Introduction. In: *Wetlands of the World: Inventory, Ecology and Management* (ed. by D. Whigham, D. Dykyjova and S. Hejny), pp. 1–31. Kluwer, Dordrecht.

Denny, P. 1993b. Wetlands of Africa: Eastern Africa. In: *Wetlands of the World: Inventory, Ecology and Management* (ed. by D. Whigham, D. Dykyjova and S. Hejny), pp. 32–46. Kluwer, Dordrecht.

Denny, P. 1993c. Wetlands of Africa: wetland use and conservation. In: *Wetlands of the World: Inventory, Ecology and Management* (ed. by D. Whigham, D. Dykyjova and S. Hejny), pp. 111–128. Kluwer, Dordrecht.

Desrochers, A., Rochefort, L. and Savard, J.P.L. 1998. Avian recolonization of eastern Canadian bogs after peat mining. *Canadian Journal of Zoology* **76**, 989–997.

Devito, K.J., Hill, A.R. and Dillon, P.J. 1999. Episodic sulphate export from wetlands in acidified headwater catchments: prediction at the landscape scale. *Biogeochemistry* **44**, 187–203.

Devito, K.J., Waddington, J.M. and Branfireun, B.A. 1997. Flow reversals in peatlands influenced by local groundwater systems. *Hydrological Processes* **11**, 103–110.

Dickinson, C.H. 1983. Micro-organisms in peatlands. In: *Mires: Swamp, Bog, Fen and Moor. Ecosytems of the World 4A* (ed. by A.J.P. Gore), pp. 225–245. Elsevier, Amsterdam.

Dickson, J.H. 1973. *Bryophytes of the Pleistocene.* Cambridge University Press.

Dickson, J.H. 1986. Bryophyte analysis. In: *Handbook of Holocene Palaeoecology and Palaeohydrology* (ed. by B.E. Berglund), pp. 627–643. Wiley, Chichester.

Dierssen, K. 1982. *Die wichtigsten Pflanzengesellschaften der Moore NW-Europas [The Major Plant Communities of Northwest European Mires].* Conservatoire et Jardin Botaniques, Geneva.

Dierssen, K., Twenhoven, F.L., Lutt, S. and Wagner, C. 1990. Drohende Vernichtung der Deckenmoore in Nordschottland [The imminent destruction of the blanket bogs in north Scotland]. *Telma* **20**, 329–334.

Dimbleby, G.W. 1985. *The Palynology of Archaeological Sites.* Academic Press, London.

Dise, N.B., Gorham, E. and Verry, E.S. 1993. Environmental factors controlling methane emissions from peatlands in northern Minnesota. *Journal of Geophysical Research – Atmospheres* **98**, 10583–10594.

Dobson, A.T. 1979. Mire types of New Zealand. *Proceedings of an International Symposium on Classification of Peat and Peatlands,* Hyytiälä, pp. 82–94. International Peat Society, Helsinki.

Doyle, G.J. (ed.) 1990. *Ecology and Conservation of Irish Peatlands.* Royal Irish Academy, Dublin.

Doyle, T. and Dowding, P. 1990. Decomposition and aspects of the physical environment in the surface layers of Mongan bog. In: *Ecology and Conservation of Irish Peatlands* (ed. by G.J. Doyle), pp. 163–171. Royal Irish Academy, Dublin.

Drake, H.L., Aumen, N.G., Kuhner, C., Wagner, C., Griesshammer, A. and Schmittroth, M. 1996. Anaerobic microflora of Everglades sediments – effects of nutrients on population profiles and activities. *Applied and Environmental Microbiology* **62**, 486–493.

Drexler, J.Z., Bedford, B.L., DeGaetano, A. and Siegel, D.I. 1999a. Quantification of the water budget and nutrient loading in a small peatland. *Journal of the American Water Resources Association* **35**, 753–769.

Drexler, J.Z., Bedford, B.L., Scognamiglio, R. and Siegel, D.I. 1999b. Fine-scale characteristics of groundwater flow in a peatland. *Hydrological Processes* **13**, 1341–1359.

Duever, M.J. 1984. Environmental factors controlling plant communities of the Big Cypress Swamp. In: *Environments of South Florida: Present and Past* (ed. by P.J. Gleason), pp. 127–137. Miami Geological Society, Coral Gables, Florida.

Dugmore, A.J., Larsen, G. and Newton, A.J. 1995. 7 Tephra isochrones in Scotland. *Holocene* **5**, 257–266.

Dugmore, A.J., Newton, A.J., Edwards, K.J., Larsen, G., Blackford, J.J. and Cook, G.T. 1996. Long-distance marker horizons from small-scale eruptions: British tephra deposits from the AD 1510 eruption of Hekla, Iceland. *Journal of Quaternary Science* **11**, 511–516.

Dumayne, L. 1993. Invader or native? Vegetation clearance in northern Britain during Romano-British time. *Vegetation History and Archaeobotany* **2**, 29–36.

Dupont, L.M. 1986. Temperature and rainfall variation in the Holocene based on comparative palaeoecology and isotope geology of a hummock and a hollow (Boutangerveen, The Netherlands). *Review of Palaeobotany and Palynology* **48**, 71–159.

Du Rietz, G.E. 1954. Die Mineralbodenwasserzeigerense als Grundlager einer natürlichen Zweigliederung der nord und mitteleuropäischen Moore. *Vegetatio* **5**, 571–585.

Du Rietz, G.E. 1957. *Linné Som Myrforskare (Linnaeus as a Paludologist).* Uppsala Universitets Årsskrift 5, Acta Universitatis Uppsaliensis, Uppsala.

Dwyer, R.B. and Mitchell, F.J.G. 1997. Investigation of the environmental impact of remote volcanic activity on North Mayo, Ireland, during the mid-Holocene. *Holocene* **7**, 113–118.

Edwards, K.J. 1983. Quaternary palynology: multiple profile studies and pollen variability. *Progress in Physical Geography* **7**, 587–609.

Edwards, K.J. and Hirons, K.R. 1982. Date of blanket peat initiation and rates of spread – a problem in research design. *Quaternary Newsletter* **36**, 32–37.

Edwards, K.J., Whittington, G. and Hirons, K.R. 1995. The relationship between fire and long-term wet heath development in South Uist, Outer Hebrides, Scotland. In: *Heaths and Moorland: Cultural Landscapes* (ed. by D.B.A. Thompson, A.J. Hester and M.B. Usher), pp. 240–248. Scottish Natural Heritage, Edinburgh.

Eggelsmann, R., Heathwaite, A.L., Grosse-Brauckmann, G., Küster, E., Naucke, W., Schuch, M. and Schweickle, V. 1993. Physical processes and properties of mires. In: *Mires: Process, Exploitation and Conservation* (ed. by A.L.

Heathwaite and Kh. Göttlich), pp. 171–262. Wiley, Chichester.

Ehrenfeld, J.G. 1995. Microsite differences in surface substrate characteristics in *Chamaecyparis* swamps of the New Jersey pinelands. *Wetlands* **15**, 183–189.

El-Daoushy, F., Tolonen, K. and Rosenberg, R. 1982. ^{210}Pb and moss-increment dating of two Finnish *Sphagnum* hummocks. *Nature* **296**, 429–431.

Erdtman, G. 1934. Über die Verwendung von Essigsaurenhydrid bei Pollenuntersuchungen. *Svensk Botanisk Tidskrift* **28**, 354–358.

Eriksson, H. 1991. Sources and sinks of carbon dioxide in Sweden. *Ambio* **20**, 146–150.

Etherington, J.R. 1983. *Wetland Ecology – Studies in Biology No. 154*. Edward Arnold.

Eurola, S. 1962. Über die regionale Einteilung der sudfinnischen Moore. *Annales Botanica Societas Vanama* **33**, 1–243.

Eurola, S. 1975. Snow and ground frost conditions of some Finnish mire types. *Annales Botanica Fennica* **12**, 1–16.

Eurola, S., Hicks, S. and Kaakinen, E. 1984. Key to Finnish mire types. In: *European Mires* (ed. by P.D. Moore), pp. 11–117. Academic Press, London.

Evans, M.G., Burt, T.P., Holden, J. and Adamson, J.K. 1999. Runoff generation and water table fluctuations in blanket peat: evidence from UK data spanning the dry summer of 1995. *Journal of Hydrology* **221**, 141–160.

Faegri, K. and Iversen, J. 1975. *Textbook of Pollen Analysis* (3rd edn). Hafner, New York.

Fan, S.H. 1987. Applications of remote-sensing techniques in peat resource investigations in Zoige. *Acta Geologica Sinica*, **61**, 274.

Farrimond, P. and Flanagan, R.L. 1996. Lipid stratigraphy of a Flandrian peat bed (Northumberland, UK) – Comparison with the pollen record. *Holocene* **6**, 69–74.

Faure, H., Adams, J.M., Debenay, J.P., Fauredenard, L., Grant, D.R., Pirazzoli, P.A., Thomassin, B., Velichko, A.A. and Zazo, C. 1996. Carbon storage and continental land surface change since the Last Glacial Maximum. *Quaternary Science Reviews* **15**, 843–849.

Fenton, J.H.C. 1980. The rate of peat accumulation in Antarctic moss banks. *Journal of Ecology* **68**, 211–228.

Ferland, C. and Rochefort, L. 1997. Restoration techniques for *Sphagnum*-dominated peatlands. *Canadian Journal of Botany* **75**, 1110–1118.

Ficken, K.J., Barber, K.E. and Eglinton, G. 1998. Lipid biomarker, delta C-13 and plant macrofossil stratigraphy of a Scottish montane peat bog over the last two millennia. *Organic Geochemistry* **28**, 217–237.

Finlayson, M. and von Oertzen, I. 1993. Wetlands of Australia – Northern (tropical) Australia. In: *Wetlands of the World: Inventory, Ecology and Management* (ed. by D. Whigham, D. Dykyjova and S. Hejny), pp. 195–243. Kluwer, Dordrecht.

Finnamore, A.T. and Marshall, S.A. 1994. Terrestrial arthropods of peatlands, with particular reference to Canada. *Memoirs of the Entomological Society of Canada* **169**.

Flower, R.J., Politov, S.V., Rippey, B., Rose, N.L., Appleby, P.G. and Stevenson, A.C. 1997. Sedimentary records of the extent and impact of atmospheric contamination from a remote Siberian highland lake. *Holocene* **7**, 161–173.

Forrest, G.I. and Smith, R.A.H. 1975. The productivity of a range of blanket bog vegetation types in the northern Pennines. *Journal of Ecology* **63**, 173–202.

Fossitt, J.A. 1996. Late Quaternary vegetation history of the Western Isles of Scotland. *New Phytologist* **132**, 171–196.

Foster, A.P. and Procter, D.A. 1995. The occurrence of some scarce East Anglian fen invertebrates in relation to vegetation management. In: *Restoration of Temperate Wetlands* (ed. by B.D. Wheeler, S.C. Shaw, W.J. Fojt and R.A. Robertson), pp. 223–240. Wiley, Chichester.

Foster, D.R. 1984. The dynamics of *Sphagnum* in forest and peatland communities in southeastern Labrador, Canada. *Arctic* **37**, 133–140.

Foster, D.R. and Fritz, S.C. 1987. Mire development, pool formation and landscape processes on patterned fens in Dalarna, Central Sweden. *Journal of Ecology* **75**, 409–437.

Foster, D.R. and King, G.A. 1984. Landscape features, vegetation and developmental history of a patterned fen in south eastern Labrador, Canada. *Journal of Ecology* **72**, 115–143.

Foster, D.R., King, G.A., Glaser, P.H. and Wright, H.E. 1983. Origin of string patterns in boreal peatlands. *Nature* **306**, 256–258.

Foster, D.R. and Wright, H.E. 1990. Role of ecosystem development and climate change in bog formation in central Sweden. *Ecology* **71**, 450–463.

Foster, D.R., Wright, H.E., Thelaus, M. and King, G.A. 1988. Bog development and landform dynamics in central Sweden and south-eastern Labrador, Canada. *Journal of Ecology* **76**, 1164–1185.

Fox, A.D. and Bell, M.C. 1994. Breeding bird communities and environmental variable correlates of Scottish peatland wetlands. *Hydrobiologia* **280**, 297–307.

Francis, I.S. 1990. Blanket peat erosion in a mid-Wales catchment during 2 drought years. *Earth Surface Processes and Landforms* **15**, 445–456.

Franzen, L.G. 1994. Are wetlands the key to the ice-age cycle enigma? *Ambio* **23**, 300–308.

Franzen, L.G., Chen, D.L. and Klinger, L.F. 1996. Principles for a climate regulation mechanism during the late Phanerozoic era, based on carbon fixation in peat-forming wetlands. *Ambio* **25**, 435–442.

Frenzel, B. 1983. Mires – repositories of climatic information or self-perpetuating ecosystems? In: *Mires: Swamp, Bog, Fen and Moor. General Studies. Ecosystems of the World 4A*. (ed. by A.J.P. Gore), pp. 35–65. Elsevier, Amsterdam.

Froggatt, P.C. and Lowe, D.J. 1990. A review of late Quaternary silicic and some other tephra formations from New Zealand: their stratigraphy, nomenclature, distribution, volume and age. *New Zealand Journal of Geology and Geophysics* **33**, 89–109.

Fuchsman, C.H. 1986. *Peat and Water: Aspects of Water Retention and Dewatering in Peat*. Elsevier, London.

Futyma, R.P. and Miller, N.G. 1986. Stratigraphy and genesis of the Lake-16 Peatland, northern Michigan. *Canadian Journal of Botany* **64**, 3008–3019.

Gaither, J.C. 1994. Understory avifauna of a Bornean peat swamp forest – is it depauperate? *Wilson Bulletin* **106**, 381–390.

Gajewski, K., Vance, R., Sawada, M., Fung, I., Gignac, L.D., Halsey, L., John, J., Maisongrande, P., Mandell, P., Mudie, P.J., Richard, P.J.H., Sherin, R.A.G., Soroko, J. and Vitt, D. 2000. The climate of North America and adjacent ocean waters ca. 6 ka. *Canadian Journal of Earth Sciences* **37**, 661–681.

Galatowitsch, S.M. and van der Walk, A.G. 1995. Natural revegetation during restoration of wetlands in the Southern Prairie Pothole region of North America. In: *Restoration of Temperate Wetlands* (ed. by B.D. Wheeler, S.C. Shaw, W.J. Fojt and R.A. Robertson), pp. 129–142. Wiley, Chichester.

Games, L.M. and Hayes, J.M. 1976. On the mechanisms of CO_2 and CH_4 production in natural anaerobic environments. In: *Environmental Biogeochemistry* Volume 1: *Carbon, Nitrogen, Phosphorous, Sulfur and Selenium Cycles* (ed. by J.O. Nriagu), pp. 51–73. Ann Arbor, Michigan.

Gauthier, G., Rochefort, L. and Reed, A. 1996. The exploitation of wetland ecosystems by herbivores. *Geoscience Canada* **23**, 253–259.

Gear, A.J. and Huntley, B. 1991. Rapid changes in the range limits of Scots pine 4000 years ago. *Science* **241**, 544–547.

Geikie, J. 1866. On the buried forests and peat mosses of Scotland and the changes of climate which they indicate. *Transactions of the Royal Society of Edinburgh* **24**, 363–384.

Gibson, N. and Kirkpatrick, J.B. 1992. Dynamics of a Tasmanian cushion heath community. *Journal of Vegetation Science*, **3**, 647–654.

Gignac, L.D. 1992. Niche structure, resource partitioning, and species interactions of mire bryophytes relative to climatic and ecological gradients in western Canada. *Bryologist* **95**, 406–418.

Gignac, L.D. 1994. Peatland species preferences – an overview of our current knowledge-base. *Wetlands* **14**, 216–222.

Gignac, L.D., Vitt, D.H. and Bayley, S.E. 1991a. Bryophyte response surfaces along ecological and climatic gradients. *Vegetatio* **93**, 29–45.

Gignac, L.D., Vitt, D.H., Zoltai, S.C. and Bayley, S.E. 1991b. Bryophyte response surfaces along climatic, chemical, and physical gradients in peatlands of western Canada. *Nova Hedwigia* **53**, 27–71.

Gilbert, D., Amblard, C., Bourdier, G. and Francez, A.J. 1998. Short-term effect of nitrogen enrichment on the microbial communities of a peatland. *Hydrobiologia* **374**, 111–119.

Giller, K.E. and Wheeler, B.D. 1986a. Past peat cutting and present vegetation patterns in an undrained fen in the Norfolk broadland. *Journal of Ecology* **74**, 219–247.

Giller, K.E. and Wheeler, B.D. 1986b. Peat and peat water chemistry of a flood-plain fen in broadland, Norfolk, UK. *Freshwater Biology* **16**, 99–114.

Gilman, K. 1994. *Hydrology and Wetland Conservation*. Wiley, Chichester.

Gilman, K. and Newson, M.D. 1980. Soil pipes and pipeflow – a hydrological study in upland Wales. *British Geomorphological Research Group Research Monograph* **1**, 1–110.

Gilvear, D.J., Andrews, R., Tellam, J.H., Lloyd, J.W. and Lerner, D.N. 1993. Quantification of the water balance and hydrogeological processes in the vicinity of a small groundwater-fed wetland, East Anglia, UK. *Journal of Hydrology* **144**, 311–334.

Gilvear, D.J., Tellam, J.H., Lloyd, J.W. and Lerner, D.N. 1992. Comparison of the hydrodynamics of three minero-trophic mires in East Anglia, England. In: *Peatland Ecosystems and Man: an Impact Assessment* (ed. by O.M. Bragg, P.D. Hulme, H.A.P. Ingram and R.A. Robertson), pp. 104–109. International Peat Society, Finland.

Gilvear, D. and Watson, A. 1995. The use of remotely sensed imagery for mapping wetland water table depths: Insh Marshes, Scotland. In: *Hydrology and Hydrochemistry of British Wetlands* (ed. by J. Hughes and L. Heathwaite), pp. 419–430. Wiley, Chichester.

Glaser, P.H. 1987a. The development of streamlined bog islands in the continental interior of North America. *Arctic and Alpine Research* **19**, 402–413.

Glaser, P.H. 1987b. The ecology of patterned boreal peatlands of northern Minnesota: a community profile. *US Fish and Wildlife Service Report* **85(7.14)**, 1–98.

Glaser, P.H. 1989. Detecting biotic and hydrogeochemical processes in large peat basins with Landsat TM imagery. *Remote Sensing of the Environment* **28**, 109–119.

Glaser, P.H. 1992a. Peat landforms. In: *The Patterned Peatlands of Minnesota* (ed. by H.E. Wright, B.A. Coffin and N.E. Aaseng), pp. 3–14. University of Minneapolis Press.

Glaser, P.H. 1992b. Ecological development of patterned peatlands. In: *The Patterned Peatlands of Minnesota* (ed. by H.E. Wright, B.A. Coffin and N.E. Aaseng), pp. 27–42. University of Minneapolis Press.

Glaser, P.H. 1992c. Raised bogs in eastern North America – regional controls for species richness and floristic assemblages. *Journal of Ecology* **80**, 535–554.

Glaser, P.H. 1998. The distribution and origin of mire pools. In: *Patterned Mires and Mire Pools* (ed. by V. Standen, J.H. Tallis and R. Meade), pp. 4–25. British Ecological Society, London.

Glaser, P.H., Bennett, P.C., Siegel, D.I. and Romanowicz, E.A. 1996. Palaeo-reversals in groundwater flow and peatland development at Lost River, Minnesota, USA. *Holocene* **6**, 413–421.

Glaser, P.H. and Janssens, J.A. 1986. Raised bogs in eastern North America: transitions in landforms and gross stratigraphy. *Canadian Journal of Botany*, **64**, 395–415.

Glaser, P.H., Janssens, J.A. and Siegel, D.I. 1990. The response of vegetation to chemical and hydrological gradients in the Lost River peatland, northern Minnesota. *Journal of Ecology* **78**, 1021–1048.

Glaser, P.H., Siegel, D.I., Romanowicz, E.A. and Shen, Y.P. 1997. Regional linkages between raised bogs and the climate, groundwater, and landscape of north-western Minnesota. *Journal of Ecology* **85**, 3–16.

Glob, P.V. 1969. *The Bog People*. Faber and Faber, London.

Glooschenko, W.A., Roulet, N.T., Barrie, L.A., Schiff, H.I. and Mcadie, H.G. 1994. The northern wetlands study (NOWES) – An overview. *Journal of Geophysical Research – Atmospheres* **99**, 1423–1428.

Glooschenko, W.A., Tarnocai, C., Zoltai, S. and Glooschenko, V. 1993. Wetlands of Canada and Greenland. In: *Wetlands of the World: Inventory, Ecology and Management* (ed. by D. Whigham, D. Dykyjova and S. Hejny), pp. 415–514. Dordrecht, Kluwer.

Goode, D.A. and Lindsay, R.A. 1979. The peatland vegetation of Lewis. *Proceedings of the Royal Society of Edinburgh* **77B**, 279–293.

Gopal, B. and Krishnamurthy, K. 1993. Wetlands of South Asia. In: *Wetlands of the World: Inventory, Ecology and Management* (ed. by D. Whigham, D. Dykyjova and S. Hejny), pp. 345–414. Kluwer, Dordrecht.

Gore, A.J.P. 1983a. *Ecosystems of the World 4A. Mires, Swamp, Bog, Fen and Moor: General Studies*. Elsevier, Amsterdam.

Gore, A.J.P. 1983b. *Ecosystems of the World 4B. Mires, Swamp, Bog, Fen and Moor: Regional Studies*. Elsevier, Amsterdam.

Gore, A.J.P. 1983c. Introduction. In: *Mires: Swamp, Bog, Fen and Moor. Ecosystems of the World 4A: General Studies* (ed. by A.J.P. Gore), pp. 1–34. Elsevier, Amsterdam.

Gore, A.J.P. and Urquhart, C. 1966. The effects of waterlogging on the growth of *Molinia caerulea* and *Eriophorum vaginatum*. *Journal of Ecology* **54**, 617–634.

Gorham, E. 1953. A note on the acidity and base status of raised and blanket bogs. *Journal of Ecology* **41**, 153–156.

Gorham, E. 1991. Northern peatlands: role in the carbon cycle and probable responses to climatic warming. *Ecological Applications* **1**, 182–195.

Gorham, E., Bayley, S.E. and Schindler, D.W. 1984. Ecological effects of acid deposition upon peatlands – a neglected field in acid-rain research. *Canadian Journal of Fisheries and Aquatic Sciences* **41**, 1256–1268.

Gorham, E., Eisenreich, S.J., Ford, J. and Santelmann, M.V. 1985. The chemistry of bog waters. In: *Chemical Processes in Lakes* (ed. by W. Stumm) pp. 339–363, New York, Wiley.

Gorham, E. and Janssens, J.A. 1992. Concepts of fen and bog re-examined in relation to bryophyte cover and the acidity of surface waters. *Acta Societatis Botanicorum Poloniae* **61**, 7–20.

Gorozhankina, SM. 1993. Structural and morphological analysis of the wetland cover of western Siberia using space photographs. *Soviet Journal of Remote Sensing*, **10**, 887–898.

Gorres, M. and Frenzel, B. 1997. Ash and metal concentrations in peat bogs as indicators of anthropogenic activity. *Water Air and Soil Pollution* **100**, 355–365.

Gosselink, J.G. and Turner, R.E. 1978. The role of hydrology in freshwater wetland systems. In: *Freshwater Wetlands* (ed. by R.I. Good, D.F. Whigham and R.L. Simpson), pp. 63–78. Academic Press, New York.

Göttlich, Kh. 1977. *Öko-Hydrologische Untersuchungen an südwestdeutschen Niedermoor-Standorten unter der Einwirkung kulturtechnischer Eingriffe 1961–1973*. Schr.-R. Kuratorium f. Kulturbauwesen, H.30 (English abstract), Verlag Parey, Hamburg-Berlin.

Göttlich, Kh., Richard, K.H., Kuntze, H., Eggelsmann, R., Günther, J., Eichelsdörfer, D. and Briemle, G. 1993. Mire utilisation. In: *Mires: Process, Exploitation and Conservation* (ed. by A.L. Heathwaite and Kh. Göttlich), pp. 325–415. Wiley, Chichester.

Graf, K. 1981. Palynological investigations of 2 post-glacial peat bogs near the boundary of Bolivia and Peru. *Journal of Biogeography* **8**, 353–368.

Graniero, P.A. and Price, J.S. 1999a. Distribution of bog and heath in a Newfoundland blanket bog complex: topographic limits on the hydrological processes governing blanket bog development. *Hydrology and Earth System Sciences* **3**, 223–231.

Graniero, P.A. and Price, J.S. 1999b. The importance of topographic factors on the distribution of bog and heath in a Newfoundland blanket bog complex. *Catena* **36**, 233–254.

Grant, S.A., Bolton, G.R. and Torvell, L. 1985. The responses of blanket bog vegetation to controlled grazing by hill sheep. *Journal of Applied Ecology* **22**, 739–751.

Grant, S.A., Lamb, W.I.C., Kerr, C.D. and Bolton, G.R. 1976. The utilization of blanket bog vegetation by grazing sheep. *Journal of Applied Ecology* **13**, 857–869.

Greatrex, P.A. 1983. Interpretation of macrofossil assemblages from surface sampling of macroscopic plant remains in mire communities. *Journal of Ecology* **71**, 773–791.

Green, B.H. and Pearson, M.C. 1977. The ecology of Wybunbury Moss, Cheshire. II. Post-glacial history and the formation of the Cheshire mere and mire landscape. *Journal of Ecology* **65**, 793–814.

Green, D.G. and Dolman, G.S. 1988. Fine resolution pollen analysis. *Journal of Biogeography* **15**, 685–701.

Green, R.E. and Robins, M. 1993. The decline of the ornithological importance of the Somerset Levels and Moors, England and changes in the management of water levels. *Biological Conservation* **66**, 95–106.

Groot, A. *Silvicultural consequences of forest harvesting on peatlands: site damage and slash conditions*. Report 0-X-384, 1–20. Great Lakes Forest Research Centre, Canadian Forestry Service, Environment Canada.

Groot, A. 1995. Harvesting method affects survival of black spruce advance growth. *Northern Journal of Applied Forestry* **12**, 8–11.

Groot, A. 1997. Uneven-aged silviculture for peatland black spruce. Emmingham, W. H. 107-120. Carvallis, Oregon. *Proceedings of the IUFRO Inter-disciplinary Uneven-aged Management Symposium*. 1 September 1997.

Groot, A. 1998. Physical effects of site disturbance on peatlands. *Canadian Journal of Soil Science* **78**, 45–50.

Groot, A. and Adams, M.J. 1994. Direct seeding black spruce on peatlands: fifth year results. *The Forestry Chronicle* **70**, 585–592.

Grootjans, A. and van Diggelen, R. 1995. Assessing the restoration prospects of degraded fens. In: *Restoration of Temperate Wetlands* (ed. by B.D. Wheeler, S.C. Shaw, W.J. Fojt and R.A. Robertson), pp. 73–90. Wiley, Chichester.

Grospietsch, T. 1967. Die Rhizopodenanalyse der Moore und ihre Anwendungsmöglichkeit. In: *Pflanzensoziologie und Palynologie* (ed. by R. Tuxen), pp. 181–192, W. Junk, Dordrecht.

Grosse-Brauckmann, G. 1986. Analysis of vegetative plant macrofossils. In: *Handbook of Holocene Palaeoecology and Palaeohydrology* (ed. by B.E. Berglund), pp. 591–618, Wiley, Chichester.

Grosvernier, P., Matthey, Y. and Buttler, A. 1997. Growth potential of three *Sphagnum* species in relation to water table level and peat properties with implications for their restoration in cut-over bogs. *Journal of Applied Ecology* **34**, 471–483.

Grünig, A. 1997. Surveying and monitoring of mires in Switzerland. In: *Conserving Peatlands* (ed. by L. Parkyn, R.E. Stoneman and H.A.P. Ingram), pp. 217–227. CAB International, Wallingford.

Gunderson, L.H. and Loftus, W.T. 1993. The Everglades. In: *Biodiversity of the South Eastern United States: Lowland Terrestrial Communities* (ed. by W.H. Martin, S.G. Boyce and A.C.E. Echternacht), pp. 199–255. Wiley, New York.

Gunnarsson, U., Rydin, H. and Sjörs, H. 2000. Diversity and pH changes after 50 years on the boreal mire Skattlosbergs Stormosse, Central Sweden. *Journal of Vegetation Science* **11**, 277–286.

Gurnell, A.M. 1998. The hydrogeomorphological effects of beaver dam building activity. *Progress in Physical Geography* **22**, 167–189.

Halsey, L.A., Vitt, D.H. and Bauer, I.E. 1998. Peatland initiation during the Holocene in continental western Canada. *Climatic Change* **40**, 315–342.

Halsey, L.A., Vitt, D.H. and Trew, D.O. 1997. Influence of peatlands on the acidity of lakes in northeastern Alberta, Canada. *Water Air and Soil Pollution* **96**, 17–38.

Hammond, R.F., van der Krogt, G. and Osinga, T. 1990. Vegetation and water-tables on two raised bog remnants in County Kildare. In: *Ecology and Conservation of Irish Peatlands* (ed. by G.J. Doyle), pp. 121–134. Royal Irish Academy, Dublin.

Hannerz, M. and Hanell, B. 1997. Effects on the flora in Norway spruce forests following clearcutting and shelterwood cutting. *Forest Ecology and Management* **90**, 29–49.

Haraguchi, A. 1995. Seasonal changes in oxygen consumption rate and redox property of floating peat in a pond in central Japan. *Wetlands* **15**, 242–246.

Hardjowigeno, S. 1997. Suitability of Indonesian peat soils for agricultural development. In: *Biodiversity and Sustainability of Tropical Peatlands* (ed. by J.O. Reiley and S.E. Page), pp. 327–333. Cardigan, Samara Publishing.

Hardwick, M.E. and Giberson, D.J. 1996. Aquatic insect populations in transplanted and natural populations of the purple pitcher plant, *Sarracenia purpurea*, on Prince Edward Island, Canada. *Canadian Journal of Zoology* **74**, 1956–1963.

Harvey, H.J. and Meredith, T.C. 1981. Ecological studies of *Peucedanum palustre* and their implications for conservation management at Wicken Fen, Cambridgeshire. In: *The Biological Aspects of Rare Plant Conservation* (ed. by H. Synge), pp. 365–377. Wiley, London.

Hatcher, P.G. and Spiker, E.C. 1988. Selective biodegradation of plant biomolecules. In: *Humic Substances and Their Role in the Environment* (ed. by F.H. Frimmel and R.F. Christman), pp. 59–74. Wiley, Chichester.

Hatcher, P.G., Spiker, E.C. and Orem, W.H. 1986. Organic geochemical studies of the humification process in low-moor peat. In: *Peat and Water* (ed. by C.H. Fuchsman), pp. 195–213, Elsevier, London.

Heal, O.W. 1964. Distribution of testate amoebae in northern England. *Journal of the Linnean Society* **44**, 369–382.

Heathwaite, A.L. 1990. The effect of drainage on nutrient release from fen peat and its implications for water quality – a laboratory simulation. *Water Air and Soil Pollution* **49**, 159–173.

Heathwaite, A.L. 1991. Solute transfer from drained fen peat. *Water Air and Soil Pollution* **55**, 379–395.

Heathwaite, A.L. 1992. Disappearing peat. *Geography Review* **5**, 26–31.

Heathwaite, L. 1995. Problems in the hydrological management of cut-over raised mires, with special reference to Thorne Moors, South Yorkshire. In: *Restoration of Temperate Wetlands* (ed. by B.D. Wheeler, S.C. Shaw, W.J. Fojt and R.A. Robertson), pp. 315–329. Wiley, Chichester.

Heathwaite, A.L., Eggelsmann, R., Göttlich, Kh. and Kaule, G. 1993a. Ecohydrology, mire drainage and mire conservation. In: *Mires: Process, Exploitation and Conservation* (ed. by A.L. Heathwaite and Kh. Göttlich), pp. 417–484. Wiley, Chichester.

Heathwaite, A.L. and Göttlich, Kh. (eds.) 1993. *Mires: Process, Exploitation and Conservation*. Wiley, Chichester.

Heathwaite, A.L., Göttlich, Kh., Burmeister, E.-G., Kaule, G. and Grospietsch, Th. 1993b. Mires: definitions and form. In: *Mires: Process, Exploitation and Conservation* (ed. by A.L. Heathwaite and Kh. Göttlich), pp. 1–75. Wiley, Chichester.

Heikkinen, K. 1994. Organic-matter, iron and nutrient transport and nature of dissolved organic-matter in the drainage-basin of a boreal humic river in northern Finland. *Science of the Total Environment* **152**, 81–89.

Heinselman, M.L. 1975. Boreal peatlands in relation to environment. In: *Coupling of Land and Water Systems* (ed. by A.D. Hasler), pp. 93–103. Springer-Verlag, Berlin.

Hemond, H.F. 1980. Biogeochemistry of Thoreau's Bog, Concord, Massachussetts. *Ecological Monographs* **50**, 507–526.

Hendon, D., Charman, D.J. and Kent, M. 2001. Comparisons of the palaeohydrological record derived from testate amoebae analysis from peatlands in northern England: within-site variability, between-site comparability and palaeoclimatic implications. *Holocene* **11**, 127–148.

Hilbert, D.W., Roulet, N. and Moore, T. 2000. Modelling and analysis of peatlands as dynamical systems. *Journal of Ecology* **88**, 230–242.

Hilton, J. and Spezzano, P. 1994. An investigation of possible processes of radiocesium release from organic upland soils to water bodies. *Water Research* **28**, 975–983.

Himberg, K.K. and Pakarinen, P. 1994. Atmospheric PCB deposition in Finland during 1970s and 1980s on the basis of concentrations in ombrotrophic peat mosses (*Sphagnum*). *Chemosphere* **29**, 431–440.

Hodder, A.P.W. and de Lange, P.J. 1991. Dissolution and depletion of ferromagnesian minerals from Holocene tephra layers in an acid bog, New Zealand and implications for tephra correlation. *Journal of Quaternary Science* **6**, 195–208.

Hofstetter, R.H. 1983. Wetlands in the United States. In: *Ecosystems of the World 4B. Mires: Swamp, Bog, Fen and Moor: Regional Studies* (ed. by A.J.P. Gore), pp. 201–244. Elsevier, Amsterdam.

Hogg, E.H. 1993. Decay potential of hummock and hollow *Sphagnum* peats at different depths in a Swedish raised bog. *Oikos* **66**, 269–278.

Hökkä, H. and Groot, A. 1999. Recent studies on black spruce management on peatlands in northern Ontario. A literature review. *Suo* **50**, 35–43.

Holmes, P.R., Boyce, D.C. and Reed, D.K. 1993a. The ground beetle (Coleoptera, Carabidae) fauna of Welsh peatland biotopes – factors influencing the distribution of ground beetles and conservation implications. *Biological Conservation* **63**, 153–161.

Holmes, P.R., Fowles, A.P., Boyce, D.C. and Reed, D.K. 1993b. The ground beetle (Coleoptera, Carabidae) fauna of Welsh peatland biotopes – species assemblages in relation to peatland habitats and management. *Biological Conservation* **65**, 61–67.

Hong, Y.T., Jiang, H.B., Liu, T.S., Zhou, L.P., Beer, J., Li, H.D., Leng, X.T., Hong, B. and Qin, X.G. 2000. Response of climate to solar forcing recorded in a 6000-year delta O-18 time-series of Chinese peat cellulose. *Holocene* **10**, 1–7.

Hope, D., Billett, M.F. and Cresser, M.S. 1997. Exports of organic carbon in two river systems in NE Scotland. *Journal of Hydrology* **193**, 61–82.

Houghton, J.T., Meiro Filho, L.G., Callander, B.A., Harris, N., Kattenburg, A. and Maskell, K. (eds.) 1996. *Climate Change 1995: the Science of Climate Change.* Cambridge University Press.

Howard-Williams, C. 1985. Cycling and retention of nitrogen and phosphorous in wetlands: a theoretical and applied perspective. *Freshwater Biology* **15**, 391–431.

Hu, F.S. and Davis, R.B. 1995. Postglacial development of a Maine bog and paleoenvironmental implications. *Canadian Journal of Botany* **73**, 638–649.

Hughes, J. and Heathwaite, A.L. 1995a. Hydrology and hydrochemistry of British wetlands. In: *Hydrology and Hydrochemistry of British Wetlands* (ed. by J. Hughes and A.L. Heathwaite), pp. 1–8. Wiley, Chichester.

Hughes, J. and Heathwaite, A.L. (eds.) 1995b. *Hydrology and Hydrochemistry of British Wetlands.* Wiley, Chichester.

Hughes, P.D.M. 2000. A reappraisal of the mechanisms leading to ombrotrophy in British raised mires. *Ecology Letters* **3**, 7–9.

Hughes, P.D.M., Kenward, H.K., Hall, A.R. and Large, F.D. 2000a. A high-resolution record of mire development and climatic change spanning the Late-glacial–Holocene boundary at Church Moss, Davenham (Cheshire, England). *Journal of Quaternary Science* **15**, 697–724.

Hughes, P.D.M., Mauquoy, D., Barber, K.E. and Langdon, P.G. 2000b. Mire-development pathways and palaeo-climatic records from a full Holocene peat archive at Walton Moss, Cumbria, England. *Holocene* **10**, 465–479.

Hulme, P.D. 1980. The classification of Scottish peatlands. *Scottish Geographical Magazine* **96**, 46–50.

Hulme, P.D. 1986. The origin and development of wet hollows and pools on Craigeazle mire, south west Scotland. *International Peat Journal* **1**, 15–28.

Hulme, P.D. 1994. A paleobotanical study of paludifying pine forest on the island of Hailuoto, northern Finland. *New Phytologist* **126**, 153–162.

Hulme, P.D. and Blyth, A.W. 1985. Observations on the erosion of blanket peat in Yell, Shetland. *Geografiska Annaler* **67A**, 119–122.

Hunt, J.B. and Hill, P.G. 1993. Tephra geochemistry: a discussion of some persistent analytical problems. *Holocene* **3**, 271–278.

Huntley, B. 1991. Historical lessons for the future. In: *The Scientific Management of Temperate Communities for Conservation* (ed. by I.F. Spellerberg), pp. 473–503. Blackwell, Oxford.

Huntley, B. and Prentice, I.C. 1988. July temperatures in Europe from pollen data, 6000 years before present. *Science* **241**, 687–690.

Huntley, B. and Prentice, I.C. 1993. Holocene vegetation and climates of northern Europe. In: *Global Climates since the Last Glacial Maximum* (ed. by H.E. Wright), pp. 136–168, University of Minesota Press, Minneapolis.

Huopalainen, M., Tuittila, E.S., Vanha-Majamaa, I., Nousiainen, H., Laine, J. and Vasander, H. 2001. Effects of long-term aerial pollution on soil seed banks in drained pine mires in southern Finland. *Water Air and Soil Pollution* **125**, 69–79.

Hutchin, P.R., Press, M.C., Lee, J.A. and Ashenden, T.W. 1995. Elevated concentrations of CO_2 may double methane emissions from mires. *Global Change Biology* **1**, 125–128.

Iivanainen, E., Sallantaus, T., Katila, M.L. and Martikainen, P.J. 1999. Mycobacteria in runoff waters from natural and drained peatlands. *Journal of Environmental Quality* **28**, 1226–1234.

Ilomets, M. 1984. The cyclical nature of the development of bogs. In: *Estonia: Nature, Man, Economy*, pp. 68–77.

Estonian and Geographical society, Academy of Sciences of the Estonian SSR, Tallinn.

Ilomets, M., Animgi, J. and Kallas, R. 1995. *Estonian Peatlands*. Ministry of Environment, Republic of Estonia, Tallinn.

IMCG 2001. *Global Action Plan for Peatlands (GAPP) – Draft Document*. http://www.imcg.net.

Immirzi, C.P., Maltby, E. and Clymo, R.S. 1992. *The global status of peatlands and their role in carbon cycling*. A report for Friends of the Earth by the Wetland Ecosystems Research Group, Dept Geography, University of Exeter, pp. 1–145. FoE, London.

Ingram, H.A.P. 1978. Soil layers in mires: function and terminology. *Journal of Soil Science* **29**, 224–227.

Ingram, H.A.P. 1982. Size and shape in raised mire ecosystems: a geophysical model. *Nature* **297**, 300–303.

Ingram, H.A.P. 1983. Hydrology. In: *Mires: Swamp, Bog, Fen and Moor. General Studies. Ecosystems of the World 4A: General Studies* (ed. by A.J.P. Gore), pp. 67–158. Elsevier, Amsterdam.

Ingram, H.A.P. 1987. Ecohydrology of Scottish peatlands. *Transactions of the Royal Society of Edinburgh: Earth Sciences* **78**, 287–296.

Ingram, H.A.P. 1992. Introduction to the ecohydrology of mires in the context of cultural perturbation. In: *Peatland Ecosystems and Man: an Impact Assessment* (ed. by O.M. Bragg, P.D. Hulme, H.A.P. Ingram and R.A. Robertson), pp. 67–93. International Peat Society, Finland.

Ivanov, K.E. 1981. *Water Movement in Mirelands*. Translated from the Russian by A. Thompson and H.A.P. Ingram. Academic Press, London.

Ivarsson, H. and Jansson, M. 1994. Regional variation of dissolved organic-matter in running waters in central northern Sweden. *Hydrobiologia* **286**, 37–51.

Jacob, J.S. and Hallmark, C.T. 1996. Holocene stratigraphy of Cobweb Swamp, a Maya wetland in northern Belize. *Geological Society of America Bulletin* **108**, 883–891.

Jacobson, G.L., Almquist-Jacobson, H. and Winne, J.C. 1991. Conservation of rare plant habitat – insights from the recent history of vegetation and fire at Crystal Fen, northern Maine, USA. *Biological Conservation* **57**, 287–314.

Janssens, J.A. 1983. A quantitative method for stratigraphic analysis of bryophytes in Holocene peat. *Journal of Ecology* **71**, 189–196.

Janssens, J.A. 1988. Fossil bryophytes and paleoenvironmental reconstruction of peatlands. In: *Methods in Bryology* (ed. by J.M. Glime), pp. 299–306. Proceedings of the Bryological Methods Workshop, Mainz Hattori Botanical Laboratory, Nichinan, Japan.

Janssens, J.A. 1989. Methods in Quaternary ecology 11. Bryophytes. *Geoscience Canada* **17**, 13–24.

Jasinski, J.P.P., Warner, B.G., Andreev, A.A., Aravena, R., Gilbert, S.E., Zeeb, B.A., Smol, J.P. and Velichko, A.A. 1998. Holocene environmental history of a peatland in the Lena River valley, Siberia. *Canadian Journal of Earth Sciences* **35**, 637–648.

Jeglum, J.K., Haavisto, V.F. and Groot, A. 1982. Peatland forestry in Ontario: an overview. In: *Symposium 82 – a Symposium on Peat and Peatlands* (ed. by J.D. Sheppard, J. Musial and T.E. Tibbetts), pp. 127–167. Shippagan, New Brunswick.

John, D.M., Leveque, C. and Newton, L.E. 1993. Wetlands of Africa: Western Africa. In: *Wetlands of the World: Inventory, Ecology and Management* (ed. by D. Whigham, D. Dykyjova and S. Hejny), pp. 47–78. Kluwer, Dordrecht.

Johnson, E.A. 1977a. A multivariate analysis of the niches of plant populations in raised bogs. I. Niche dimensions. *Canadian Journal of Botany* **55**, 1201–1210.

Johnson, E.A. 1977b. A multivariate analysis of the niches of plant populations in raised bogs. II Niche width and overlap. *Canadian Journal of Botany* **55**, 1211–1220.

Johnson, L.C. and Damman, A.W.H. 1991. Species controlled *Sphagnum* decay on a south Swedish raised bog. *Oikos* **61**, 234–242.

Johnson, L.C. and Damman, A.W.H. 1993. Decay and its regulation in *Sphagnum* peatlands. *Advances in Bryology* **5**, 249–296.

Johnson, L.C., Damman, A.W.H. and Malmer, N. 1990. *Sphagnum* macrostructure as an indicator of decay and compaction in peat cores from an ombrotrophic south Swedish peat-bog. *Journal of Ecology* **78**, 633–647.

Jones, C.G., Lawton, J.H. and Shachak, M. 1994. Organisms as ecosystem engineers. *Oikos* **69**, 373–386.

Jones, R.D. and Amador, J.A. 1992. Removal of total phosphorus and phosphate by peat soils of the Florida Everglades. *Canadian Journal of Fisheries and Aquatic Sciences* **49**, 577–583.

Jones, R.K., Pierpoint, G., Wickware, G.M., Jeglum, J.K., Arnap, R.W. and Bowles, J.M. 1983. *Field Guide to Forest Ecosystem Classification for the Clay belt, Site Region 3e*. Queen's Printer, Ministry of Natural Resources, Toronto.

Joosten, J.H.J. 1985. A 130 year micro- and macrofossil record from regeneration peat in former peasant peat pits in the Peel, the Netherlands: a palaeoecological study with agricultural and climatological implications. *Palaeogeography, Palaeoclimatology, Palaeoecology* **49**, 277–312.

Joosten, J.H.J. 1995. Time to regenerate: long-term perspectives of raised bog regeneration with special emphasis on palaeoecological studies. In: *Restoration of Temperate Wetlands* (ed. by B.D. Wheeler, S.C. Shaw, W.J. Fojt and R.A. Robertson), pp. 379–404. Wiley, Chichester.

Joy, J. and Pullin, A.S. 1997. The effects of flooding on the survival and behaviour of overwintering large heath butterfly *Coenonympha tullia* larvae. *Biological Conservation* **82**, 61–66.

Junk, W.J. 1983. Ecology of swamps on the middle Amazon. In: *Ecosystems of the World 4B. Mires: Swamp, Bog, Fen and Moor: Regional Studies* (ed. by A.J.P. Gore), pp. 269–294. Elsevier, Amsterdam.

Junk, W.J. 1993. Wetlands of Tropical South America. In: *Wetlands of the World: Inventory, Ecology and*

Management (ed. by D. Whigham, D. Dykyjova and S. Hejny), pp. 679–739. Dordrecht, Kluwer.

Kac, N.J. 1971. *Die Moore der Erde*. Nauka, Moscow.

Kanninen, M. and Anttila, P. (eds.) 1992. *The Finnish Research Programme on Climate Change*. Publications of the Academy of Finland 3/92, VAPK, Helsinki.

Karofeld, E. 1996. The effects of fly ash precipitation on the *Sphagnum* mosses in Niinsaare bog, NE Estonia. *Suo* **47**, 105–114.

Karofeld, E. 1998. The dynamics of the formation and development of hollows in raised bogs in Estonia. *Holocene* **8**, 697–704.

Karofeld, E. 1999. The effects of bombing and regeneration of plant cover in Konnu-Suurosoo raised bog, North Estonia. *Wetlands Ecology and Management* **6**, 253–259.

Karofeld, E. and Toom, M. 1999. Mud-bottoms in Männikjärve Bog, central Estonia. *Proceedings of the Estonian Academy of Sciences Biology Ecology* **48**, 216–235.

Katase, T. 1993. Phenolic-acids in tropical peats from peninsular Malaysia – occurrence and possible diagenetic behavior. *Soil Science* **155**, 155–165.

Keddy, P.A. 2000. *Wetland Ecology: Principles and Conservation*. Cambridge, Cambridge University Press.

Kempter, H., Gorres, M. and Frenzel, B. 1997. Ti and Pb concentrations in rainwater-fed bogs in Europe as indicators of past anthropogenic activities. *Water Air and Soil Pollution* **100**, 367–377.

Kershaw, A.P. and Bohte, A. 1997. The impact of prehistoric fires on tropical peatland forests. In: *Biodiversity and Sustainability of Tropical Peatlands* (ed. by J.O. Reiley and S.E. Page), pp. 73–80. Samara Publishing, Cardigan.

Kershaw, A.P., Reid, M. and Bulman, D. 1997. The nature and development of peatlands in Victoria, Australia. In: *Biodiversity and Sustainability of Tropical Peatlands* (ed. by J.O. Reiley and S.E. Page), pp. 81–92. Samara Publishing, Cardigan.

Keys, D. *Canadian peat harvesting and the environment*. Ottawa, North American Wetlands Conservation Council (Canada), pp. 1–29. Sustaining wetlands, issues paper no. 1992–3.

Kilian, M.R., Van der Plicht, J. and Vangeel, B. 1995. Dating raised bogs – new aspects of AMS C-14 wiggle matching, a reservoir effect and climatic-change. *Quaternary Science Reviews* **14**, 959–966.

Kilian, M.R., van Geel, B. and van der Plicht, J. 2000. C-14 AMS wiggle matching of raised bog deposits and models of peat accumulation. *Quaternary Science Reviews* **19**, 1011–1033.

King, W. 1685. Of the Bogs, and Loighs of Ireland. *Philosophical Transactions, Oxford 15*.

Kingsbury, C.M. and Moore, T.R. 1987. The freeze–thaw cycle of a subarctic fen, Northern Québec, Canada. *Arctic and Alpine Research* **19**, 289–295.

Kingston, J.C. 1982. Association and distribution of common diatoms in surface samples from northern Minnesota peatlands. *Nova Hedwigia* **73**, 333–345.

Kinzel, H. 1983. Influence of limestone, silicates and soil pH on vegetation. In: *Physiological Plant Ecology III. Responses to the Chemical and Biological Environment* (ed. by O.L. Lange, P.S. Nobel, C.B. Osmond and H. Ziegler), pp. 201–244. Springer-Verlag, Berlin.

Kirkpatrick, A.H., Scott, L. and MacDonald, A.J. 1995. Moorlands of Orkney – cultural landscapes. In: *Heaths and Moorland: Cultural Landscapes* (ed. by D.B.A. Thompson, A.J. Hester and M.B. Usher), pp. 309–311. Scottish Natural Heritage, Edinburgh.

Kivinen, E. 1977. Survey, classification and conservation of peatlands. *Bulletin of the International Peat Society* **8**, 24–25.

Kivinen, E. and Pakarinen, P. 1981. Geographical distribution of peat resources and major peatland complex types in the world. *Annales Academiae Scientorum Fennicae AIII* **132**, 1–28.

Klinger, L.F., Taylor, J.A. and Franzen, L.G. 1996. The potential role of peatland dynamics in ice-age initiation. *Quaternary Research* **45**, 89–92.

Koch-Rose, M.S., Reddy, K.R. and Chanton, J.P. 1994. Factors controlling seasonal nutrient profiles in a subtropical peatland of the Florida Everglades. *Journal of Environmental Quality* **23**, 526–533.

Koerselman, W. and Beltman, B. 1988. Evapotranspiration from fens in relations to Penman's potential free water evaporation (E_o) and pan evaporation. *Aquatic Botany* **31**, 307–320.

Koerselman, W. and Verhoeven, J.T.A. 1995. Eutrophication of fen ecosystems: external and internal nutrient sources and restoration strategies. In: *Restoration of Temperate Wetlands* (ed. by B.D. Wheeler, S.C. Shaw, W.J. Fojt and R.A. Robertson), pp. 91–112. Wiley, Chichester.

Kooijman, A.M. 1992. The decrease of rich fen bryophytes in the Netherlands. *Biological Conservation* **59**, 139–143.

Kooijman, A.M. and Bakker, C. 1994. The acidification capacity of wetland bryophytes as influenced by simulated clean and polluted rain. *Aquatic Botany* **48**, 133–144.

Korhola, A. 1994. Radiocarbon evidence for rates of lateral expansion in raised mires in southern Finland. *Quaternary Research* **42**, 299–307.

Korhola, A. 1995. Holocene climatic variations in southern Finland reconstructed from peat-initiation data. *Holocene* **5**, 43–58.

Korhola, A. 1996. Initiation of a sloping mire complex in southwestern Finland – autogenic versus allogenic controls. *Écoscience* **3**, 216–222.

Korhola, A., Alm, J., Tolonen, K. and Jungner, H. 1996. Three-dimensional reconstruction of carbon accumulation and CH_4 emission during nine millennia in a raised mire. *Journal of Quaternary Science* **11**, 161–165.

Korhola, A., Tolonen, K., Turunen, J. and Jungner, H. 1995. Estimating long-term carbon accumulation rates in boreal peatlands by radiocarbon dating. *Radiocarbon* **37**, 575–584.

Korhonen, R. and Lüttig, G.W. 1996. Peat in balneology and healthcare. In: *Global Peat Resources* (ed. by E.

Lappalainen), pp. 339–345. International Peat Society, Finland.

Kortelainen, P. 1993. Content of total organic-carbon in Finnish lakes and its relationship to catchment characteristics. *Canadian Journal of Fisheries and Aquatic Sciences* **50**, 1477–1483.

Kortelainen, P. and Saukkonen, S. 1995. Organic vs minerogenic acidity in headwater streams in Finland. *Water Air and Soil Pollution* **85**, 559–564.

Kosters, E.C., Chmura, G.L. and Bailey, A. 1987. Sedimentary and botanical factors influencing peat accumulation in the Mississippi delta. *Journal of the Geological Society* **144**, 423–434.

Kratz, T.K. 1988. A new method for estimating horizontal growth of the peat mat in basin filling peatlands. *Canadian Journal of Botany* **66**, 826–828.

Kratz, T.K. and Dewitt, C.B. 1986. Internal factors controlling peatland-lake ecosystem development. *Ecology* **67**, 100–107.

Kuhry, P. 1985. Transgression of a raised bog across a coversand ridge, originally covered with an oak-lime forest. Palaeoecological study of a Middle Holocene local vegetational succession in the Amtsven (northwest Germany). *Review of Palaeobotany and Palynology* **44**, 303–353.

Kuhry, P. 1994. The role of fire in the development of *Sphagnum* dominated peatlands in western boreal Canada. *Journal of Ecology* **82**, 899–910.

Kuhry, P. 1997. The palaeoecology of a treed bog in western boreal Canada: a study based on microfossils, macrofossils and physico-chemical properties. *Review of Palaeobotany and Palynology* **96**, 183–224.

Kulczynski, S. 1949. Peat bogs of Polesie. *Mémoires de l'Académie Polonaise des Sciences et des Lettres* **B15**.

Kullman, L. 1989. Tree-limit history during the Holocene in the Scandes Mountains, Sweden, inferred from subfossil wood. *Review of Palaeobotany and Palynology* **58**, 163–171.

Kullman, L. and Engelmark, O. 1990. A high Late Holocene tree-limit and the establishment of the spruce forest-limit – a case study in northern Sweden. *Boreas* **19**, 323–331.

Kullman, L. 1999. Early Holocene tree growth at a high elevation site in the northernmost Scandes of Sweden (Lapland): a palaeobiogeographical case study based on megafossil evidence. *Geografiska Annaler Series A – Physical Geography* **81A**, 63–74.

Kvenvolden, K.A. and Lorenson, T.D. 1993. Methane in permafrost – preliminary results from coring at Fairbanks, Alaska. *Chemosphere* **26**, 609–616.

Labadz, J.C., Burt, T.P. and Potter, A.W.R. 1991. Sediment yield and delivery in the blanket peat moorlands of the southern Pennines. *Earth Surface Processes and Landforms* **16**, 255–271.

Lafleur, P.M. 1990. Evapotranspiration from sedge-dominated wetland surfaces. *Aquatic Botany* **37**, 341–353.

Lafleur, P.M. and Roulet, N.T. 1992. A comparison of evaporation rates from 2 fens of the Hudson Bay lowland. *Aquatic Botany* **44**, 59–69.

Lag, J. 1986. Jordbunnsgrunnlaget for plantevekst pa Svalbard [Soils as a basis for plant growth in Svalbard]. *Jordundersokelsens Saertrykk* **352**, 1-26.

Laiho, R., Laine, J., Vasander, H. (eds.) 1995. *Northern Peatlands in Global Climatic Change*. Publications of the Academy of Finland 1/96, Edita, Helsinki.

Laine, J., Silvola, J., Tolonen, K., Alm, J., Nykanen, H., Vasander, H., Sallantaus, T., Savolainen, I., Sinisalo, J. and Martikainen, P.J. 1996. Effect of water-level drawdown on global climatic warming – northern peatlands. *Ambio* **25**, 179–184.

Laine, J., Vasander, H. and Laiho, R. 1995. Long-term effects of water-level drawdown on the vegetation of drained pine mires in southern Finland. *Journal of Applied Ecology* **32**, 785–802.

Lamb, H.H. 1977. *Climate: Past, Present and Future*. Volume 2. Methuen, London.

Lamers, L.P.M., Farhoush, C., Van Groenendael, J.M. and Roelofs, J.G.M. 1999. Calcareous groundwater raises bogs; the concept of ombrotrophy revisited. *Journal of Ecology* **87**, 639–648.

Lan, Y.L. and Breslin, V.T. 1999. Sedimentary records of spheroidal carbonaceous particles from fossil-fuel combustion in western Lake Ontario. *Journal of Great Lakes Research* **25**, 443–454.

Lang, K., Silvola, J., Ruuskanen, J. and Martikainen, P.J. 1995. Emissions of nitric-oxide from boreal peat soils. *Journal of Biogeography* **22**, 359–364.

Lappalainen, E. (ed.) 1996a. *Global Peat Resources*. International Peat Society, Finland.

Lappalainen, E. 1996b. General review on world peatland and peat resources. In: *Global Peat Resources* (ed. by E. Lappalainen), pp. 53–56. International Peat Society, Finland.

LaRose, S., Price, J. and Rochefort, L. 1997. Rewetting of a cutover peatland: hydrologic assessment. *Wetlands* **17**, 416–423.

Latter, P.M., Howson, G., Howard, D.M. and Scott, W.A. 1998. Long-term study of litter decomposition on a Pennine peat bog: which regression? *Oecologia* **113**, 94–103.

Lavers, C.P. and Haines-Young, R.H. 1996. Using models of bird abundance to predict the impact of current land-use and conservation policies in the flow country of Caithness and Sutherland, northern Scotland. *Biological Conservation* **75**, 71–77.

Lavers, C.P. and Haines-Young, R.H. 1997. Displacement of dunlin *Calidris alpina schinzii* by forestry in the Flow Country and an estimate of the value of moorland adjacent to plantations. *Biological Conservation* **79**, 87–90.

Lavoie, C., Elias, S.A. and Filion, L. 1997a. A 7000-year record of insect communities from a peatland environment, southern Quebec. *Écoscience* **4**, 394–403.

Lavoie, C., Elias, S.A. and Payette, S. 1997b. Holocene fossil beetles from a treeline peatland in subarctic Quebec. *Canadian Journal of Zoology* **75**, 227–236.

Lavoie, C. and Rochefort, L. 1996. The natural revegetation of a harvested peatland in southern Québec: a spatial and dendroecological analysis. *Écoscience* 3, 101–111.

Lee, J.A. and Tallis, J.H. 1973. Regional and historical aspects of lead pollution in Britain. *Nature* 245, 216–218.

Lehtonen, K. and Ketola, M. 1993. Solvent-extractable lipids of *Sphagnum*, *Carex*, bryales and *Carex*-bryales peats – content and compositional features vs peat humification. *Organic Geochemistry* 20, 363–380.

Lewis, F.J. 1905. The plant remains in the Scottish peat mosses I. *Transactions of the Royal Society of Edinburgh* 41, 699–723.

Lewis, F.J. 1906. The plant remains in the Scottish peat mosses II. *Transactions of the Royal Society of Edinburgh* 45, 335–360.

Lewis, F.J. 1907. The plant remains in the Scottish peat mosses III. *Transactions of the Royal Society of Edinburgh* 46, 33–70.

Lewis, F.J. 1911. The plant remains in the Scottish peat mosses IV. *Transactions of the Royal Society of Edinburgh* 47, 793–833.

Lewis-Smith, R.I. and Clymo, R.S. 1984. An extraordinary peat forming community on the Falkland Islands. *Nature* 309, 617–620.

Lightowlers, D. 1988. A poisoned landscape gathers no moss. *New Scientist* 5 May, 54–58.

Lillesand, T.M. and Kiefer, R.W. 2000. *Remote Sensing and Image Interpretation* (4th edn). Wiley, Chichester.

Linderholm, H.W. 1999. Climatic and anthropogenic influences on radial growth of Scots pine at Hanvedmossen, a raised peat bog in south central Sweden. *Geografiska Annaler* 81A, 75–86.

Lindholm, T. and Markkula, I. 1984. Moisture conditions in hummocks and hollows in virgin and drained sites on the raised bog Laaviosuo, southern Finland. *Annales Botanici Fennici* 21, 241–255.

Lindsay, R.A. 1977. *Glasson Moss N.N.R. and the 1976 fire.* 51. Nature Conservancy Council report. London, 51 pp.

Lindsay, R.A. 1995. *Bogs: the Ecology, Classification and Conservation of Ombrotrophic Mires.* Scottish Natural Heritage, Edinburgh.

Lindsay, R.A., Charman, D.J., Everingham, F., O'Reilly, R.M., Palmer, M.A., Rowell, T.A. and Stroud, D.A. 1988. *The Flow Country: the Peatlands of Caithness and Sutherland.* NCC, Peterborough.

Lindsay, R.A., Riggall, J. and Burd, F. 1985. The use of small scale surface patterns in the classification of British peatlands. *Aquilo (Seriales Botanica)* 21, 69–79.

Lloyd, J.W. and Tellam, J.H. 1995. Groundwater-fed wetlands in the UK. In: *Hydrology and Hydrochemistry of British Wetlands* (ed. by J. Hughes and A.L. Heathwaite), pp. 39–61. Wiley, Chichester.

Lowe, D.J. 1985. Application of impulse radar to continuous profiling of tephra-bearing lake sediments and peats: an initial evaluation. *New Zealand Journal of Geology and Geophysics* 28, 667–674.

Lowe, D.J. 1988. Stratigraphy, age, composition and correlation of late Quaternary tephras interbedded with organic sediments in Waikato lakes, North Island, New Zealand. *New Zealand Journal of Geology and Geophysics* 31, 125–165.

Lowe, D.J. and Hogg, A.G. 1986. Tephrostratigraphy and chronology of the Kaipo Lagoon, an 11,500 year old montane peat bog in Urewera National Park, New Zealand. *Journal of The Royal Society of New Zealand* 16, 25–41.

Lowe, D.J., Newnham, R.H. and Ward, C.M. 1999. Stratigraphy and chronology of a 15 ka sequence of multi-sourced silicic tephras in a montane peat bog, eastern North Island, New Zealand. *New Zealand Journal of Geology and Geophysics* 42, 565–579.

Lowe, J.J. (ed.) 1991. Radiocarbon dating: recent applications and future potential. *Quaternary Proceedings* 1, Quaternary Research Association, Cambridge.

Lowe, J.J. and Walker, M.J.C. 1997. *Reconstructing Quaternary Environments.* Longman, Harlow.

Lucas, R.E. 1982. *Organic Soils (Histosols). Formation Distribution, Physical and Chemical Properties and Management for Crop Production.* Research Report 435 (Farm Science). Michigan State University.

Luken, J.O. and Billings, W.D. 1985. Succession and biomass allocation as controlled by *Sphagnum* in an Alaskan peatland. *Canadian Journal of Botany* 63, 1500–1507.

Maas, D. and Schopp-Guth, A. 1995. Seed banks in fen areas and their potential use in restoration ecology. In: *Restoration of Temperate Wetlands* (ed. by B.D. Wheeler, S.C. Shaw, W.J. Fojt and R.A. Robertson), pp. 189–206. Wiley, Chichester.

Macdonald, S.E. and Yin, F.Y. 1999. Factors influencing size inequality in peatland black spruce and tamarack: evidence from post-drainage release growth. *Journal of Ecology* 87, 404–412.

MacDonell, M.R. and Groot, A. 1996. *Uneven-aged Silviculture for Peatland Second-growth Black Spruce: Biological Feasibility.* NODA/NFP Technical report TR–36. Natural Resources Canada, Canadian Forest Service, Sault Ste Marie.

MacDonell, M.R. and Groot, A. 1997. Harvesting peatland black spruce: impacts on advance growth and site disturbance. *The Forestry Chronicle* 73, 249–255.

McGee, E. and Bradshaw, R. 1990. Erosion of high-level blanket peat. In: *Ecology and Conservation of Irish Peatlands* (ed. by G. Doyle), pp. 109–120. Royal Irish Academy, Dublin.

McGlone, M.S. and Wilmshurst, J.M. 1999. A Holocene record of climate, vegetation change and peat bog development, east Otago, South Island, New Zealand. *Journal of Quaternary Science* 14, 239–254.

McGlone, M.S., Mark, A.F. and Bell, D. 1995. Late Pleistocene and Holocene vegetation history, Central Otago, South Island, New Zealand. *Journal of The Royal Society of New Zealand* 25, 1–22.

McGlone, M.S., Moar, N.T. and Meurk, C.D. 1997. Growth and vegetation history of alpine mires on the Old Man

Range, Central Otago, New Zealand. *Arctic and Alpine Research* **29**, 32–44.

Mackay, A.W. and Tallis, J.H. 1996. Summit-type blanket mire erosion in the Forest of Bowland, Lancashire, UK: predisposing factors and implications for conservation. *Biological Conservation* **76**, 31–44.

McKenzie, C., Schiff, S., Aravena, R., Kelly, C. and Louis, V.S. 1998. Effect of temperature on production of CH_4 and CO_2 from peat in a natural and flooded boreal forest wetland. *Climatic Change* **40**, 247–266.

McNamara, J.P., Siegel, D.I., Glaser, P.H. and Beck, R.M. 1992. Hydrogeologic controls on peatland development in the Malloryville wetland, New York (USA). *Journal of Hydrology* **140**, 279–296.

Macphail, M.K., Pemberton, M. and Jacobson, G. 1999. Peat mounds of southwest Tasmania: possible origins. *Australian Journal of Earth Sciences* **46**, 667–677.

McTiernan, K.B., Garnett, M.H., Mauquoy, D., Ineson, P. and Couteaux, M.M. 1998. Use of near-infrared reflectance spectroscopy (NIRS) in palaeoecological studies of peat. *Holocene* **8**, 729–740.

Madden, B. and Doyle, G.J. 1990. Primary production on Mongan bog. In: *Ecology and Conservation of Irish Peatlands* (ed. by G.J. Doyle), pp. 147–161. Royal Irish Academy, Dublin.

Magnusson, B. and Stewart, J.M. 1987. Effects of disturbances along hydroelectrical transmission corridors through peatlands in northern Manitoba, Canada. *Arctic and Alpine Research* **19**, 470–478.

Mahaney, W.C. 1984. *Quaternary Dating Methods*. Elsevier, Amsterdam.

Mäkilä, M. 1997. Holocene lateral expansion, peat growth and carbon accumulation on Haukkasuo, a raised bog in southeastern Finland. *Boreas* **26**, 1–14.

Mallik, A.U. 1989. Small-scale plant succession towards fen on floating mat of a *Typha* marsh in Atlantic Canada. *Canadian Journal of Botany* **67**, 1309–1316.

Mallik, A.U., Gimingham, C.H. and Rahman, A.A. 1984. Ecological effects of heather burning. I Water infiltration, moisture retention and porosity of the surface soil. *Journal of Ecology* **72**, 767–776.

Malmer, N. 1986. Vegetational gradients in relation to environmental conditions in north western European mires. *Canadian Journal of Botany* **64**, 375–383.

Maltby, E. 1986. *Waterlogged Wealth – Why Waste the World's Wet Places?* Earthscan, London.

Mark, A.F., Johnson, P.N., Dickinson, K.J.M. and McGlone, M.S. 1995. Southern Hemisphere patterned mires, with emphasis on southern New Zealand. *Journal of The Royal Society of New Zealand* **25**, 23–54.

Markon, C.J. and Derksen, D.V. 1994. Identification of tundra land-cover near Teshekpuk Lake, Alaska using spot satellite data. *Arctic* **47**, 222–231.

Martikainen, P.J., Nykanen, H., Crill, P. and Silvola, J. 1993. Effect of a lowered water-table on nitrous-oxide fluxes from northern peatlands. *Nature* **366**, 51–53.

Martin, N.J., Siwasin, J. and Holding, A.J. 1982. The bacterial population of a blanket peat. *Journal of Applied Bacteriology* **53**, 35–48.

Masing, V. 1982. The plant cover of Estonian bogs: a structural analysis. In: *Peatland Ecosystems* (ed. by T. Frey *et al.*) Academy of Sciences of the Estonian SSR, Tallinn.

Matthews, E. and Fung, I. 1987. Methane emissions from natural wetlands: global distribution, area and environmental characteristics of sources. *Global Biogeochemical Cycles* **1**, 61–86.

Matthews, J.A., Dahl, S.O., Berrisford, M.S. and Nesje, A. 1997. Cyclic development and thermokarstic degradation of palsas in the mid-Alpine zone at Leirpullan, Dovrefjell, southern Norway. *Permafrost and Periglacial Processes* **8**, 107–122.

Matthews, S.B. 1991. An assessment of bison habitat in the Mills Mink Lakes area, Northwest-Territories, using Landsat Thematic Mapper data. *Arctic* **44**, 75–80.

Mawby, F.J. 1995. Efects of damming peat cuttings on Glasson Moss and Wedholme Flow, two lowland raised bogs in Northwest England. In: *Restoration of Temperate Wetlands* (ed. by B.D. Wheeler, S.C. Shaw, W.J. Fojt and R.A. Robertson), pp. 349–357. Wiley, Chichester.

Meade, R. 1992. Some early changes following the rewetting of a vegetated cutover peatland surface at Danes Moss, Cheshire, UK, and their relevance to conservation management. *Biological Conservation* **61**, 31–40.

Meadows, M.E. 1988. Late Quaternary peat accumulation in Southern Africa. *Catena* **15**, 459–472.

Mets, L. 1982. Changes in a bog pool complex during an observation period of 17 years. *Estonian Contributions to the International Biological Program* **9**, 128–134.

Middeldorp, A.A. 1982. Pollen concentration as a basis for indirect dating and quantifying net organic and fungal production in a peat bog ecosystem. *Review of Palaeobotany and Palynology* **37**, 225.

Mighall, T.M. and Chambers, F.M. 1995. Holocene vegetation history and human impact at Bryn-y-Castell, Snowdonia, North Wales. *New Phytologist* **130**, 299–321.

Miller, B.B. and Bajc, A.F. 1990. Non-marine molluscs. In: *Methods in Quaternary Ecology:* Geoscience Canada Reprint Series No. 5 (ed. by B.G. Warner), pp. 101–112.

Miller, N.G. and Futyma, R.P. 1987. Palaeohydrological implications of Holocene peatland development in northern Michigan. *Quaternary Research* **27**, 297–311.

Mitchell, E.A.D., van der Knaap, W.O., van Leeuwen, J.F.N. Buttler, A., Warner, B.G. and Gobat, J.M. 2001. The palaeoecological history of the Praz-Rodet bog (Swiss Jura) based on pollen, plant macrofossils and testate amoebae (Protozoa). *Holocene* **11**, 65–80.

Mitsch, W.J. and Gosselink, J.G. 2000. *Wetlands*, 3rd edn. Wiley, New York.

Mitsch, W.J. and Wilson, R.F. 1996. Improving the success of wetland creation and restoration with know-how, time, and self-design. *Ecological Applications* **6**, 77–83.

Moen, A. 1985. Classification of mires for conservation purposes in Norway. *Aquilo Seriales Botanica* **21**, 95–100.

Moen, A. 1995. Vegetational changes in boreal rich fens induced by haymaking; management plant for the Sølendet Nature Reserve. In: *Restoration of Temperate Wetlands* (ed. by B.D. Wheeler, S.C. Shaw, W.J. Fojt and R.A. Robertson), pp. 167–181. Wiley, Chichester.

Money, R.P. 1995. Restoration of cut-over peatlands: the role of hydrology in determining vegetation quality. In: *Hydrology and Hydrochemistry of British Wetlands* (ed. by J. Hughes and A.L. Heathwaite), pp. 383–400. Wiley, Chichester.

Moore, P.D. 1973. The influence of prehistoric cultures upon the initiation and spread of blanket bog in upland Wales. *Nature* **241**, 350–353.

Moore, P.D. 1975. Origin of blanket mires. *Nature* **256**, 267–269.

Moore, P.D. 1977. Stratigraphy and pollen analysis of Claish Moss, Scotland: significance for the origin of surface pools and forest history. *Journal of Ecology* **65**, 375–397.

Moore, P.D. 1982a. Fire: catastrophic or creative force? *Impact of Science on Society* **32**, 5–14.

Moore, P.D. 1982b. Sub surface formation of charcoal: an unlikely event in peat. *Quaternary Newsletter* **38**, 13–14.

Moore, P.D. (ed.) 1984a. *European Mires*. Academic Press, London.

Moore, P.D. 1984b. The classification of mires: an introduction. In: *European Mires* (ed. by P.D. Moore), pp. 1–10. Academic Press, London.

Moore, P.D. 1993. The origin of blanket mire, revisited. In: *Climate Change and Human Impact on the Landscape* (ed. by F.M. Chambers), pp. 217–224. Chapman and Hall, London.

Moore, P.D. 1997. Bog standards in Minnesota. *Nature* **386**, 655–657.

Moore, P.D. and Bellamy, D.J. 1973. *Peatlands*. Elek Science, London.

Moore, P.D., Merryfield, D.L. and Price, M.D.R. 1984. The vegetation and development of blanket mires. In: *European Mires* (ed. by P.D. Moore). Academic Press, London.

Moore, P.D., Webb, J.A. and Collinson, M.E. 1991. *Pollen Analysis*. Blackwell Scientific, Oxford.

Moore, T.R., Heyes, A. and Roulet, N.T. 1994. Methane emissions from wetlands, southern Hudson-Bay lowland. *Journal of Geophysical Research – Atmospheres* **99**, 1455–1467.

Moore, T.R., Roulet, N.T. and Waddington, J.M. 1998. Uncertainty in predicting the effect of climatic change on the carbon cycling of Canadian peatlands. *Climatic Change* **40**, 229–245.

Morley, R.J. 1981. Development and vegetation dynamics of a lowland ombrogenous peat swamp in Kalimantan-Tengah, Indonesia. *Journal of Biogeography* **8**, 383.

Muscutt, A.D., Reynolds, B. and Wheater, H.S. 1993. Sources and controls of aluminum in storm runoff from a headwater catchment in mid-Wales. *Journal of Hydrology* **142**, 409–425.

Mutka, K. 1996. Environmental use of peat. In: *Global Peat Resources* (ed. by E. Lappalainen), pp. 335–338. International Peat Society, Finland.

National Wetlands Working Group 1988. *Wetlands of Canada*. Environment Canada and Polyscience Publications, Montreal, Canada.

Naucke, W., Heathwaite, A.L., Eggelsmann, R. and Schuch, M. 1993. Mire chemistry. In: *Mires: Process, Exploitation and Conservation* (ed. by A.L. Heathwaite and Kh. Göttlich), pp. 263–309. Wiley, Chichester.

Neckles, H.A., Murkin, H.R. and Cooper, J.A. 1990. Influences of seasonal flooding on macroinvertebrate abundance in wetland habitats. *Freshwater Biology*, **23**, 311–322.

Neuhäusl, R. 1975. *Hochmoore am Teich Velke Darko*. Vegetace CSSR, A9, Academia, Prague.

Neuzil, S.G. 1997. Onset and rate of carbon accumulation in four domed ombrogenous peat deposits, Indonesia. In: *Biodiversity and Sustainability of Tropical Peatlands* (ed. by J.O. Reiley and S.E. Page), pp. 55–72. Samara Publishing, Cardigan.

Newnham, R.M., Delange, P.J. and Lowe, D.J. 1995. Holocene vegetation, climate and history of a raised bog complex, northern New Zealand based on palynology, plant macrofossils and tephrochronology. *Holocene* **5**, 267–282.

Newnham, R.M., Lowe, J.J. and Williams, P.W. 1999. Quaternary environmental change in New Zealand: a review. *Progress in Physical Geography* **23**, 567–610.

Ng, P.K.L., Tay, J.B. and Lim, K.K.P. 1994. Diversity and conservation of blackwater fishes in peninsular Malaysia, particularly in the North Selangor peat swamp forest. *Hydrobiologia* **285**, 203–218.

Nicholson, B.J. and Vitt, D.H. 1990. The paleoecology of a peatland complex in continental western Canada. *Canadian Journal of Botany* **68**, 121–138.

Nicholson, I.A., Robertson, R.A. and Robinson, M. 1989. The effects of drainage on the hydrology of a peat bog. *International Peat Journal* **3**, 59–83.

Nilssen, E. and Vorren, K.D. 1991. Peat humification and climate history. *Norsk Geologisk Tidsskrift* **71**, 215–217.

Nilsson, M., Baath, E. and Soderstrom, B. 1992. The microfungal communities of a mixed mire in northern Sweden. *Canadian Journal of Botany* **70**, 272–276.

Nilsson, M. and Bohlin, E. 1993. Methane and carbon dioxide concentrations in bogs and fens – with special reference to the effects of the botanical composition of the peat. *Journal of Ecology* **81**, 615–625.

Nilsson, T. and Lundin, L. 1996. Effects of drainage and wood ash fertilization on water chemistry at a cutover peatland. *Hydrobiologia* **335**, 3–18.

Noble, M.G., Lawrence, D.B. and Streveler, G.P. 1984. *Sphagnum* invasion beneath an evergreen forest canopy in southeastern Alaska. *Bryologist* **87**, 119–127.

Nordbakken, J.F. 1996. Plant niches along the water-table gradient on an ombrotrophic mire expanse. *Ecography* **19**, 114–121.

Nordbakken, J.F. 2000. Fine-scale persistence of boreal bog plants. *Journal of Vegetation Science* **11**, 269–276.

North American Wetlands Conservation Council (Canada) 1996. *Global Mire and Peatland Conservation: Proceedings of an International Workshop.* Report No. 96-1. North American Wetlands Conservation Council (Canada). Ottawa. Also available at http://www.imcg.net/docum/brisbane.htm.

Notohadiprawiro, T. 1997. Twenty-five years experience in peatland development for agriculture in Indonesia. In: *Biodiversity and Sustainability of Tropical Peatlands* (ed. by J.O. Reiley and S.E. Page), pp. 301—309. Samara Publishing, Cardigan.

Nykanen, H., Alm, J., Lang, K., Silvola, J. and Martikainen, P.J. 1995. Emissions of CH_4, N_2O and CO_2 from a virgin fen and a fen drained for grassland in Finland. *Journal of Biogeography* **22**, 351–357.

Nyrönen, T. 1996. Peat production. In: *Global Peat Resources* (ed. by E. Lappalainen), pp. 315–318. International Peat Society, Finland.

O'Connell, M. 1990. Origins of Irish lowland blanket bog. In: *Ecology and Conservation of Irish Peatlands* (ed. by G. Doyle), pp. 49–71. Royal Irish Academy, Dublin.

Oechel, W.C., Hastings, S.J., Vourlitis, G., Jenkins, M., Riechers, G. and Grulke, N. 1993. Recent change of arctic tundra ecosystems from a net carbon dioxide sink to a source. *Nature* **361**, 520–523.

Ogilvie, J.F. 1990. Forestry and wetland – conservation of the Border Mires. *Scottish Forestry* **44**, 10–18.

Ohlson, M. and Dahlberg, B. 1991. Rate of peat increment in hummock and lawn communities on Swedish mires during the last 150 years. *Oikos* **61**, 369–378.

Okland, R.H. and Ohlson, M. 1998. Age–depth relationships in Scandinavian surface peat: a quantitative analysis. *Oikos* **82**, 29–36.

Oldfield, F., Appleby, P.G., Canbray, R.S., Eakins, J.D., Barber, K.E., Battarbee, R.W., Pearson, G.R. and Williams, J. M. 1979. ^{210}Ob, ^{137}Cs and ^{239}Pu profiles in ombrotrophic peat. *Oikos* **33**, 40–45.

Oldfield, F., Richardson, N. and Appleby, P.G. 1995. Radiometric dating (Pb-210, Cs-137, Am-241) of recent ombrotrophic peat accumulation and evidence for changes in mass-balance. *Holocene* **5**, 141–148.

Oldfield, F. and Statham, D.C. 1963. Pollen analytical data from Erswick Tarn and Ellerside Moss. *New Phytologist* **62**, 53-66.

Olmsted, I. 1993. Wetlands of Mexico. In: *Wetlands of the World: Inventory, Ecology and Management* (ed. by D. Whigham, D. Dykyjova and S. Hejny), pp. 637–677. Kluwer, Dordrecht.

Osborne, P.L. 1993. Wetlands of Papua New Guinea. In: *Wetlands of the World: Inventory, Ecology and Management* (ed. by D. Whigham, D. Dykyjova and S. Hejny), pp. 305–344. Kluwer, Dordrecht.

Ovenden, L. 1982. Vegetation history of a polygonal peatland, northern Yukon. *Boreas* **11**, 209–224.

Ovenden, L. 1990. Peat accumulation in northern wetlands. *Quaternary Research* **33**, 377–386.

Page, S.E., Rieley, J.O., Shotyk, O.W. and Weiss, D. 1999. Interdependence of peat and vegetation in a tropical peat swamp forest. *Philosophical Transactions of the Royal Society of London* **B354**, 1885–1897.

Painter, T.J. 1991. Lindow man, Tollund man and other peat-bog bodies – the preservative and antimicrobial action of sphagnum, a reactive glycuronoglycan with tanning and sequestering properties. *Carbohydrate Polymers* **15**, 123–142.

Päivänen, J. 1996. Forestry use of peatlands. In: *Global Peat Resources* (ed. by E. Lappalainen), pp. 311–314. International Peat Society, Finland.

Pakarinen, P. and Tolonen, K. 1977a. Distribution of lead in *Sphagnum fuscum* profiles in Finland. *Oikos* **28**, 69–73.

Pakarinen, P. and Tolonen, K. 1977b. On the growth-rate and dating of surface peat. *Suo* **28**, 19–24.

Parkyn, L. and Stoneman, R. 1997. The Scottish raised bog land cover survey. In: *Conserving Peatlands* (ed. by L. Parkyn, R.E. Stoneman and H.A.P. Ingram) pp. 192–203. CAB International, Wallingford.

Patterson III, W.A., Edwards, K.J. and Maguire, D.J. 1987. Microscopic charcoal as a fossil indicator of fire. *Quaternary Science Reviews* **6**, 3–23.

Payette, S. 1984. Peat inception and climatic change in northern Quebec. In: *Climatic Change on a Yearly to Millennial Basis* (ed. by N.A. Morner and W. Karlen) pp. 173–179. Reidel, the Netherlands.

Payette, S. 1988. Late Holocene development of subarctic ombrotrophic peatlands: allogenic and autogenic succession. *Ecology* **69**, 516–531.

Payette, S., Gauthier, L. and Grenier, I. 1986. Dating ice-wedge growth in subarctic peatlands following deforestation. *Nature* **322**, 724–727.

Pearce, F. 1994. Peat bogs hold bulk of Britain's carbon. *New Scientist* **144**, 6.

Pearsall, W.H. 1956. Two blanket bogs in Sutherland. *Journal of Ecology* **44**, 493–516.

Peteet, D., Andreev, A., Bardeen, W. and Mistretta, F. 1998. Long-term Arctic peatland dynamics, vegetation and climate history of the Pur-Taz region, Western Siberia. *Boreas* **27**, 115–126.

Petterson, R. 1997. Peatland conservation versus a wise use: an International Peat Society view. In: *Biodiversity and Sustainability of Tropical Peatlands* (ed. by J.O. Reiley and S.E. Page), pp. 31–33. Samara Publishing, Cardigan.

Pfadenhauer, J. and Klötzli, F. 1996. Restoration experiments in middle European wet terrestrial ecosystems: an overview. *Vegetatio* **126**, 101–115.

Pfadenhauer, J., Schneekloth, H., Schneider, R. and Schneider, S. 1993. Mire distribution. In: *Mires: Process, Exploitation and Conservation* (ed. by A.L. Heathwaite and Kh. Göttlich), pp. 77–121. Wiley, Chichester.

Phillips, V.D. 1998. Peatswamp ecology and sustainable development in Borneo. *Biodiversity and Conservation* **7**, 651–671.

Pilcher, J.R., Hall, V.A. and McCormac, F.G. 1995. Dates of Holocene icelandic volcanic-eruptions from tephra layers in Irish peats. *Holocene* **5**, 103–110.

Pirozynski, K.A. 1990. Fungi. In: *Methods in Quaternary Ecology:* Geoscience Canada Reprint Series No. 5 (ed. by B.G. Warner), pp. 15-22.

Pisano, E. 1983. The Magellanic tundra complex. In: *Ecosystems of the World 4B. Mires: Swamp, Bog, Fen and Moor* (ed. by A.J.P. Gore), pp. 295–329. Elsever, Amsterdam.

Pitkänen, A., Turunen, J. and Tolonen, K. 1999. The role of fire in the carbon dynamics of a mire, eastern Finland. *Holocene* **9**, 453–462.

Pons, L.J. 1992. Holocene peat formation in the lower parts of The Netherlands. In: *Fens and Bogs in the Netherlands: Vegetation, History, Nutrient Dynamics and Conservation* (ed. by J.T.A. Verhoeven), pp. 7–79. Kluwer, Dordrecht.

Post, W.M., Emanuel, W.R., Zinkel, P.J. and Strangenberger, A.G. 1982. Soil carbon pools and world life zones. *Nature* **298**, 156–159.

Poulin, M., Rochefort, L. and Desrochers, A. 1999. Conservation of bog plant species assemblages: assessing the role of natural remnants in mined sites. *Applied Vegetation Science* **2**, 169–180.

Price, J.S. 1991. Evaporation from a blanket bog in a foggy coastal environment. *Boundary Layer Meteorology* **57**, 391–406.

Price, J.S. 1992. Blanket bog in Newfoundland.2. Hydrological processes. *Journal of Hydrology* **135**, 103–119.

Price, J.S. 1994. Sources and sinks of sea-salt in a Newfoundland blanket bog. *Hydrological Processes* **8**, 167–177.

Price, J.S. 1996. Hydrology and microclimate of a partly restored cutover bog, Quebec. *Hydrological Processes* **10**, 1263–1272.

Price, J. 1997. Soil moisture, water tension, and water table relationships in a managed cutover bog. *Journal of Hydrology* **202**, 21–32.

Price, J., Rochefort, L. and Quinty, F. 1998. Energy and moisture considerations on cutover peatlands: surface microtopography, mulch cover and *Sphagnum* regeneration. *Ecological Engineering* **10**, 293–312.

Price, J.S. and Waddington, J.M. 2000. Advances in Canadian wetland hydrology and biogeochemistry. *Hydrological Processes* **14**, 1579–1589.

Price, M.D.R. and Moore, P.D. 1984. Pollen dispersion in the hills of Wales: a pollen shed hypothesis. *Pollen et Spores* **26**, 127–136.

Proctor, M.C.F. 1992. Regional and local variation in the chemical composition of ombrogenous mire waters in Britain and Ireland. *Journal of Ecology* **80**, 719–736.

Proctor, M.C.F. 1994. Seasonal and shorter-term changes in surface water chemistry on four English ombrogenous bogs. *Journal of Ecology* **82**, 597–610.

Proctor, M.C.F. and Maltby, E. 1998. Relations between acid atmospheric deposition and the surface pH of some ombrotrophic bogs in Britain. *Journal of Ecology* **86**, 329–340.

Pugh, J.C. 1975. *Surveying for Field Scientists*. Methuen, London.

Pulteney, C. 1988. Mire vegetation of the Blackdown Hills. *Nature in Devon* **9**, 31–36.

Punning, J.M., Koff, T., Ilomets, M. and Jogi, J. 1995. The relative influence of local, extra-local, and regional factors on organic sedimentation in the Vallamae kettle hole, Estonia. *Boreas* **24**, 65–80.

Pyatt, D.G. 1993. Multipurpose forests on peatland. *Biodiversity and Conservation* **2**, 548–555.

Rabassa, Coronato, A. and Roig, C. 1996. The peat bogs of Tierra del Fuego, Argentina. In: *Global Peat Resources* (ed. by E. Lappalainen), pp. 261–266. International Peat Society, Finland.

Rackham, O. 1986. *The History of the Countryside*. Dent, London.

Railton, J.B. and Sparling, J.H. 1973. Preliminary studies on the ecology of palsa mounds in northern Ontario. *Canadian Journal of Botany* **51**, 1037–1044.

Rawes, M. 1983. Changes in two high altitude blanket bogs after the cessation of sheep grazing. *Journal of Ecology* **71**, 219–235.

Rawes, M. and Hobbs, R. 1979. Management of semi-natural blanket bog in the northern Pennines. *Jouurnal of Ecology* **67**, 789–807.

Reddy, K.R. and D'Angelo, E.M. 1994. Soil processes regulating water quality in wetlands. In: *Global Wetlands: Old World and New* (ed. by W.J. Mitsch), pp. 309–324. Elsevier, Amsterdam.

Reeve, A.S., Siegel, D.I. and Glaser, P.H. 1996. Geochemical controls on peatland pore-water from the Hudson Bay lowland – a multivariate statistical approach. *Journal of Hydrology* **181**, 285–304.

Reid, E., Mortimer, G.N., Thompson, D.B.A. and Lindsay, R.A. 1994. Blanket bogs in Great Britain: an assessment of large scale pattern and distribution using remote sensing and GIS. In: *Large Scale Ecology and Conservation Biology*. (ed. by P.J. Edwards, R.M. May and N.R. Webb) pp. 229–246. Blackwell Scientific, Oxford.

Reid, E., Ross, S.Y., Thompson, D.B.A. and Lindsay, R.A. 1997. From *Sphagnum* to satellite: towards a comprehensive inventory of the blanket mires of Scotland. In: *Conserving Peatlands* (ed. by L. Parkyn, R.E. Stoneman and H.A.P. Ingram) pp. 204–216. CAB International, Wallingford.

Reiley, J.O. and Page, S.E. (eds.) 1997. *Biodiversity and Sustainability of Tropical Peatlands*. Samara Publishing, Cardigan.

Reiley, J.O., Sieffermann, R.G. and Page, S.E. 1992. The origin, development, present status and importance of the lowland peat swamp forests of Borneo. *Suo* **43**, 241–244.

Reinikainen, A., Lindholm, T. and Vasander, H. 1984. Ecological variation of mire site types in the small kettle-hole mire Heinisuo, southern Finland. *Annales Botanica Fennici* **21**, 79–101.

Reynolds, B. and Pomeroy, A.B. 1988. Hydrochemistry of chloride in an upland catchment in mid-Wales. *Journal of Hydrology* **99**, 19–32.

Reynolds, J.D. 1990. Ecological relationships of peatland invertebrates. In: *Ecology and Conservation of Irish Peatlands* (ed. by G.J. Doyle), pp. 135–143. Royal Irish Academy, Dublin.

Richardson, C.J. 1983. Vanishing wastelands or valuable wetlands. *Bioscience* **33**, 626–633.

Robert, E.C., Rochefort, L. and Garneau, M. 1999. Natural revegetation of two block-cut mined peatlands in eastern Canada. *Canadian Journal of Botany* **77**, 447–459.

Roberts, C.N., 1998. *Holocene*. Blackwell, Oxford.

Robinson, D. 1984. The estimation of the charcoal content of sediments: a comparison of methods on peat sections from the Island of Arran. *Circaea* **2**, 121–128.

Robinson, D. 1987. Investigations into the Aukhorn peat mounds, Keiss, Caithness – pollen, plant macrofossil and charcoal analyses. *New Phytologist* **106**, 185–200.

Rochefort, L. 2000. New frontiers in bryology and lichenology – *Sphagnum* – a keystone genus in habitat restoration. *Bryologist* **103**, 503–508.

Rochefort, L. and Bastien, D.F. 1998. Reintroduction of *Sphagnum* into harvested peatlands: evaluation of various methods for protection against desiccation. *Écoscience* **5**, 117–127.

Rochefort, L. and Vitt, D.H. 1988. Effects of simulated acid rain on *Tomenthypnum nitens* and *Scorpidium scorpiodes* in a rich fen. *Bryologist* **91**, 121–129.

Rochefort, L., Vitt, D.H. and Bayley, S.E. 1990. Growth, production and decomposition dynamics of *Sphagnum* under natural and experimentally acidified conditions. *Ecology* **71**, 1986–2000.

Rodhe, H. and Malmer, N. 1997. Comments on an article by Franzen *et al.* 1996. Principles for a climate regulation mechanism during the Late Phanerozoic era, based on carbon fixation in peat-forming wetlands. *Ambio* **25**, 435–442. *Ambio* **26**, 187.

Rodwell, J.S. 1991. *British Plant Communities*, Volume 2: *Mires and Heaths*. Cambridge University Press, Cambridge.

Romanov, V.V. 1968. *Hydrophysics of Bogs*. Israel Program for Scientific Translation, Jerusalem.

Rose, N.L., Harlock, S. and Appleby, P.G. 1999. The spatial and temporal distributions of spheroidal carbonaceous fly-ash particles (SCP) in the sediment records of European mountain lakes. *Water Air and Soil Pollution* **113**, 1–32.

Rose, N.L., Harlock, S., Appleby, P.G. and Battarbee, R.W. 1995. Dating of recent lake sediments in the United Kingdom and Ireland using spheroidal carbonaceous particle (SCP) concentration profiles. *Holocene* **5**, 328–335.

Ross, S.M. 1995. Overview of the hydrochemistry and solute processes in British wetlands. In: *Hydrology and Hydrochemistry of British Wetlands* (ed. by J. Hughes and A.L. Heathwaite), pp. 133–181. Wiley, Chichester.

Roulet, N. 1990. The hydrological role of peat-covered wetlands. *Canadian Geographer* **34**, 82–83.

Roulet, N., Moore, T., Bubier, J. and Lafleur, P. 1992. Northern fens – methane flux and climatic change. *Tellus Series B–Chemical and Physical Meteorology* **44**, 100–105.

Rowell, T.A. 1986. The history of drainage at Wicken Fen, Cambridgeshire, England, and its relevance to conservation. *Biological Conservation* **35**, 111–142.

Rowell, T.A. 1988. *The Peatland Management Handbook*. Research and survey in nature conservation No. 14. Nature Conservancy Council, Peterborough.

Rowell, T.A., Guarino, L. and Harvey, H.J. 1985. The experimental management of Wicken Fen, Cambridgeshire. *Journal of Applied Ecology* **22**, 217–227.

Roy, V., Bernier, P.Y., Plamondon, A.P. and Ruel, J.C. 1999. Effect of drainage and microtopography in forested wetlands on the microenvironment and growth of planted black spruce seedlings. *Canadian Journal of Forest Research* **29**, 563–574.

Roy, V., Ruel, J.C. and Plamondon, A.P. 2000. Establishment, growth and survival of natural regeneration after clearcutting and drainage on forested wetlands. *Forest Ecology and Management* **129**, 253–267.

Royal Society for Nature Conservation 1990. *The Peat Report*. 1–28. Royal Society for Nature Conservation, Nettleham.

Ruhland, K., Smol, J.P., Jasinski, J.P.P. and Warner, B.G. 2000. Response of diatoms and other siliceous indicators to the developmental history of a peatland in the Tiksi forest, Siberia, Russia. *Arctic, Antarctic and Alpine Research* **32**, 167–178.

Runge, M. 1983. Physiology and ecology of nitrogen nutrition. In: *Physiological Plant Ecology III. Responses to the Chemical and Biological Environment* (ed. by O.L. Lange, P.S. Nobel, C.B. Osmond and H. Ziegler), pp. 163–200. Springer-Verlag, Berlin.

Ruuhijarvi, R. 1983. The Finnish mire types and their regional distribution. In: *Ecosystems of the World 4B. Mires: Swamp, Bog, Fen and Moor* (ed. by A.J.P. Gore), pp. 47–67. Elsevier, Amsterdam.

Rybnicek, K. 1984. The vegetation and develoment of central European mires. In: *European Mires* (ed. by P.D. Moore), pp. 177–201. London, Academic Press.

Rycroft, D.W., Williams, D.J.A. and Ingram, H.A.P. 1975a. The transmission of water through peat. I Review. *Journal of Ecology* **63**, 535–556.

Rycroft, D.W., Williams, D.J.A. and Ingram, H.A.P. 1975b. The transmission of water through peat. II Field experiments. *Journal of Ecology* **63**, 557–568.

Saarinen, T., Tolonen, K. and Vasander, H. 1992. Use of ^{14}C labelling to measure below-ground biomass of mire plants. *Suo* **43**, 245–247.

Saarnio, S., Alm, J., Martikainen, P.J. and Silvola, J. 1998. Effects of raised CO_2 on potential CH_4 production and oxidation in, and CH_4 emission from, a boreal mire. *Journal of Ecology* **86**, 261–268.

Sagot, C. and Rochefort, L. 1996. *Sphagnum* desiccation tolerance. *Cryptogamie Bryologie Lichenologie* **17**, 171–183.

Salonen, V. 1987. Relationship between the seed rain and the establishment of vegetation in two areas abandoned after peat harvesting. *Holarctic Ecology* **10**, 171–174.

Schauffler, M., Jacobson, G.L., Pugh, A.L. and Norton, S.A. 1996. Influence of vegetational structure on capture of salt and nutrient aerosols in a Maine peatland. *Ecological Applications* **6**, 263–268.

Schiff, S., Aravena, R., Mewhinney, E., Elgood, R., Warner, B., Dillon, P. and Trumbore, S. 1998. Precambrian shield wetlands: hydrologic control of the sources and export of dissolved organic matter. *Climatic Change* **40**, 167–188.

Schiller, C.L. and Hastie, D.R. 1994. Exchange of nitrous oxide within the Hudson Bay Lowland. *Journal of Geophysical Research – Atmospheres* **99**, 1573–1588.

Schmilewsksi, G.K. 1996. Horticultural use of peat. In: *Global Peat Resources* (ed. by E. Lappalainen), pp. 327–334. International Peat Society, Finland.

Schouwenaars, J.M. 1992. Hydrological charcacteristics of bog relicts in the Engbertsdijksvenen after peat-cutting and rewetting. In: *Peatland Ecosystems and Man: an Impact Assessment* (ed. by O.M. Bragg, P.D. Hulme, H.A.P. Ingram and R.A. Robertson), pp. 125–132. Finland, International Peat Society.

Schouwenaars, J.M. 1993. Hydrological differences between bogs and bog-relicts and consequences for bog restoration. *Hydrobiologia* **265**, 217–224.

Schouwenaars, J.M. 1995. The selection of internal and external water management options for bog restoration. In: *Restoration of Temperate Wetlands* (ed. by B.D. Wheeler, S.C. Shaw, W.J. Fojt and R.A. Robertson), pp. 331–346. Wiley, Chichester.

Segerstrom, U., Bradshaw, R. and Hornberg, G.A.B. 1994. Disturbance history of a swamp forest refuge in northern Sweden. *Biological Conservation* **68**, 189–196.

Sellers, P., Hall, F., Margolis, H., Kelly, B., Baldocchi, D., Denhartog, G., Cihlar, J., Ryan, M.G., Goodison, B., Crill, P., Ranson, K.J., Lettenmaier, D. and Wickland, D.E. 1994. The Boreal Ecosystem-Atmosphere Study (BOREAS) – an overview and early results from the 1994 field year. *Bulletin of the American Meteorological Society* **76**, 1549–1577.

Seppälä, M. and Koutaniemi, L. 1985. Formation of a string and pool topography as expressed by morphology, stratigraphy and current processes on a mire in Kuusamo, Finland. *Boreas* **14**, 287–309.

Sernander, R. 1908. On the evidence of postglacial changes of climate furnished by the peat-mosses of Northern Europe. *Geologiska Föreningens i Stockholm Förhandlingar* **30**, 467–478.

Shore, J.S., Bartley, D.D. and Harkness, D.D. 1995. Problems encountered with the ^{14}C dating of peat. *Quaternary Science Reviews* **14**, 373–383.

Shotyk, W. 1988. Review of the inorganic geochemistry of peats and peatland waters. *Earth Science Reviews* **25**, 95–176.

Shotyk, W., Cheburkin, A.K., Appleby, P.G., Fankhauser, A. and Kramers, J.D. 1997. Lead in three peat bog profiles,

Jura Mountains, Switzerland: enrichment factors, isotopic composition, and chronology of atmospheric deposition. *Water Air and Soil Pollution* **100**, 297–310.

Shotyk, W., Weiss, D., Appleby, P.G., Cheburkin, A.K., Frei, R., Gloor, M., Kramers, J.D., Reese, S. and van der Knaap, W.O. 1998. History of atmospheric lead deposition since 12,370 C-14 yr BP from a peat bog, Jura Mountains, Switzerland. *Science* **281**, 1635–1640.

Shurpali, N.J., Verma, S.B., Kim, J. and Arkebauer, T.J. 1995. Carbon dioxide exchange in a peatland ecosystem. *Journal of Geophysical Research–Atmospheres* **100**, 14319–14326.

Siegel, D.I. 1983. Groundwater and the evolution of patterned mires, glacial Lake Agassiz peatlands, northern Minnesota. *Journal of Ecology* **71**, 913–921.

Siegel, D.I. 1992. Groundwater hydrology. In: *The Patterned Peatlands of Minnesota* (ed. by H.E. Wright, B.A. Coffin and N.E. Aaseng), pp. 163–172. University of Minneapolis Press, Minneapolis.

Siegel, D.I. and Glaser, P.H. 1987. Groundwater flow in a bog-fen complex, Lost River Peatland, northern Minnesota. *Journal of Ecology* **75**, 743–754.

Siegel, D.I., Reeve, A.S., Glaser, P.H. and Romanowicz, E.A. 1995. Climate-driven flushing of pore water in peatlands. *Nature* **374**, 531–533.

Sikora, L.J. and Keeney, D.R. 1983. Further aspects of soil chemistry under anaerobic conditions. In: *Mires: Swamp, Bog, Fen and Moor. Ecosystems of the World 4A* (ed. by A.J.P. Gore), pp. 247–256. Elsevier, Amsterdam.

Silins, U. and Rothwell, R.L. 1998. Forest peatland drainage and subsidence affect soil water retention and transport properties in an Alberta peatland. *Soil Science Society of America Journal* **62**, 1048–1056.

Silvola, J. 1985. CO_2 dependence of photosynthesis in certain forest and peat mosses and simulated photosynthesis at various actual and hypothetical CO_2 concentrations. *Lindbergia* **11**, 86–93.

Silvola, J. 1986. Carbon dioxide dynamics in mires reclaimed for forestry in eastern Finland. *Annales Botanica Fennici* **23**, 59–67.

Silvola, J. and Aaltonen, H. 1984. Water content and photosynthesis in the peat mosses *Sphagnum fuscum* and *S. angustifolium*. *Annales Botanica Fennici* **21**, 1–6.

Silvola, J. and Hanski, I. 1979. Carbon accumulation in a raised bog. *Oecologia* **37**, 285–295.

Silvola, J. and Heikkinen, S. 1978. CO_2 exchange in the *Empetrum nigrum–Sphagnum fuscum* community. *Oecologia* **37**, 273–283.

Simola, H. 1983. Limnological effects of peatland drainage and fertilization as reflected in the varved sediment of a deep lake. *Hydrobiologia* **106**, 43–58.

Simola, H. and Lodenius, M. 1982. Recent increase in mercury sedimentation in a forest lake attributable to peatland drainage. *Bulletin of Environmental Contamination and Toxicology* **29**, 298–305.

Sjörs, H. 1959. Forest and peatland at Hawley Lake, northern Ontario. *National Museum of Canada Bulletin* **171**, 1–31.

Sjörs, H. 1983. Mires of Sweden. In: *Ecosystems of the World 4B. Mires: Swamp, Bog, Fen and Moor* (ed. by A.J.P. Gore), pp. 69–94. Elsevier, Amsterdam.

Sjörs, H. 1985. A comparison between mires of southern Alaska and Fennoscandia. *Aquilo Seria Botanica* **21**, 89–94.

Skiba, U., Cresser, M.S., Derwent, R.G. and Futty, D.W. 1989. Peat acidification in Scotland. *Nature* **337**, 68–69.

Slack, N.G. and Hallingback, T. 1992. Community and species responses to environmental gradients in suboceanic mires of the west Swedish coast. *Annales Botanici Fennici* **29**, 269–293.

Sleigh, M.A. 1989. *Protozoa and Other Protists*, 4th edn. Edward Arnold, London.

Smart, P.J., Wheeler, B.D. and Willis, A.J. 1986. Plants and peat cuttings: historical ecology of a much exploited peatland – Thorne Waste, Yorkshire, UK. *New Phytologist* **104**, 731–748.

Smart, P.J., Wheeler, B.D. and Willis, A.J. 1989. Revegetation of peat excavations in a derelict raised bog. *New Phytologist* **111**, 733–748.

Smart, P.L. 1991. Uranium series dating. In: *Quaternary Dating Methods – a User's Guide* (ed. by P.L. Smart and P.D. Frances), pp. 45–83. Technical Guide No. 4. Quaternary Research Association, Cambridge.

Smart, P.L. and Frances, P.D. 1991. *Quaternary Dating Methods – a User's Guide*. Technical Guide No. 4, Quaternary Research Association, Cambridge.

Smit, R., Bragg, O.M. and Ingram, H.A.P. 1999. Area separation of streamflow in an upland catchment with partial peat cover. *Journal of Hydrology* **219**, 46–55.

Smith, A.G. and Cloutman, E.W. 1988. Reconstruction of Holocene vegetation history in three dimensions at Waun-Fignen-Felen, an upland site in South Wales. *Philosophical Transactions of the Royal Society of London* **B322**, 159–219.

Smith, A.G. and Green, C.A. 1995. Topogenous peat development and late-flandrian vegetation history at a site in upland South Wales. *Holocene* **5**, 172–183.

Smith, A.G. and Morgan, L.A. 1989. A succession to ombrotrophic bog in the Gwent Levels, and its demise: a Welsh parallel to the peats of the Somerset Levels. *New Phytologist* **112**, 145–167.

Smith, R.S. and Charman, D.J. 1988. The vegetation of upland mires within conifer plantations in Northumberland, northern England. *Journal of Appplied Ecology* **25**, 579–594.

Smith, R.S., Lunn, A.G. and Newson, M.D. 1995. The Border Mires in Kielder Forest – a review of their ecology and conservation management. *Forest Ecology and Management* **79**, 47–61.

Smith, R.T. and Taylor, J.A. 1989. Biopedological processes in the inception of peat formation. *International Peat Journal* **3**, 1–24.

Smol, J.P. 1990. Freshwater algae. In: *Methods in Quaternary Ecology*: Geoscience Canada Reprint Series No. 5 (ed. by B.G. Warner), pp. 3–14.

Solem, T. 1986. Age, origin and development of blanket mires in Sor-Trondelag, central Norway. *Boreas* **15**, 101–115.

Solem, T. 1989. Blanket mire formation at Haramsoy, More og Romsdal, Western Norway. *Boreas* **18**, 221–235.

Sorensen, K.W. 1993. Indonesian peat swamp forests and their role as a carbon sink. *Chemosphere* **27**, 1065–1082.

Soro, A., Sundberg, S. and Hakan, R. 1999. Species diversity, niche metrics and species associations in harvested and undisturbed bogs. *Journal of Vegetation Science* **10**, 549–560.

Sparling, J.H. 1962. Occurrence of *Schoenus nigricans* L. in the blanket bogs of western Ireland. *Nature* **195**, 723–724.

Speranza, A., Hanke, J., van Geel, B. and Fanta, J. 2000. Late-holocene human impact and peat development in the Cerna Hora bog, Krkonose Mountains, Czech Republic. *Holocene* **10**, 575–585.

Standen, V., Tallis, J.H. and Meade, R. (eds.) 1998. *Patterned Mires and Mire Pools*. British Ecological Society, London.

Stead, I.M., Bourke, J.B., Brothwell, D. (eds.) 1986. *Lindow Man: the Body in the Bog*. British Museum Publications, London.

Steindorsson, S. 1975. Studies on the mire vegetation of Iceland. *Societas Scientarum Islandica* **41**, 1–226.

Steinnes, E. 1997. Trace element profiles in ombrogenous peat cores from Norway: evidence of long range atmospheric transport. *Water Air and Soil Pollution* **100**, 405–413.

Stevenson, A.C. and Moore, P.D. 1988. Studies in the vegetational history of SW Spain. IV Palynological investigations of a valley mire at El Acebron, Huelva. *Journal of Biogeography* **15**, 339–361.

Stewart, C. and Fergusson, J.E. 1994. The use of peat in the historical monitoring of trace metals in the atmosphere. *Environmental Pollution* **86**, 243–249.

Stoneman, R. and Brooks, S. (eds.) 1997. *Conserving Bogs: the Management Handbook*. The Stationery Office, Edinburgh.

Stove, G.C. 1983. The current use of remote-sensing data in peat, soil, land-cover and crop inventories in Scotland. *Philosophical Transactions of the Royal Society of London Series A – Mathematical and Physical Sciences*, **309**, 271–281.

Stove, G.C. and Hulme, P.D. 1980. *International Journal of Remote Sensing*, **1**, 319–344.

Stroud, D.A., Reed, T.M., Pienkowski, M.W. and Lindsay, R.A. 1987. *Birds, Bogs and Forestry*. Nature Conservancy Council, Peterborough.

Stuiver, M., Reimer, P.J., Bard, E., Beck, J.W., Burr, G.S., Hughen, K.A., Kromer, B., McCormac, G., van der Plicht, J. and Spurk, M. 1998. INTCAL 98 radiocarbon age calibration, 24000–0 cal BP. *Radiocarbon* **40**, 1041–1083.

Sundberg, S. and Rydin, H. 2000. Experimental evidence for a persistent spore bank in *Sphagnum*. *New Phytologist* **148**, 105–116.

Sundh, I., Nilsson, M. and Borga, P. 1997. Variation in microbial community structure in two boreal peatlands as determined by analysis of phospholipid fatty acid profiles. *Applied and Environmental Microbiology* **63**, 1476–1482.

Sundström, E. 1992a. *Effects of Removal of the Paperpot Container on Growth and Development of Black Spruce Seedlings on a Drained Peatland Clearcut in Northeastern Ontario.* NEST Technical Report TR-01, Forestry Canada, Sault Ste Marie, Ontario.

Sundström, E. 1992b. *Five-year Growth Response in Drained and Fertilised Black Spruce Peatlands. I Permanent Growth Plot Analysis.* 1–19. NEST Technical Report TR-02, Forestry Canada, Sault Ste Marie, Ontario.

Sundström, E. and Jeglum, J.K. 1992. *Five-year Growth Response in Drained and Fertilised Black Spruce Peatlands. II Stem Analysis.* NEST Technical Report Tr-03, Forestry Canada, Sault Ste Marie, Ontario.

Sundström, E., Magnusson, T. and Hanell, B. 2000. Nutrient conditions in drained peatlands along a north–south climatic gradient in Sweden. *Forest Ecology and Management* **126**, 149–161.

Svensson, B.M. 1995. Competition between *Sphagnum fuscum* and *Drosera rotundifolia* – a case of ecosystem engineering. *Oikos* **74**, 205–212.

Svensson, G. 1988a. Bog development and environmental conditions as shown by the stratigraphy of Store Mosse in southern Sweden. *Boreas* **17**, 89–111.

Svensson, G. 1988b. Fossil plant communities and regeneration patterns on a raised bog in south Sweden. *Journal of Ecology* **76**, 41–59.

Szumigalski, A.R. and Bayley, S.E. 1996. Decomposition along a bog to rich fen gradient in central Alberta, Canada. *Canadian Journal of Botany* **74**, 573–581.

Tallis, J.H. 1964. Studies on southern Pennine peats: II. The pattern of erosion. *Journal of Ecology* **52**, 333–344.

Tallis, J.H. 1965. Studies on southern pennine peats: IV. Evidence of recent erosion. *Journal of Ecology* **53**, 509–520.

Tallis, J.H. 1975. Tree remains in southern Pennine peats. *Nature* **256**, 482–484.

Tallis, J.H. 1983. Changes in wetland communities. In: *Mires: Swamp, Bog, Fen and Moor. General Studies. Ecosystems of the World 4A.* (ed. by A.J.P. Gore), pp. 311–347. Amsterdam, Elsevier.

Tallis, J.H. 1985. Mass movement and erosion of a southern Pennine blanket peat. *Journal of Ecology* **73**, 283–316.

Tallis, J.H. 1987. Fire and flood at Holme Moss: erosion processes in an upland blanket mire. *Journal of Ecology* **75**, 1099–1129.

Tallis, J.H. 1991. Forest and moorland in the south Pennine uplands in the mid-Flandrian period. 3. The spread of moorland local regional and national. *Journal of Ecology* **79**, 401–415.

Tallis, J.H. 1994. Pool-and-hummock patterning in a southern Pennine blanket mire II. The formation and erosion of the pool system. *Journal of Ecology* **82**, 789–803.

Tallis, J.H. 1995a. Blanket mires in the upland landscape. In: *Restoration of Temperate Wetlands* (ed. by B.D. Wheeler, S.C. Shaw, W.J. Fojt and R.A. Robertson), pp. 495–508. Wiley, Chichester.

Tallis, J.H. 1995b. Climate and erosion signals in British blanket peats – the significance of *Racomitrium lanuginosum* remains. *Journal of Ecology* **83**, 1021–1030.

Tallis, J.H. and Livett, E.A. 1994. Pool-and-hummock patterning in a southern Pennine blanket mire I. Stratigraphic profiles for the last 2800 years. *Journal of Ecology* **82**, 775–788.

Tansley, A.G. 1939. *The British Islands and Their Vegetation.* Cambridge University Press.

Tarnocai, C., Adams, G.D., Glooschenko, W.A., Grondin, P., Hirvonen, H.E., Lynch-Stewart, P., Mills, G.F., Oswald, E.T., Pollett, F.C., Rubec, C.D.A. and Wells, E. 1988. The Canadian wetland classification system. In: *Wetlands of Canada* (ed. by National Wetlands Working Group), pp. 413–427. Polyscience, Montreal, Canada.

Tauber, H. 1965. Differential pollen dispersal and the interpretation of pollen diagrams. *Danm. Geol. Unders. Ser. II* **89**, 1–69.

Taylor, J.A. and Smith, R.T. 1980. The role of pedogenic factors in the initiation of peat formation and in the classification of mires. *Proceedings of the 6th International Peat Congress*, pp. 109–118. International Peat Society, Duluth, USA.

Thompson, K. and Hamilton, A.C. 1983. Peatlands and swamps of the African continent. In: *Ecosystems of the World 4B. Mires: Swamp, Bog, Fen and Moor* (ed. by A.J.P. Gore), pp. 331–373. Amsterdam, Elsevier.

Thorhallsdottir, T.E. 1994. Effects of changes in groundwater level on palsas in central Iceland. *Geografiska Annaler Series A – Physical Geography* **76**, 161–167.

Thormann, M.N. and Bayley, S.E. 1997a. Aboveground plant production and nutrient content of the vegetation in six peatlands in Alberta, Canada. *Plant Ecology* **131**, 1–16.

Thormann, M.N. and Bayley, S.E. 1997b. Decomposition along a moderate-rich fen–marsh peatland gradient in boreal Alberta, Canada. *Wetlands* **17**, 123–137.

Thormann, M.N., Currah, R.S. and Bayley, S.E. 1999. The mycorrhizal status of the dominant vegetation along a peatland gradient in southern boreal Alberta, Canada. *Wetlands* **19**, 438–450.

Todd, P.A., McAdam, J.H. and Montgomery, W.I. 1995. The ecological effects of mechansied peat extraction on blanket bogs in Northern Ireland. In: *Heaths and Moorland: Cultural Landscapes* (ed. by D.B.A. Thompson, A.J. Hester and M.B. Usher), pp. 312–318. Scottish Natural Heritage, Edinburgh.

Tolonen, K. 1966. Stratigraphic and rhizopod analyses on an old raised bog, Varrassuo, in Hollola, South Finland. *Annales Botanici Fennici* **3**, 147–166.

Tolonen, K. 1986a. Charred particle analysis. In: *Handbook of Holocene Palaeoecology and Palaeohydrology* (ed. by B.E. Berglund), pp. 485–496, Wiley, Chichester.

Tolonen, K. 1986b. Rhizopod analysis. In: *Handbook of Holocene Palaeoecology and Palaeohydrology* (ed. by B.E. Berglund), pp. 645–666. Wiley, Chichester.

Tolonen, K., Warner, B.G. and Vasander, H. 1994. Ecology of testaceans (Protozoa, Rhizopoda) in mires in southern Finland: II. Multivariate analysis. *Archiv für Protistenkunde* **144**, 97–112.

Troels-Smith, J. 1955. Karakterisering af lose jordater (Characterisation of unconsolidated sediments). *Danmarks Geologiske Undersogelse IV series* **3(10)**, 1–73.

Trudgill, S.T. 1988. *Soil and Vegetation Systems.* Oxford University Press.

Turetsky, M.R. and Wieder, R.K. 1999. Boreal bog *Sphagnum* refixes soil-produced and respired (CO_2)-C-14. *Écoscience* **6**, 587–591.

Turner, J., Innes, J.B. and Simmons, I.G. 1993. Spatial diversity in the mid-Flandrian vegetation history of North Gill, North Yorkshire. *New Phytologist* **123**, 599–647.

Turner, R.C., Scaife, R.G. (eds.) 1995. *Bog Bodies: New Discoveries and New Perspectives.* British Museum Press, London.

Tyler, C. 1984. Calcareous fens in south Sweden. Previous uses, effects of management and management recommendations. *Biological Conservation* **30**, 69–89.

Tyler, S.C. 1991. The global methane budget. In: *Microbial Production and Consumption of Greenhouse Gases: Methane, Nitrogen Oxides and Halomethanes* (ed. by J.E. Rogers and W.B. Whitman), pp. 7–38. American Society for Microbiology, Washington, DC.

Ugolini, F.C. and Mann, D.H. 1979. Biopedological origin of peatlands in southeast Alaska. *Nature* **281**, 366–368.

Umeda, Y. and Inoue, T. 1996. Peatlands of Japan. In: *Global Peat Resources* (ed. by E. Lappalainen), pp. 179–182. Finland, International Peat Society.

Urban, N., Eisenreich, S.J., Grigal, D.F. and Schwarcz, H.P. 1990. Mobility and diagenesis of Pb and Pb-210 in peat. *Geochimica et Cosmochimica Acta* **54**, 3329–3346.

Urban, N.R., Verry, E.S. and Eisenreich, S.J. 1995. Retention and mobility of cations in a small peatland – trends and mechanisms. *Water Air and Soil Pollution* **79**, 201–224.

Usher, M.B. 1986. Wildlife conservation evaluation: attributes, criteria and values. In: *Wildlife Conservation Evaluation* (ed. by M.B. Usher), pp. 3–44. Chapman and Hall, London.

Vaisanen, R.A. and Rauhala, P. 1983. Succession of land bird communities on large areas of peatland drained for forestry. *Annales Zoologici Fennici* **20**, 115–127.

Vaithiyanathan, P. and Richardson, C.J. 1997. Nutrient profiles in the Everglades: examination along the eutrophication gradient. *Science of the Total Environment* **205**, 81–95.

Vaithiyanathan, P. and Richardson, C.J. 1999. Macrophyte species changes in the Everglades: examination along a eutrophication gradient. *Journal of Environmental Quality* **28**, 1347–1358.

Valentine, D.W., Holland, E.A. and Schimel, D.S. 1994. Ecosystem and physiological controls over methane production in northern wetlands. *Journal of Geophysical Research – Atmospheres* **99**, 1563–1571.

Van Breemen, N. 1995. How *Sphagnum* bogs down other plants. *Trends in Ecology and Evolution* **10**, 270–275.

Van Den Bogaard, C., Dorfler, W., Sandgren, P. and Schmincke, H.U. 1994. Correlating the Holocene records: Icelandic tephra found in Schleswig-Holstein (Northern Germany). *Naturwissenchaften* **81**, 554–556.

Van der Molen, P.C. and Hoekstra, S.P. 1988. A palaeoecological study of a hummock–hollow complex from Engbertsdijksveen in the Netherlands. *Review of Palaeobotany and Palynology* **56**, 213–274.

Van der Molen, P.C., Schalkoort, M. and Smit, R. 1994. Vegetation and ecology of hummock–hollow complexes on an Irish raised bog. *Proceedings of the Royal Irish Academy* **94B**, 145–175.

Van der Molen, P.C. and Wijmstra, T.A. 1994. The thermal regime of hummock–hollow complexes on Clara bog, Co. Offaly. *Proceedings of the Royal Irish Academy* **94B**, 209–221.

Van der Wijk, A., Mook, G. and Ivanovich, M. 1988. Correction for environmental Th-230 in U/Th disequilibrium dating of peat. *Science of the Total Environment* **70**, 19–40.

van Diggelen, R., Grootjans, A.P., Kemmers, R.H., Kooijman, A.M., Succow, M., Devries, N.P.J. and Vanwirdum, G. 1991. Hydro-ecological analysis of the fen system Lieper Posse, Eastern Germany. *Journal of Vegetation Science* **2**, 465–476.

van Diggelen, R., Molenaar, W.J. and Kooijman, A.M. 1996. Vegetation succession in a floating mire in relation to management and hydrology. *Journal of Vegetation Science* **7**, 809–820.

van Geel, B. 1978. A palaeoecological study of Holocene peat bog sections in Germany and the Netherlands. *Review of Palaeobotany and Palynology* **25**, 1–120.

van Geel, B. 1986. Application of fungal and algal remains and other microfossils in palynological analyses. In: *Handbook of Holocene Palaeoecolgy and Paleohydrology* (ed. by B.E. Berglund), pp. 497–505. Wiley, London.

van Geel, B. 1989. Holocene raised bog deposits in the Netherlands as geochemical archives of prehistoric aerosols. *Acta Botanica Neerlandica* **38**, 467–476.

Vardy, S.R., Warner, B.G. and Aravena, R. 1997. Holocene climate effects on the development of a peatland on the Tuktoyaktuk Peninsula, northwest Territories. *Quaternary Research* **47**, 90–104.

Vardy, S.R., Warner, B.G. and Aravena, R. 1998. Holocene climate and the development of a subarctic peatland near Inuvik, Northwest territories, Canada. *Climatic Change* **40**, 285–313.

Vardy, S.R., Warner, B.G., Turunen, J. and Aravena, R. 2000. Carbon accumulation in permafrost peatlands in the Northwest Territories and Nunavut, Canada. *Holocene* **10**, 273–280.

Varjo, J. 1996. Controlling continuously updated forest data by satellite remote-sensing. *International Journal of Remote Sensing*, **17**, 43–67.

Vasander, H. 1984. Effect of forest amelioration on diversity in an ombrotrophic bog. *Annales Botanica Fennici* **21**, 7–15.

Vasander, H., Leivo, A. and Tanninen, T. 1992. Rehabilitation of a drained peatland area in the Seitseminen National Park in southern Finland. In: *Peatland Ecosystems and Man: an Impact Assessment* (ed. by O.M. Bragg, P.D. Hulme, H.A.P. Ingram and R.A. Robertson), pp. 381–387. International Peat Society, Finland.

Verhoeven, J.T.A. (ed.) 1992. *Fens and Bogs in the Netherlands: Vegetation, History, Nutrient Dynamics and Conservation.* Kluwer, Dordrecht.

Verhoeven, J.T.A. and Liefveld, W.M. 1997. The ecological significance of organochemical compounds in *Sphagnum. Acta Botanica Neerlandica* 46, 117–130.

Verhoeven, J.T.A., Maltby, E. and Schmitz, M.B. 1990. Nitrogen and phosphorous mineralization in fens and bogs. *Journal of Ecology* 78, 713–726.

Verry, E.S. and Timmons, D.R. 1982. Waterborne nutrient flow through an upland-peatland watershed in Minnesota. *Ecology* 63, 1456–1467.

Viereck, L.A. 1983. The effects of fire in black spruce ecosystems in Alaska and northern Canada. In: *The Role of Fire in Northern Circumpolar Ecosystems* (ed. by R.W. Wein and D.A. MacLean), pp. 201–220. Wiley, New York.

Villagrán, C. 1988. Expansion of Magellanic moorland during the Late Pleistocene: palynological evidence from northern Isla de Chiloé, Chile. *Quaternary Research* 30, 304–314.

Vitt, D.H., Achuff, P. and Andrus, R.E. 1975. The vegetation and chemical properties of patterned fens in the Swan Hills, north central Alberta. *Canadian Journal of Botany* 53, 2776–2795.

Vitt, D.H., Bayley, S.E. and Jin, T.L. 1995. Seasonal-variation in water chemistry over a bog-rich fen gradient in continental western Canada. *Canadian Journal of Fisheries and Aquatic Sciences* 52, 587–606.

Vitt, D.H. and Chee, W.L. 1990. The relationships of vegetation to surface water chemistry to peat chemistry in fens of Alberta, Canada. *Vegetatio,* 89, 87–106.

Vitt, D.H., Halsey, L.A., Bauer, I.E. and Campbell, C. 2000. Spatial and temporal trends in carbon storage of peatlands of continental western Canada through the Holocene. *Canadian Journal of Earth Sciences* 37, 683–693.

Vogelmann, J.E. and Moss, D.M. 1993. Spectral reflectance measurements in the genus *Sphagnum. Remote Sensing of Environment,* 45, 273–279.

Von Post, L. 1924, *Das genetische System der organogenen Bildungen Schwedens.* Comité International de Pédologie IV Commission No. 22.

Von Post, L. and Sernander, R. 1910. *Pflanzen-physiognomische Studien auf Torfmooren in Nrken.* XI International Geological Congress: Excursion Guide No. 14 (A7), 1–48. Stockholm.

Waddington, J.M., Griffis, T.J. and Rouse, W.R. 1998. Northern Canadian wetlands: net ecosystem CO_2 exchange and climatic change. *Climatic Change* 40, 267–275.

Waddington, J.M. and Roulet, N.T. 1997. Groundwater flow and dissolved carbon movement in a boreal peatland. *Journal of Hydrology* 191, 122–138.

Waddington, J.M., Roulet, N.T. and Swanson, R.V. 1996. Water-table control of CH_4 emission enhancement by vascular plants in boreal peatlands. *Journal of Geophysical Research –Atmospheres* 101, 22775–22785.

Wahlen, M., Tanaka, N., Henry, B., Deck, B., Zeglen, J., Vogel, J.S., Southon, J.S., Fairbanks, R. and Broecker, W. 1989. Carbon-14 in methane sources and in atmospheric methane: the contribution from fossil carbon. *Science* 245, 286–290.

Wahren, C.H., Williams, R.J. and Papst, W.A. 1999. Alpine and subalpine wetland vegetation on the Bogong High Plains, south-eastern Australia. *Australian Journal of Botany* 47, 165–188.

Walbridge, M.R. 1994. Plant community composition and surface-water chemistry of fen peatlands in West Virginia's Appalachian plateau. *Water Air and Soil Pollution* 77, 247–269.

Walker, D. 1970. Direction and rate in some British postglacial hydroseres. In: *Studies in the Vegetational History of the British Isles* (ed. by D. Walker and R.G. West), pp. 117–139. Cambridge University Press, Cambridge.

Walker, D. and Walker, P.M. 1961. Stratigraphic evidence of regeneration in some Irish bogs. *Journal of Ecology* 49, 169–185.

Wallace, H.L., Good, J.E.G. and Williams, T.G. 1992. The effects of afforestation on upland plant communities: an application of the British National Vegetation Classification. *Journal of Applied Ecology* 29, 180–194.

Wallen, B., Falkengren, U. and Malmer, N. 1988. Biomass, productivity and relative rate of photosynthesis of *Sphagnum* at different water levels on a south Swedish peat bog. *Holarctic Ecology* 11, 70–76.

Wardenaar, E.C.P. 1987. A new hand tool for cutting peat profiles. *Canadian Journal of Botany* 65, 1772–1773.

Warner, B.G. 1987. Abundance and diversity of testate amoebae (Rhizopoda, Testacea) in *Sphagnum* peatlands in southwestern Ontario, Canada. *Archiv für Protistenkunde* 133, 173–189.

Warner, B.G. 1988a. Methods in Quaternary ecology 5. Testate amoebae (Protozoa). *Geoscience Canada* 15, 251–260.

Warner, B.G. 1988b. Methods in Quaternary Ecology 3. Plant macrofossils. *Geoscience Canada* 15, 121–129.

Warner, B.G. 1990a. Other fossils. In: *Methods in Quaternary Ecology.* Geoscience Canada Reprint Series No. 5 (ed. by B.G. Warner), pp. 149–162.

Warner, B.G. (ed.) 1990b. *Methods in Quaternary Ecology.* Geoscience Canada Reprint Series No. 5, Geological Association of Canada, St John's, Newfoundland.

Warner, B.G. 1993. Palaeoecology of floating bogs and landscape change in the Great Lakes drainage basin of North America. In: *Climate Change and Human Impact on the Landscape* (ed. by F.M. Chambers), pp. 237–245. Chapman and Hall, London.

Warner, B.G. 1996. Vertical gradients in peatlands. In: *Wetlands: Environmental Gradients, Boundaries and Buffers* (ed. by G. Mulamoottil, B.G. Warner and

E.A. McBean), pp. 45–65. CRC Press, Boca Raton, Florida.

Warner, B.G. 2001. Peat. In: *Encyclopedia of Sediments and Sedimentary Rocks* (ed. by G.V. Middleton), in press. Kluwer Academic Publishers, Dordrecht.

Warner, B.G. and Bunting, M.J. 1996. Indicators of rapid environmental change in northern peatlands. In: *Geoindicators: Assessing Rapid Environmental Changes in Earth Systems* (ed. by A.R. Berger and W.J. Iams), pp. 235–246. A.A. Balkema, Rotterdam.

Warner, B.G. and Buteau, P. 2000. The early peat industry in Canada 1864–1945. *Geoscience Canada* **27**, 57–66.

Warner, B.G. and Charman, D.J. 1994. Holocene changes on a peatland in northwestern Ontario interpreted from testate amebas (Protozoa) analysis. *Boreas* **23**, 270–279.

Warner, B.G., Clymo, R.S. and Tolonen, K. 1993. Implications of peat accumulation at Point Escuminac, New Brunswick. *Quaternary Research* **39**, 245–248.

Warner, B.G., Kubiw, H.J. and Hanf, K.I. 1989. An anthropogenic cause for quaking mire formation in southwestern Ontario. *Nature* **340**, 380–384.

Warner, B.G., Kubiw, H.J. and Karrow, P.F. 1991a. Origin of a postglacial kettle-fill sequence near Georgetown, Ontario. *Canadian Journal of Earth Sciences* **28**, 1965.

Warner, B.G., Nobes, D.C. and Theimer, B.D. 1990. An application of ground penetrating radar to peat stratigraphy of Ellice Swamp, southwestern Ontario. *Canadian Journal of Earth Science* **27**, 932–938.

Warner, B.G., Tolonen, K. and Tolonen, M. 1991b. A postglacial history of vegetation and bog formation at Point Escuminac, New Brunswick. *Canadian Journal of Earth Science* **28**, 1572–1582.

Wassen, M.J. 1996. A comparison of fens in natural and artificial landscapes. *Vegetatio* **126**, 5–26.

Wassen, M.J., Barendregt, A., Palczynski, A., Desmidt, J.T. and de Mars, H. 1990. The relationship between fen vegetaton gradients, groundwater-flow and flooding in an undrained valley mire at Biebrza, Poland. *Journal of Ecology* **78**, 1106–1122.

Wasylikowa, K. 1986. Analysis of fossil fruits and seeds. In: *Handbook of Holocene Palaeoecology and Palaeohydrology* (ed. by B.E. Berglund), pp. 571–590. Wiley, Chichester.

Waton, P.V. and Barber, K.E. 1987. Rimsmoor, Dorset: biostratigraphy and chronology of an infilled doline. In: *Wessex and the Isle of Wight Field Guide* (ed. by K.E. Barber), pp. 75–80. Quaternary Research Association, Cambridge.

Webb, R.S. and Overpeck, J.T. 1993. Carbon reserves released? *Nature* **361**, 497–498.

Webb, R.S. and Webb III, T. 1988. Rates of sediment accumulation in pollen cores from small lakes and mires of eastern North America. *Quaternary Research* **30**, 284–297.

Wein, R.W. 1983. Fire behaviour and ecological effects in organic terrain. In: *The Role of Fire in Northern Circumpolar Ecosystems* (ed. by R.W. Wein and D.A. MacLean), pp. 81–95. Wiley, New York.

Wein, R.W. and MacLean, D.A. 1983. An overview of fire in northern ecosystems. In: *The Role of Fire in Northern*

Circumpolar Ecosystems (ed. by R.W. Wein and D.A. MacLean), pp. 1–18. Wiley, New York.

Wells, C.E., Hodgkinson, D. and Huckerby, E. 2000. Evidence for the possible role of beaver (*Castor fiber*) in the prehistoric ontogenesis of a mire in northwest England, UK. *Holocene* **10**, 503–508.

Wells, C.E., Huckerby, E. and Hall, V. 1997. Mid- and late-Holocene vegetation histroy and tephra studies at Fenton Cottage, Lancashire, UK. *Vegetation History and Archaeobotany* **6**, 153–166.

Wells, C.E. and Wheeler, B.D. 1999. Evidence for possible climatic forcing of late-Holocene vegetation changes in Norfolk Broadland floodplain mires, UK. *Holocene* **9**, 595–608.

Wells, E.D. and Hirvonen, H.E. 1988. Wetlands of Atlantic Canada. In: *Wetlands of Canada* (ed. by National Wetlands Working Group), pp. 249–303. Polyscience, Montreal, Canada.

Wells, E.D. and Pollett, F.C. 1983. Peatlands. In: *Biogeography and Ecology of the Island of Newfoundland* (ed. by G.R. South), pp. 207–265. W. Junk, The Hague.

Wells, E.D. and Williams, B.L. 1996. Effects of drainage, tilling and pK-fertilization on bulk-density, total N, P, K, Ca and Fe and net N-mineralization in 2 peatland forestry sites in Newfoundland, Canada. *Forest Ecology and Management* **84**, 97–108.

Wheeler, A.J. 1992. Vegetational sussession, acidification and allogenic events recorded in Flandrian peat deposits from an isolated Fenland embayment. *New Phytologist* **122**, 745–756.

Wheeler, B.D. 1984. British fens: a review. In: *European Mires* (ed. by P.D. Moore), pp. 237–282. Academic Press, London.

Wheeler, B.D. 1995. Introduction: restoration and wetlands. In: *Restoration of Temperate Wetlands* (ed. by B.D. Wheeler, S.C. Shaw, W.J. Fojt and R.A. Robertson), pp. 1–18. Wiley, Chichester.

Wheeler, B.D. 1996. Extracts from a speech at the International Peat Society conference on peat industry and environment, Parnu, Estonia, September 1995. *International Peat Society Bulletin* **27**, 29–31.

Wheeler, B.D. 1999. Water and plants in freshwater wetlands. In: *Ecohydrology* (ed. by A.J. Baird and R.L. Wilby), pp. 127–180. Routledge, London.

Wheeler, B.D. and Proctor, M.C.F. 2000. Ecological gradients, subdivisions and terminology of north-west European mires. *Journal of Ecology* **88**, 187–203.

Wheeler, B.D. and Shaw, S.C. 1995a. A focus on fens – controls on the composition of fen vegetation in relation to restoration. In: *Restoration of Temperate Wetlands* (ed. by B.D. Wheeler, S.C. Shaw, W.J. Fojt and R.A. Robertson), pp. 49–72. Wiley, Chichester.

Wheeler, B.D. and Shaw, S.C. 1995b. *Restoration of Damaged Peatlands*. Department of the Environment, HMSO, London.

Wheeler, B.D., Shaw, S.C. and Cook, R.E.D. 1992. Phytometric assessment of the fertility of undrained rich-fen soils. *Journal of Applied Ecology* **29**, 466–475.

Wheeler, B.D., Shaw, S.C., Fojt, W.J. and Robertson, R.A. (eds.) 1995. *Restoration of Temperate Wetlands*. Wiley, Chichester.

Whinam, J. and Buxton, R. 1997. *Sphagnum* peatlands of Australasia: an assessment of harvesting sustainability. *Biological Conservation* **82**, 21–29.

White, J.W.C., Ciais, P., Figge, R.A., Kenny, R. and Markgraf, V. 1994. A high-resolution record of atmospheric CO_2 content from carbon isotopes in peat. *Nature* **367**, 153–156.

Whiting, G.J. and Chanton, J.P. 1993. Primary production control of methane emission from wetlands. *Nature* **364**, 794–795.

Widjaja-Ahdi, I.P.G. 1997. Developing tropical peatlands for agriculture. In: *Biodiversity and Sustainability of Tropical Peatlands* (ed. by J.O. Reiley and S.E. Page), pp. 293–300. Samara Publishing, Cardigan.

Wieder, R.K., Yavitt, J.B. and Lang, G.E. 1990. Methane production and sulfate reduction in two Appalachian peatlands. *Biogeochemistry* **10**, 81–104.

Wilen, B.O. and Tiner, R.W. 1993. Wetlands of the United States. In: *Wetlands of the World: Inventory, Ecology and Management* (ed. by D. Whigham, D. Dykyjova and S. Hejny), pp. 515–636. Kluwer, Dordrecht.

Wilkie, N., Thompson, P. and Russell, N. 1997. The conservation and restoration of active blanket bog in Caithness and Sutherland. In: *Blanket Mire Degradation: Causes, Consequences and Challenges* (ed. by J.H. Tallis, R. Meade and P.D. Hulme), pp. 183–188. Macaulay Land Use Research Institute, Aberdeen.

Williams, M.A.J., Dunkerley, D.L., De Deckker, P., Kershaw, A.P. and Chapell, J. 1998. *Quaternary Environments*. Arnold, London.

Wilmshurst, J.M. and McGlone, M.S. 1996. Forest disturbance in the central North Island, New Zealand, following the 1850 BP Taupo eruption. *Holocene* **6**, 399–411.

Wilson, J.B. and Agnew, A.D.Q. 1992. Positive feedback switches in plant communities. *Advances in Ecological Research* **23**, 263–336.

Wind-Mulder, H.L., Rochefort, L. and Vitt, D.H. 1996. Water and peat chemistry comparisons of natural and post-harvested peatlands across Canada and their relevance to peatland restoration. *Ecological Engineering* **7**, 161–181.

Winkler, M.G. 1985. Charcoal analysis for palaeoenvironmental interpretation: a chemical assay. *Quaternary Research* **23**, 313–326.

Winkler, M.G. 1988. Effect of climate on development of two *Sphagnum* bogs in south-central Wisconsin. *Ecology* **69**, 1032–1043.

Winkler, M.G. and Dewitt, C.B. 1985. Environmental impacts of peat mining in the United States – documentation for wetland conservation. *Environmental Conservation* **12**, 317–330.

Winston, R.B. 1994. Models of the geomorphology, hydrology, and development of domed peat bodies. *Geological Society of America Bulletin*, **106**, 1594–1604.

Winston, R.B. 1996. Models of the geomorphology, hydrology, and development of domed peat bodies (vol. 106, pg. 1594, 1994). *Geological Society of America Bulletin*, **108**, 126.

Wolfe, B.B., Edwards, T.W.D., Aravena, R., Forman, S.L., Warner, B.G., Velichko, A.A. and MacDonald, G.M. 2000. Holocene paleohydrology and paleoclimate at treeline, north-central Russia, inferred from oxygen isotope records in lake sediment cellulose. *Quaternary Research* **53**, 319–329.

Woodin, S.J. and Lee, J.A. 1987. The fate of some components of acidic deposition in ombrotrophic mires. *Environmental Pollution* **45**, 61–72.

Woodland, W.A., Charman, D.J. and Sims, P.C. 1998. Quantitative estimates of water tables and soil moisture in Holocene peatlands from testate amoebae. *Holocene* **8**, 261–273.

Wösten, J.H.M., Ismail, A.B. and van Wijk, A.L.M. 1997. Peat subsidence and its practical implications: a case study in Malaysia. *Geoderma* **78**, 25–36.

Wright, H.E., Coffin, B.A. and Aaseng, N.E. (eds.) 1992. *The Patterned Peatlands of Minnesota*. University of Minneapolis Press, Minneapolis.

Wright, H.E., Mann, D.H. and Glaser, P.P.H. 1984. Piston cores for peat and lake sediments. *Ecology* **65**, 657–659.

Xie, S., Nott, C.J., Avsejs, L.A., Volders, F., Maddy, D., Chambers, F.M., Gledhill, A., Carter, J.F. and Evershed, R.P. 2000. Palaeoclimate records in compound-specifid delta D values of a lipid biomarker in ombrotrophic peat. *Organic Geochemistry* **31**, 1053–1057.

Xuehui, M. and Yan, H. 1996. Peat and peatlands in China. In: *Global Peat Resources* (ed. by E. Lappalainen), pp. 163–168. International Peat Society, Finland.

Yalden, P.E. and Yalden, D.W. 1988. The level of recreational pressure on blanket bog in the Peak District National Park, England. *Biological Conservation* **44**, 213–227.

Yavitt, J.B. 2000. Carbon dynamics in Appalachian peatlands of West Virginia and western Maryland. *Water Air and Soil Pollution* **77**, 271–290.

Yavitt, J.B., Angell, L.L., Fahey, T.J., Cirmo, C.P. and Driscoll, C.T. 1992. Methane fluxes, concentrations, and production in 2 Adirondack beaver impoundments. *Limnology and Oceanography* **37**, 1057–1066.

Yavitt, J.B., Lang, G.E. and Sexstone, A.J. 1990. Methane fluxes in wetland and forest soils, beaver ponds, and low-order streams of a temperate forest ecosystem. *Journal of Geophysical Research –Atmospheres* **95**, 22463–22474.

Yonebayashi, C. and Minaki, M. 1997. Late Quaternary vegetation and climatic history of eastern Nepal. *Journal of Biogeography* **24**, 837–843.

Yoshikawa, S., Yamaguchi, S. and Hata, A. 2000. Paleolimnological investigation of recent acidity changes in Sawanoike Pond, Kyoto, Japan. *Journal of Paleolimnology* **23**, 285–304.

Younger, P.L. and Mchugh, M. 1995. Peat development, sand cones and paleohydrogeology of a spring-fed mire in east Yorkshire, UK. *Holocene* **5**, 59–67.

Zoltai, S.C. 1988. Wetland environments and classification. In: *Wetlands of Canada* (ed. by National Wetlands Working Group), pp. 3–26. Polyscience, Montreal.

Zoltai, S.C., Tarnocai, C., Mills, G.F. and Veldhuis, H. 1988a. Wetlands of subarctic Canada. In: *Wetlands of Canada* (ed. by National Wetlands Working Group), pp. 57–96. Polyscience, Montreal.

Zoltai, S.C., Taylor, S., Jeglum, J.K., Mills, G.F. and Johnson, J.D. 1988b. Wetlands of boreal Canada. In: *Wetlands of Canada* (ed. by National Wetlands Working Group), pp. 97–154. Polyscience, Montreal.

Zoltai, S.C. and Vitt, D.H. 1990. Holocene climatic change and the distribution of peatlands in western interior Canada. *Quaternary Research* **33**, 231–240.

Zurek, S. 1984. Distribution and character of European mires [in German]. *Telma* **14**, 113–125.

Index

General notes on index:

Page numbers in **bold** type indicate a figure or table.

The prefixes 'peatland' and 'peat' (and similar general related terms) have been avoided wherever possible in the index to avoid too many entries under these broad headings. For example, 'Peat extraction' is found in the index under 'Extraction'.

Geographical terms are entered only where a significant example is mentioned and the location is especially relevant. Locations in the USA and Canada are grouped by state or province. Some notable sites are listed under their site names. Species names are also only indexed where a significant example occurs.